Nonlinear Meta-Optics

Multidisciplinary and Applied Optics

Series Editor
Vasudevan Lakshminarayanan
University of Waterloo,
Ontario, Canada

Quantum Mechanics of Charged Particle Beam Optics: Understanding Devices from Electron Microscopes to Particle Accelerators
Ramaswamy Jagannathan, Sameen Ahmed Khan

Understanding Optics with Python
Vasudevan Lakshminarayanan, Hassen Ghalila, Ahmed Ammar, L. Srinivasa Varadharajan

Nonlinear Meta-Optics
Edited by Costantino De Angelis, Giuseppe Leo, Dragomir N. Neshev

For more information about this series, please visit: https://www.crcpress.com/Multidisciplinary-and-Applied-Optics/book-series/CRCMULAPPOPT

Nonlinear Meta-Optics

Edited by
Costantino De Angelis
Giuseppe Leo
Dragomir N. Neshev

CRC Press
Taylor & Francis Group
Boca Raton London New York

CRC Press is an imprint of the
Taylor & Francis Group, an **informa** business

First edition published 2020
by CRC Press
6000 Broken Sound Parkway NW, Suite 300, Boca Raton, FL 33487-2742

and by CRC Press
4 Park Square, Milton Park, Abingdon, Oxon OX14 4RN

First issued in paperback 2023

Publisher's Note
The publisher has gone to great lengths to ensure the quality of this reprint but points out that some imperfections in the original copies may be apparent.

ISBN-13: 978-1-138-57654-4 (hbk)
ISBN-13: 978-1-03-265309-9 (pbk)
ISBN-13: 978-1-351-26976-6 (ebk)

DOI: 10.1201/b22515

Typeset in Times
by Lumina Datamatics Limited

Contents

Preface

Optical metasurfaces are arrays of optical antennas, with sub-wavelength size and separation. The metasurfaces represent an original concept of flat optics with no classical analogs. They allow for the ultimate miniaturization of optical components, as well as the enabler of new functionalities not possible to date. In the past two decades, the optical properties of metasurfaces have been intensely studied in the linear regime, with either metallic or amorphous dielectric nanostructures.

Until that time, the venerable branch of physics and engineering dealing with light propagation in free space had not qualitatively deviated from the paradigms established back in the seventeenth century by Newton, Huygens, and Fresnel. Even after Maxwell's theory of classical electromagnetism in the nineteenth century and the development of Fourier optics, lasers, and quantum optics in the second half of twentieth century, molding light stayed mostly based on glass components of variable optical thickness that focus, deflect or disperse light beams by exploiting the accumulation of phase along different optical paths. This state of things has been revolutionized by the breakthrough brought about by flat optics, with its promise of replacing bulky and difficult-to-align assemblies of traditional optical components with stacks of nanostructured thin films.

A few years after the development of linear flat optics, the sub-wavelength physics of its nonlinear counterpart has gained increasing attention, with frequency conversion effects observed first in the hot spots associated to localized plasmon resonances in metal nanoantennas and then in association to Mie-type multipolar resonances in dielectric nanostructures. This transition to the nano-world has marked an important turning point for 60-year-old nonlinear optics, with the role of phase matching being replaced by that of near-field resonances occurring in dissipative open nanostructures in a non-Hermitian regime.

The aim of this text is to provide a comprehensive overview of the main aspects associated with this new branch of nonlinear optics, which we propose to name as "Nonlinear Meta-Optics". Our ambition is to elucidate the foundations and the potentials of such multidisciplinary domain, thanks to the valuable worldwide contributions of some of its main actors in the recent past. Emphasis is placed on both theoretical foundations and experimental demonstrations, spanning from the different modeling and technological approaches to the design and fabrication of radically new nonlinear devices.

As a consequence, this book naturally starts by placing its core message in the broader context of linear optical metasurfaces and flat optics as described in Chapter 1. Based on the nanofabrication materials and protocols illustrated in Chapter 3, several types of nonlinear nanostructures are presented: from those relying on resonant intersubband transitions (Chapter 2) to those associated with nonlinear plasmonic metasurfaces (Chapter 4). While the nonlinear optical response of the latter is directly dictated by geometry, and thus enables relatively intuitive design techniques, the much higher nonlinear efficiencies of Mie-resonant nanostructures in $\chi^{(2)}$ and $\chi^{(3)}$ semiconductors (Chapter 9) often come at the price of involved tensor physics in the near field. Of course, as is already the case for the more established domain of integrated optics, two of the most exciting perspectives of nonlinear optics at the nanoscale are given by the ultrafast and quantum regimes (Chapters 5 and 6), with their promise of both exciting fundamental research and impactful applications like high-frequency data processing, secure communications, and ultimate metrology. Thus, it is the sole reason to place these two chapters in the proper context, by devoting Chapter 7 to nonlinear photonic crystals and Chapter 8 to nonlinear and quantum effects in microcavities. Finally, since we believe in the potential of top-down/bottom-up hybrid structures for future developments of nonlinear metasurfaces, this book ends with a chapter on these perspectives (Chapter 10).

As the present text results from 10 independent contributions, we have not avoided some repetitions and stitches between chapters. After all, in our opinion, this can increase its readability in a non-sequential way, following the specific needs of each reader.

We hope that our efforts will be useful for the increasing number of scientists, engineers, and graduate students attracted by this fascinating new field of modern optics.

Finally, we wish to thank all the contributors for taking time to write and revise their chapters, and for in-depth and friendly discussions. Our last word is a big thank to our families, for their infinite patience and understanding.

Costantino De Angelis
Giuseppe Leo
Dragomir N. Neshev
Brescia, Paris, Canberra

Editors

Costantino De Angelis received the Laurea Degree Summa Cum Laude in Electronic Engineering and a PhD in Telecommunications from the University of Padova, in 1989 and 1993, respectively. Since 1998, he has been at the University of Brescia where he is a full professor of electromagnetic fields. He is an OSA Fellow and has been responsible for several university research contracts in the last 20 years within Europe, the United States, and Italy. His technical interests are in optical antennas and nanophotonics. He is the author of over 150 peer-reviewed scientific journal articles.

Giuseppe Leo received his Laurea Degree Summa Cum Laude in Electronic Engineering at La Sapienza University (Rome) in 1990 and his PhD degree in Physics at Orsay University (Paris) in 2001. He was at Rome-3 University from 1992 to 2004. He has been a full professor in physics at Paris Diderot University since 2004, and in charge of the Nonlinear devices group at MPQ Laboratory since 2006. His research areas include nonlinear optics, micro- and nano-photonics and optoelectronics, with a focus on AlGaAs platform. He has coordinated several research programs and coauthored over 110 peer-reviewed journal articles and 4 patents.

Dragomir N. Neshev is a Professor in Physics at the Australian National University (ANU) and the Director of the Australian Research Council Centre of Excellence for Transformative Meta-Optical Systems (TMOS). He received his PhD degree from Sofia University, Bulgaria, in 1999. Since then he has worked in the field of nonlinear optics at several research centers around the world. He joined ANU in 2002. He is the recipient of several awards, including a Queen Elizabeth II Fellowship, an Australian Research Fellowship, a Marie Curie Individual Fellowship, and the Academic Award for Best Young Scientist of Sofia University. His activities span over several branches of optics, including nonlinear periodic structures, singular optics, plasmonics, and photonic metamaterials. He has coauthored 200 publications in international peer-reviewed scientific journals.

Contributors

Andrea Alù
Photonics Initiative
Advanced Science Research Center
City College of New York
New York, New York

J. Azaña
INRS-EMT
Varennes, Quebec, Canada

Mikhail A. Belkin
University of Texas at Austin
Austin, Texas

S. Bharadwaj
INRS-EMT
Varennes, Quebec, Canada

Igal Brener
Sandia National Laboratories
Albuquerque, New Mexico
and
Center for Integrated
    Nanotechnologies
Sandia National Laboratories
Albuquerque, New Mexico

Luca Carletti
Department of Information Engineering
University of Brescia
Brescia, Italy

L. Caspani
University of Strathclyde
Glasgow, United Kingdom

Michele Celebrano
Department of Physics
Politecnico di Milano
Milano, Italy

P. Colman
Thales Research and Technology
    France
Palaiseau, France

S. Combrié
Thales Research and Technology
    France
Palaiseau, France

Costantino De Angelis
Department of Information Engineering
University of Brescia
and
National Institute of Optics – CNR
Brescia, Italy

A. De Rossi
Thales Research and Technology
    France
Palaiseau, France

Aloyse Degiron
Laboratoire MPQ
Université de Paris - CNRS
Paris, France

Marco Finazzi
Department of Physics
Politecnico di Milano
Milano, Italy

B. Fischer
INRS-EMT
Varennes, Quebec, Canada

Patrice Genevet
Université Côte d'Azur, CNRS
CRHEA
Valbonne, France

Lavinia Ghirardini
Department of Physics
Politecnico di Milano
Milano, Italy

Carlo Gigli
Laboratoire MPQ
Université de Paris - CNRS
Paris, France

V. F. Gili
Institute of Applied Physics
Abbe Center of Photonics
Friedrich-Schiller-Universität Jena,
    Germany

Juan Sebastian Gomez-Diaz
University of California
Davis, California

Sébastien Héron
Université Côte d'Azur, CNRS
CRHEA
Valbonne, France

M. Islam
INRS-EMT
Varennes, Quebec, Canada

C. Jagadish
Department of Electronic Materials
    Engineering
Research School of Physics
The Australian National University
Canberra, Australian Capital Territory,
    Australia

Alexey V. Krasavin
Department of Physics
London Centre for Nanotechnology
King's College London
London, United Kingdom

Alex Krasnok
University of Texas at Austin
Austin, Texas

M. Kues
Hannover Center for Optical
    Technologies
Leibniz University Hannover
Hanover, Germany

and

INRS-EMT
Varennes, Quebec, Canada

Giuseppe Leo
Laboratoire MPQ
Université de Paris - CNRS
Paris, France

Andrea Locatelli
Department of Information
    Engineering
University of Brescia
and
National Institute of Optics – CNR
Brescia, Italy

Giuseppe Marino
Laboratoire MPQ
Université de Paris - CNRS
Paris, France

A. Martin
Thales Research and Technology
    France
Palaiseau, France

Alexander E. Minovich
Department of Physics
London Centre for Nanotechnology
King's College London
London, United Kingdom

Andrey Miroshnichenko
School of Engineering and Information
  Technology
University of New South Wales
Canberra, Australian Capital Territory,
  Australia

G. Moille
Thales Research and Technology France
Palaiseau, France

R. Morandotti
INRS-EMT
Varennes, Quebec, Canada

and

University of Electronic Science and
  Technology of China
Chengdu, China

D. J. Moss
Center for Microphotonics
Swinburne University of Technology
Hawthorn, Australia

Dragomir N. Neshev
Nonlinear Physics Centre
Research School of Physics
The Australian National University
Canberra, Australian Capital Territory,
  Australia

Alexander N. Poddubny
Nonlinear Physics Centre
Research School of Physics
The Australian National University
Canberra, Australian Capital Territory,
  Australia

and

ITMO University
and
Ioffe Institute
Saint Petersburg, Russia

M. Rahmani
Nonlinear Physics Centre
Research School of Physics
The Australian National University
Canberra, Australian Capital Territory,
  Australia

C. Reimer
Hyperlight Corporation
Cambridge, Massachusetts

and

INRS-EMT
Varennes, Quebec, Canada

Davide Rocco
Department of Information Engineering
University of Brescia
Brescia, Italy

L. Romero Cortés
INRS-EMT
Varennes, Quebec, Canada

P. Roztocki
INRS-EMT
Varennes, Quebec, Canada

Giovanni Sartorello
Department of Physics
London Centre for Nanotechnology
King's College London
London, United Kingdom

and

School of Applied and Engineering
  Physics
Cornell University
Ithaca, New York

S. Sciara
INRS-EMT
Varennes, Quebec, Canada

Maxim R. Shcherbakov
School of Applied and Engineering
    Physics
Cornell University
Ithaca, New York

and

Faculty of Physics
Lomonosov Moscow State University
Moscow, Russia

Gennady Shvets
School of Applied and Engineering
    Physics
Cornell University
Ithaca, New York

Andrey A. Sukhorukov
Nonlinear Physics Centre
Research School of Physics
The Australian National University
Canberra, Australian Capital Territory,
    Australia

Mykhailo Tymchenko
University of Texas at Austin
Austin, Texas

Polina P. Vabishchevich
Sandia National Laboratories
Albuquerque, New Mexico

and

Center for Integrated
    Nanotechnologies
Sandia National Laboratories
Albuquerque, New Mexico

Lei Xu
School of Engineering and Information
    Technology
University of New South Wales
Canberra, Australian Capital Territory,
    Australia

Anatoly V. Zayats
Department of Physics
London Centre for Nanotechnology
King's College London
London, United Kingdom

Y. Zhang
INRS-EMT
Varennes, Quebec, Canada

1 Metasurfaces, Then and Now

Sébastien Héron and Patrice Genevet

CONTENTS

1.1 INTRODUCTION

The general function of most optical devices can be described as the modification of the wavefront of light by altering its phase, amplitude, and polarization in a desired manner. Conventional optical elements, such as lenses, liquid crystal, planar phased arrays, and vortex elements fabricated using grayscale lithography, are modifying the phase of the wavefront using "propagation effect," $\varphi = kr$. This limits the longitudinal dimension of the optical elements to be several orders thicker than the wavelength, $l > \lambda$. The class of optical components that alter the phase of light waves includes lenses, prisms, spiral phase plates, axicons, and more generally spatial light modulators (SLMs), which are able to behave such many of these components by means of a dynamically tunable spatial phase response. A second class of optical components such as waveplates utilizes bulk birefringent crystals with optical anisotropy to change the polarization of light. A third class of optical components such as gratings and holograms is based on diffractive optics, where diffracted in-phase spherical waves departing from different parts of the components interfere in the far-field to produce the desired optical patterns. All of these components shape optical wavefronts using the propagation effect: the change in phase and polarization is gradually accumulated during light propagation. This approach is generalized in transformation optics [1] which utilizes metamaterials to engineer the spatial distribution of refractive indices and therefore bend light in unusual ways, achieving remarkable phenomena such as negative refraction, subwavelength-focusing, and cloaking [2–7].

It is possible to break away from our reliance on the propagation effect, and attain new degrees of freedom in molding the wavefront and in optical design by introducing abrupt phase changes over the scale of the wavelength (dubbed "phase discontinuities") into the optical path. A novel approach to achieve "phase jumps" is to rely on the abrupt phase change $\Delta\varphi(x, y)$ associated with the light scattering process across a large number of resonant optical scatterers, assembled in suitable arrays. Interacting with a simple resonator, such as a plasmonic nanoparticle, the phase of a re-emitted electromagnetic (EM) wave is shifted from that of the excitation by an amount ranging from 0 to π as the wavelength is swept across the resonance. To obtain full 2π-phase coverage, several techniques have been proposed and are discussed in this chapter. To avoid diffraction effects, the phase elements arrangement should satisfy two requirements: subwavelength thickness of the array and subwavelength separation of the scatterers, i.e., the array must form a metasurface. Numerous optical functions have already been fulfilled by such interfaces [3,8–15] and based on those results, one can appreciate the potential of metasurfaces for going beyond simple wavefront control, for example their implication in nonlinear optics and quantum optics. This introduction chapter revisits the notion of metasurface, and also shows a brief landscape of what has been done in this exciting field of research over the last few years. For more exhaustive contents, the reader is invited to refer to these recent reviews [18–21].

The chapter is split into two main sections: the first part discusses the main physical ideas and concepts of metasurfaces, and the second presents selected recent works in the field. This way, the readers could both understand the basic physical phenomena of linear flat optics, and have an overview of how scientists melt them together to forge intuitions for designing new meta-optic components.

1.2 CONCEPTS OF FLAT OPTICS

1.2.1 From Blazed Grating to Beam Deflector

In conventional optics, light rays passing through a diffractive/refractive component experience different optical paths when propagating inside the material. The accumulated phase of light propagating along different trajectories, or rays, interferes to form the desired wavefront [22–24]. The first idea that led to smaller, and especially thinner optics, was the invention of Fresnel lenses which benefit from the modulo 2π definition of the phase. Augustin-Jean Fresnel developed this concept to reduce the weight of large lighthouse lenses. He introduced thin and flat optics, at least flat components compared to ancient devices that have the same behavior as bulky lenses. The second idea that opened Pandora's Box was the invention of Fresnel zone plates, going from ray to wave optics, from Gaussian ray tracing to Huygens-Fresnel construction. Adding all the light rays in the far-field is somehow equivalent to calculating the interference pattern of the secondary diffracted waves. Then, a seminal proposal by W. E. Kock published in 1948 discussing the possibility to locally reduce the phase velocity of light using subwavelength metallic patches showed that optical components, a lens in his paper, can be much more compact using metallic patches than refractive devices [25]. This work was followed by Stork et al. and M. W. Farn who realized grating structures with a period small compared with the wavelength of light [26,27]. For better understanding, let us discuss the working principle of a conventional blazed grating: it is nothing more than a periodic arrangement of tiny prisms disposed side by side and designed with a varying thickness such that the propagation phase spans 0–2π depending on the light path. On the one hand, the deflection angle caused by this device is governed by the grating law essentially due to the periodicity of the phase. On the other hand, the prism-like phase gradient leads to concentration of the light flux into one selected diffraction order corresponding to the specular transmission direction. The saw-tooth phase gradient given by the blazed grating can be reproduced considering its spatial phase retardation. This observation is essentially the starting point of the concept of metasurfaces. To address spatial phase retardation at subwavelength scale, it is necessary to consider the response of nanoscale photonic structures, also called nanophotonics devices. These nano-structures provide ways to control the amplitude, the polarization and the phase of scattered fields with subwavelength resolution. Lately, the concept of high-contrast subwavelength (HCG) dielectric structures led to the development of ultrathin optical components [28–31]. By varying the material composition, i.e., by variation of the duty cycle of the grating, one can create artificial materials with unconventional properties [32–39]. Through tailoring the diffracting properties at the subwavelength scale using spatially varying TiO_2 nano-structures, Lalanne et al. [40–42] created such phased blazed grating devices as seen in Figure 1.1. This example is of one of the first metasurfaces: it comprises nano-elements that have subwavelength spacing, and that sample the desired phase gradient realizing tilted wavefront in agreement with the Huygens-Fresnel construction.

The field rapidly started to grow thanks to the advance in nanofabrication. Metallic lift-off procedures enabled the development of plasmonic antennas with various shapes, such as V-shaped antennas, to create almost any wavefront at will with the help of their resonant properties. To properly address the light wavefront with high efficiency, the nano-antennas should provide any phase shift between 0 and 2π without modifying the light intensity.

1.2.2 METASURFACES BASED ON V-SHAPED NANO-ANTENNAS

Miniaturizing photonic devices requires control of the phase of light waves over the full 2π range within the smallest possible distance. Plasmonic vectorial interfaces are able to control phase over the full 2π range and have small footprint in both the

FIGURE 1.1 Scanning-electron micrograph of the blazed binary subwavelength grating. The horizontal period (along the x axis) is 1.9 mm, and the period in the perpendicular direction (y axis) is equal to the sampling period (380 nm). The maximum pillar aspect ratio is 4.6. (From Lalanne, P. et al., *Optics Letters* 23, 1081–1083, 1998. With permission of Optical Society of America.)

longitudinal and lateral directions without the "paraxial" limitation (that can exceed the angular range of the reflected light). This is achieved by using V-shaped plasmonic antennas that transfer energy from a linearly polarized incident beam to the orthogonal polarization with tailorable phase shifts.

In metallic nano-structures, resonant scattering of light by oscillating free electrons at the surface occurs due to a resonant electronic-electromagnetic oscillation known as localized plasmonic resonance (LPR). These LPR have previously been extensively studied for creating high electromagnetic field enhancement, and have important application in chemical and biological sensing. They can be described as radiating metallic antennas which geometry is shown in Figure 1.2b. To the extent that the metallic nanostructure and the corresponding LPR can be roughly described by a single harmonic oscillator, the scattering phase shift cannot exceed π. As depicted in Figure 1.2a, we are describing the LPR using a simple model discussed in Ref. [43] in which it is represented by charge q located at $x(t)$ with mass m on a spring with spring constant k driven by an harmonic incident electric field with frequency ω. This model accurately describes the near- and far-field spectral features of plasmonics resonances, including their phase response (the rigorous treatment of time harmonic radiation of thin-wire antennas involves the solution of the Electric Field Integral Equation [44]). Because of ohmic losses, the charge experiences internal damping with damping coefficient γ:

$$\frac{d^2x}{dt^2} + \frac{\gamma}{m}\frac{dx}{dt} + \frac{k}{m}x = \frac{q}{m}E_0 e^{i\omega t} + \frac{2q^2}{3mc^3}\frac{d^3x}{dt^3} \tag{1.1}$$

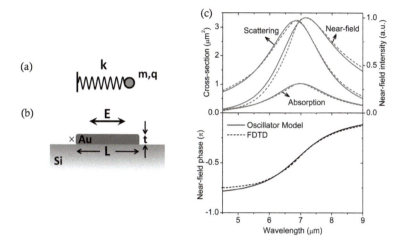

FIGURE 1.2 (a) An optical antenna can be modeled as a charged harmonic oscillator, where q is the charge and m is the inertial mass. (b) Schematics for FDTD simulations. A gold optical antenna (length $L = 1$ μm, thickness $t = 50$ nm, width $w = 130$ nm) lies on a silicon substrate and is illuminated by a normally incident plane wave polarized along the antenna axis. The cross represents a point ~4 nm away from the antenna edge where the near-field is calculated. (c) Upper panel: Scattering (σ_{scat}) and absorption (σ_{abs}) cross-sections, and near-field intensity as calculated via the oscillator model (solid curves) and FDTD simulations (dashed curves). The scattering and absorption cross-sections are defined as $\sigma_{scat}(\omega) = P_{scat}(\omega)/I_0$ and $\sigma_{abs}(\omega) = P_{abs}(\omega)/I_0$, where I_0 is the incident intensity. A total-field/scattered-field plane-wave source was employed in FDTD simulations to extract the scattered power $P_{scat}(\omega)$ and absorbed power $P_{abs}(\omega)$ of the antenna. Lower panel: Oscillator phase (solid curve) and the phase of the near-field calculated via FDTD (dashed curve). (From Yu, N. et al., *IEEE J. Sel. Top. Quantum Electron.*, 19, 4700423, 2013. © 2013 IEEE. With permission.)

In addition to internal damping which is proportional to the velocity dx/dt, the charge experiences an additional damping force due to radiation reaction which is proportional to the time derivative of the acceleration dx^3/dt. This term describes the recoil that the charge feels when it emits radiation, and is referred to as the Abraham-Lorentz force or simply the radiation reaction force [22]. By assuming harmonic motion $x(t) = x_0 e^{i\omega t}$ we can write down the steady-state solution to Eq. (1.3) as

$$x(t) = \frac{AE_0}{\left(\omega_0^2 - \omega^2\right) + i\left(\omega\Gamma_\alpha + \omega^3\Gamma_s\right)} e^{i\omega t} \qquad (1.2)$$

where we have replaced the quantities q, k, m and γ with more general oscillator parameters $A = q/m$, $\omega_0 = \sqrt{k/m}$, $\Gamma_a = \gamma/m$ and $\Gamma_s = 2q^2/3mc^3$ with Γ_a and Γ_s respectively describe the non-radiative and radiative damping mechanisms. From Eq. (1.2) one then sees that the amplitude of the oscillation is in phase with the incident field for $\omega \to 0$ and is phase delayed by π for $\omega \to \infty$, acquiring all intermediate values as the frequency of the signal is swept across the resonance, and in particular a value of $\pi/2$

when $\omega = \omega_0$ as can be seen in Figure 1.2c. So far, a resonant metallic nanorod cannot induce phase shifts higher than π. In the following two of them are assembled to create V-shaped antennas behaving as coupled oscillators.

These plasmonic vectorial interfaces composed of V-shaped antennas of various shapes allow the realization of any arbitrary 2D phase distribution. A V-antenna consists of two arms with angle Δ between them and with equal length h. Assume normal incidence to the plane of the antenna, for an arbitrary incident polarization, the modes excited in the V-antenna can be decomposed into a "symmetric" mode and an "antisymmetric" mode. The symmetric mode is driven by the projection of the incident electric field to the symmetric axis of the V-antenna (E_s in Figure 1.3b, c); this mode is characterized by identical current and charge distributions in the two antenna arms and no current is crossing the junction connecting the two arms (Figure 1.3b, c). The "antisymmetric" mode is driven by the projection of the incident electric field in the direction perpendicular to the symmetric axis of the V-antenna (E_a in Figure 1.3b, c). This mode is characterized by current and charge distributions in the two arms of the same amplitude but of opposite signs. Electric current flows across the antenna junction (Figure 1.3b, c). Figure 1.3d provides a palette of phase covering the entire 2π range. In the simplest case, one can span ~2π by drawing a straight line from one end of the symmetric mode (h~1.1 µm, Δ~0°) to the other end of the antisymmetric mode (h~0.5 µm, Δ~180°). With the constraint of equal amplitude in the scattered light, one can choose, for example, the two dotted curves in Figure 1.3d, which sample a variety of antenna geometries while covering the 2π phase delay required to fully sample the wavefront. The points chosen for experiments are then indicated by these circular markers in Figure 1.3d, e. From left

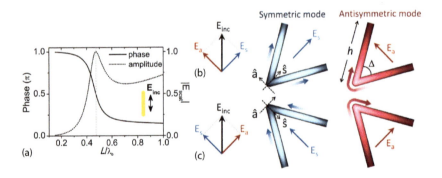

FIGURE 1.3 (a) Calculated phase and amplitude of scattered light from a straight rod antenna made of a perfect electric conductor. The vertical dashed line indicates the first-order dipolar resonance of the antenna. (b) A V-antenna supports symmetric and antisymmetric modes, which are excited, respectively, by components of the incident field along ŝ and â axes. The angle between the incident polarization and the antenna symmetry axis is 45°. The schematic current distribution is represented by colors on the antenna (blue for symmetric and red for antisymmetric mode), with brighter color representing larger currents. The direction of current flow is indicated by arrows with color gradient. (c) V-antennas corresponding to mirror images of those in (b). The components of the scattered electric field perpendicular to the incident field in (b) and (c) have a π phase difference.

(Continued)

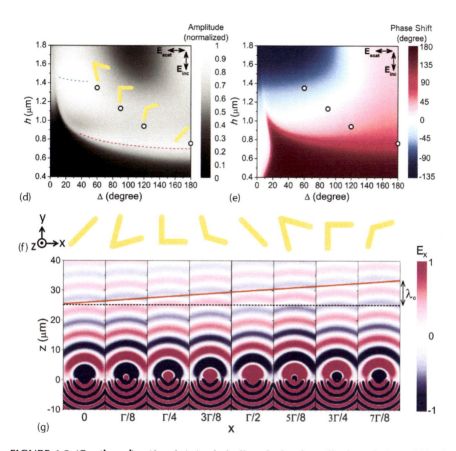

FIGURE 1.3 (Continued) (d and e) Analytically calculated amplitude and phase shift of the cross-polarized scattered light for V-antennas consisting of gold rods with a circular cross section and with various length h and angle between the rods D at $\lambda_0 = 8$ mm (20). The four circles in (d) and (e) indicate the values of h and Δ used in experiments. (f) Schematic unit cell of the plasmonic interface for demonstrating the generalized laws of reflection and refraction. (g) Finite-difference-time-domain (FDTD) simulations of the scattered electric field for the individual antennas composing the array in (f). Plots show the scattered electric field polarized in the x-direction for y-polarized plane-wave excitation at normal incidence from the substrate. (Reprinted from Yu, N. et al., *Science*, 334, 333–337, 2011. With permission from Science.)

to right these antennas have a phase difference $\pi/4$ and the total phase coverage of the four elements is π. The beauty of transferring to the orthogonal polarization direction with the scattered light is that by simply taking the mirror structure of an existing antenna (Figure 1.3f), it is able to create a new antenna whose emitted light has a flipped sign or an additional π phase shift. This way, eight V-antennas were created with an incremental phase of $\pi/4$ and 2π total phase coverage. If equally spaced, they would provide a linear phase gradient, as confirmed by 3D full-wave simulations (Figure 1.3g). A large phase coverage (~300°) can also be achieved by using arrays

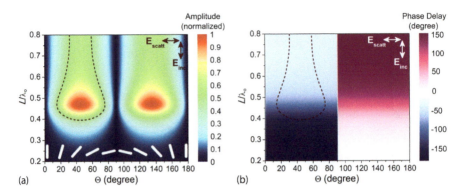

FIGURE 1.4 (a) Amplitude of the cross-polarized radiation from a straight rod antenna as a function of antenna length L and orientation angle θ. The latter is the angle between the antenna axis and the incident polarization. The dashed curve is an equal amplitude contour. Schematics of antennas of various orientations are shown at the bottom of the figure. (b) Phase of the cross-polarized antenna radiation. (Reprinted from Yu, N. et al., *Science*, 334, 333–337, 2011. With permission from Science.)

of straight antennas (see Figure 1.4). However, while offering the same phase shift span, their scattering amplitudes are substantially smaller than those of V-antennas. As a consequence of their split resonances, the V-antenna instead allows one to design an array with phase coverage of 2π and equal, yet high, scattering amplitudes for all of the array elements, leading to anomalously reflected and refracted beams of substantially higher intensities. As shall be discussed in Section 1.2.5 on Pancharatnam-Berry phase, high scattering efficiency can be obtained by considering circular polarization conversion with specifically designed single rod antennas, equivalent to what is described in Figure 1.4.

Up to this point, traditional wavefronts have been molded with the use of V-shaped antennas by translating the phase introduced by an optical component into the metasurface language. Accounting for the phase gradient imparted on the wavefront at the interface, it is possible to introduce a generalized version of Snell-Descartes law to predict the reflection and refraction angles across a phase-gradient metasurface.

1.2.3 GENERALIZED LAW OF REFRACTION

To derive the generalized laws of reflection and refraction, one can consider two approaches, (1) by deriving the reflection and refraction laws from the boundary conditions of Maxwell equations in the presence of a phase-gradient metasurface using the generalized sheet transmission conditions (GSTC) or (2) by considering the conservation of wavevector (i.e., k-vector) along the interface. In the following, we are discussing the second simple approach but we invite the readers to consult these references [45–48] that treat the more formal derivation. Importantly, the interfacial phase gradient $d\varphi/dr$ provides an effective wavevector along the interface that is imparted onto the transmitted and reflected photons. Let us consider the 2D situation in Figure 1.5, where the phase gradient $d\varphi/dx$ lies in the plane of incidence.

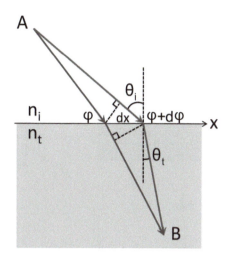

FIGURE 1.5 Schematics used to derive the two-dimensional generalized Snell's law. The interface between the two media is artificially structured to introduce a constant phase gradient $d\varphi(x)/dx$ in the plane of incidence, which serves as an effective wavevector that deflects the transmitted beam. The generalized law is a result of wavevector conservation along the interface. (Reprinted from Yu, N. et al., *Science*, 334, 333–337, 2011. With permission from Science.)

Wavevector conservation states that the sum of the tangential component of the incident wavevector, $k_0\, n_i \sin(\theta_i)$, and of $d\varphi/dx$ should equal the tangential component of the wavevector of the refracted light, $k_0\, n_t\, \sin(\theta_t)$. From this simple relation one derives:

$$n_t \sin\theta_t - n_i \sin\theta_i = \frac{1}{k_0}\frac{d\varphi}{dx} \tag{1.3}$$

The interfacial phase gradient originates from inhomogeneous plasmonic structures on the interface that provides an extra momentum to the reflected and transmitted photons. In turn, the photons exert a recoil force on the interface. Note that the generalized law can also be derived following Fermat's principle (or the principle of stationary phase). The latter states that the total phase shift $\varphi(r_s) + \int_A^B k_0 n(r)dr$ accumulated must be stationary for the actual path a light beam takes. Here the total phase shift includes the contribution due to propagation $\int_A^B k_0 n(r)dr$ and abrupt phase changes $\varphi(r_s)$ acquired when the light beam passes through, r_s being the position along the interface. Following the analogy discussed in the introduction, this law which has been discussed in several paper (see for example [2,10,15,20,49]), is essentially equivalent to the blazed grating equation.

By using the eight V-antennas described in the previous section, it is possible to create metasurfaces that imprint a linear distribution of phase shifts to the optical wavefronts. A representative fabricated sample with the highest packing density of antennas is shown in Figure 1.6a. A periodic antennas arrangement with a constant

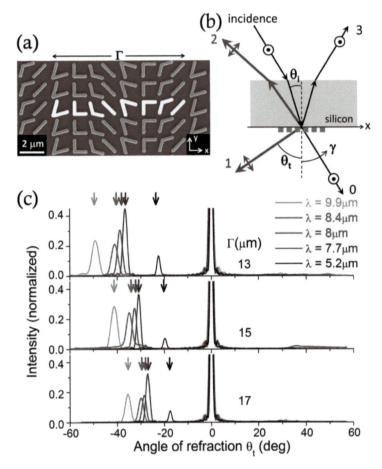

FIGURE 1.6 (a) SEM image of a metasurface consisting of a phased optical antenna array fabricated on a silicon wafer. It introduces a linear phase distribution along the interface and is used to demonstrate the generalized laws of reflection and refraction. The unit cell of the structure (highlighted) comprises eight gold V-antennas of width ~220 nm and thickness ~50 nm and it repeats with a periodicity of Γ = 11 m in the x-direction and 1.5 m in the y-direction. The antennas are designed so that their x- and y-polarized components of the scattered waves have equal amplitudes and constant phase difference π/4 between neighbors. (b) Schematic experimental setup for y-polarized excitation (electric field normal to the plane of incidence) and phase gradient in the plane of incidence. Beams "0," "1," "2," and "3" are ordinary refraction, extraordinary refraction, extraordinary reflection, and ordinary reflection, respectively; indicates the angular position of the detector. (c) Experimental far-field scans showing the ordinary and extraordinary refraction generated by metasurfaces with different interfacial phase gradients (from 2/13 to 2/17 μm) at different wavelengths (from 5.2 to 9.9 μm), given normally incident light. This broadband response is independent of the incident polarization. The scans are normalized with respect to the intensity of the ordinary beams. (From Yu, N. et al., *IEEE J. Sel. Top. Quantum Electron.*, 19, 4700423, 2013. © 2013 IEEE. With permission.)

incremental phase is used here for convenience, but is not necessary to satisfy the generalized laws. The phase gradient only needs to be constant along the interface. Figure 1.6c shows experimental far-field scans at excitation wavelengths from 5.2 to 9.9 μm. Three samples with periods $\Gamma = 13$, 15, and 17 μm were tested. For all samples and excitation wavelengths, we observed the ordinary and extraordinary refractions and negligible optical background. Samples with smaller Γ create larger phase gradients and therefore deflect extraordinary refraction into larger angles from the surface normal; smaller Γ also means a higher antenna packing density and therefore more efficient scattering of light into the extraordinary beams (Figure 1.6c). Eventually, the observed angular positions of the extraordinary refraction agree very well with the generalized law of refraction, $\theta_t = -\sin^{-1}(\lambda/\Gamma)$.

In this generalized Snell-Descartes law, the phase gradient must be continuous as a function of the position modulo 2π, meaning that the saw-tooth phase profile should be periodic. To ease fabrication, the periodic phase profiles are obtained using always the same sampling, i.e., with the same selected antennas. It is however possible to use different antennas from one period to another by simply changing their position and geometry while preserving the functionality of the optical element. This aperiodic regime has been studied and the results are discussed in the next paragraph.

1.2.4 THE APERIODIC BEAM DEFLECTOR

By disposing side-by-side antennas that are separated by random distances to create an aperiodic distribution of plasmonic nano-antennas, it is possible to obtain linear phase gradient, according that each element address the required phase retardation. With respect to periodic phase-gradient metasurfaces that are obtained by repeating a supercell unit of 2π phase ramps, randomly disposed phased antennas interfere along a given direction and cancel the higher order diffracted background signal. The aperiodic metasurfaces are experimentally realized by randomly shifting the positions of each antenna and adjusting their phase response in consequence. Repeated elements are avoided to the point that there is not obvious periodicity and thereof no direct analogy with classical blazed gratings. The presence of anomalous reflected and refracted beams demonstrates that the generalized laws derived in [2] are intimately related to the spatial phase gradient and cannot naively be interpreted in terms of periodical diffraction as for conventional gratings. The residual signals originating from residual diffraction in the periodic case, is totally suppressed using an aperiodic arrangement. It is worth noticing that the analogy between a perfectly designed blazed grating and our aperiodic interface still occurs. It can be appreciated by doing the Fourier transform of its transmission function. Nonetheless, aperiodic interfaces feature many other interesting properties among which is their robustness with respect to design and fabrication imperfections.

The scanning electron microscopy images of the structures used in our experiments are presented in Figure 1.7b. The periodic and the aperiodic metasurfaces are fabricated using electron beam lithography on double side polished intrinsic silicon substrate followed by electron beam evaporation of 5 nm of titanium adhesion layer and 45 nm of gold. The Si wafer has negligible absorption for the chosen incident light of 10.6 μm. For both designs, the width of the antennas is fixed at about 150 nm.

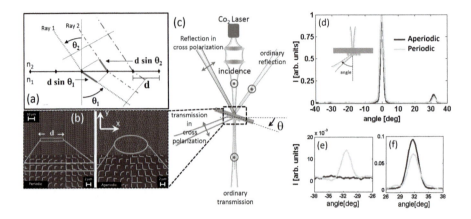

FIGURE 1.7 (a) Geometry of classical diffraction for planar wavefronts. Two parallel light rays are incident on a periodically indented interface with a period d. (b) SEM images of the periodic and the random arrangements of plasmonic V-shaped antennas. We choose our phase gradient $\partial\Phi/\partial x = 2\pi/d$ with $d = 20$ μm. When the latter is excited at normal incidence with a 10.6 μm laser wavelength, it gives rise to an anomalous refracted angle of 32°. Each of the eight elements in the period are designed to have equal scattering amplitudes and constant phase difference $\Delta\Phi = \pi/4$ between neighbors. (c) Sketch of the experimental setup used to characterize the metasurfaces. A linearly polarized CO_2 laser operating at 10.6 μm is impinging onto a silicon wafer patterned with our two different arrays of nanoantennas. A liquid-nitrogen-cooled mercury-cadmium-telluride (MCT) detector, rotating around the sample axis at approximately 15 cm, is angularly scanned to measure the far-field intensity profiles. The angular resolution is set to 0.2° and a polarizer can be inserted along the beam paths between the sample and the detector to isolate the signal generated by the antennas in co and cross state of polarization. (d) Far-field intensity profiles experimentally measured for the refracted beams for periodically (grey) and randomly (black) distributed antennas. For both, the incident beam is impinging the interface at normal incidence. Figure (d) shows the full angular scan without polarizer. Ordinary and anomalously refracted beams are observed for both designs at angles equal to −32° and 32°, respectively Figure (e,f). We experimentally checked that the latter is crossed polarized with the incident light. Figure (e) confirms that the random design suppresses the undesired residual signal originating from diffraction.

In the d-periodic case, the elements are separated along the x dimension of the interface by a fixed distance of $d/8$. Each one of them scatters light with an increasing phase delay $\Delta\varphi = \pi/4$. The periodic phase gradient is achieved by replicating the same unit cell of eight elements in a periodic manner. However, in case of the aperiodic interface, the elements are disposed randomly but the finite size of the elements imposes several constraints on the amount of spatial disorder that can be introduced along the interface. (1) We need to avoid overlapping structures. (2) We have to maintain the subwavelength packing to avoid diffraction in the far-field which would cause diffuse background signal. (3) For comparison with the periodic case, the random structure should impose the same spatial phase gradient $\partial\varphi/\partial x = 2\pi/(20$ μm$)$, keeping the same spatial density of antennas. To fulfill these requirements, the random interfaces are created by randomly shifting the position of antennas around position occupied by the periodic elements. To avoid overlap, a maximum of 10% of the inter-antennas distance is considered. To maintain the correct linear phase gradient, we choose the appropriate antenna accordingly to the local value of the phase at each antenna position. While such distribution preserves the same density of antennas and a linear uniform phase profile oriented along x exclusively, the resulting metasurface does not present any sort of periodicity whether it is along the x or the y-direction. The experimental results carried out at normal incidence for the two metasurfaces are shown in Figure 1.7. For both devices, we observe a large signal at a zero angle which corresponds to the classical transmission. In each case, a second peak at an angle of about 32°, following the generalized formula for refraction, is also observed. The residual peak appearing with the periodic design at the angle of −32°, i.e., at the angle for the −1 diffracted order which is observed at symmetric angle with respect to the anomalous beam for normal incidence, is totally suppressed using aperiodic metasurfaces. Figure 1.7e not only demonstrates generalized refraction for aperiodic surfaces but also the ability of random metasurfaces to suppress undesired and spurious diffraction effects generally resulting from periodicity.

Beyond these interesting phase manipulation properties achieved with metasurface phased arrays, it is important for real-world application to determine their performances in terms of throughput efficiency, deflection angle capabilities and reliability with respect to fabrication uncertainties and errors. So far, one can easily realize that the use of plasmonic nano-structures working through polarization conversion limits the component efficiencies. On the one hand, metallic antennas have high intrinsic non-radiative losses because of the existence of a plasma frequency and on the other hand, radiative losses also exist due to unwanted backward or forward scattered fields. To circumvent these issues, interesting scenarios, known as the Kerker conditions, have been proposed to achieve pure forward (or backward) emission. They are derived from Mie theory, adapted to antennas geometry from the canonical sphere problem.

1.2.5 MIE THEORY AND KERKER CONDITIONS

Sufficiently small spherical dielectric nanoparticles provide pronounced low-order resonances associated with the excitation of both magnetic and electric dipolar modes. Under certain conditions described in [50], the first resonance that roughly arises when the effective wavelength is of the order of the particle diameter ($\lambda_0 / n_{sp} \approx D_{sp}$), is a magnetic dipole resonance. It is worth pointing out that unlike split-ring resonator (SRR) magnetic resonances, dipolar magnetic resonances in high dielectric resonators are essentially excited by electric field and not by the magnetic contribution. In rectangular resonators, for example, the resonant wavelength can be easily tuned by changing either the geometry or the size of the scatterer. The combination of the scattering properties of both electric and magnetic resonances are of great interest for the realization of efficient planar optical metasurfaces. In order to fully appreciate the potential of "electromagnetic" scatterers for metasurfaces, we review here the work of Kerker and co-workers [51,52]. In this seminal paper, the authors considered a dielectric sphere with equal values of the relative dielectric constant and relative magnetic permeability, $\varepsilon = \mu$, and showed that it exhibits zero backscattering and no depolarization. For long considered as theoretical curiosity, particles with $\varepsilon = \mu$ feature equal electric (a_n) and magnetic (b_n) multipole coefficients such that light is scattered to destructively interfere in the backward direction. This effect is known as the first Kerker condition. Considering nanoparticles with relatively small size, on the order of the excitation wavelength, Mie's coefficients predicting the scattered light are greatly simplified and only a few coefficients remain to characterize the scattered field. Nanoparticles can be treated as dipolar particles in the sense that only the dipolar terms contribute to the scattered field. The first two coefficients of the Mie expansion can therefore be used to describe the electric and magnetic dipole polarizabilities:

$$\alpha_e = \frac{3i\varepsilon}{2k^3} a_1 \text{ and } \alpha_m = \frac{3i}{2\mu k^3} b_1 \text{ where } k = n_{sp}k_0.$$

Interference between the electric and magnetic dipoles leads to a back scattering cross section given by:

$$\sigma_s(\pi) = 4\pi k^4 \left(\left| \varepsilon^{-1}\alpha_e \right|^2 + \left| \mu\alpha_m \right|^2 \right) \left[1 + \cos(\pi - \Delta\varphi_\alpha) \right] \qquad (1.4)$$

which has a minimum when dipoles are oscillating in phase ($\Delta\varphi_\alpha = 0$) with an overall pattern similar to that of a Huygens source [53–55]. However, nano-antennas can be tailored not to just sustain resonances dominated by dipolar moments, but also to sustain higher order multipole moments. Various geometries have been investigated that exhibit such higher order multipole moments to achieve a directional response. For instance, nanoring antennas have been explored because of their promising applications, e.g., for nonlinear effects at the nanoscale and in quantum plasmonics. Since these two topics are mainly the purpose of this book, we are not providing an exhaustive literature review but a slight insight, and are referring to the following chapters of this book.

For specific parameters, plasmonic nanorings can support both an electric dipole and quadrupole mode (see Figure 1.8a, b). The geometry can be tuned to meet the generalized Kerker condition, i.e., both moments are balanced [56]. Meeting the generalized Kerker condition suppresses the nanoring's backward scattering and enhances its forward scattering. Zero backscattering occurs if the so-called generalized Kerker condition

$$p_x - \frac{\sqrt{\varepsilon_r} m_y}{c} + \frac{ik}{6} Q_{xz} = 0 \qquad (1.5)$$

is fulfilled, where p_x, m_y and Q_{xz} are dipolar electric and magnetic moments and quadrupolar electric moment respectively. This equation reduces to the well-known Kerker condition, $p_x - \frac{\sqrt{\varepsilon_r} m_y}{c} = 0$, for a vanishing quadrupole moment. To take

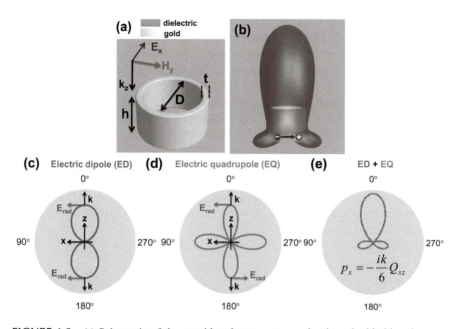

FIGURE 1.8 (a) Schematic of the considered nano-antenna that is embedded in a homogenous host medium. (b) Artistic view of the radiation pattern of the nano-antenna. (c–e) Radiation pattern for different multipole moments in the xz-plane. (c) Electric dipole moment (p_x), i.e., $|E_{far}|^2 \propto \cos^2\theta$. The arrows indicate the phase of the radiated field. (d) Electric quadrupole moment (Q_{xz}), i.e., $|E_{far}|^2 \propto \cos^2 2\theta$. (e) Superposition of electric dipole and quadrupole moments, i.e., $|E_{far}|^2 \propto (\cos\theta + \cos 2\theta)^2$, when the generalized Kerker condition is fulfilled (see Eq. (1.5) with m_y 0), i.e., $p_x = i\,k\,Q_{xz}/6$. (From Alaee, R. et al., *Opt. Lett.*, 40, 2645–2648. With permission of Optical Society of America.)

additional higher order multipole moments into account, Eq. (1.5) can be easily extended. The forward radar scattering cross section of the nano-antenna reads as

$$
\sigma_{Forward} = \lim_{r \to \infty} 4\pi r^2 \frac{\left| E_{far}\left(\varphi = 0, \theta = 0\right)\right|^2}{|E_{inc}|^2}
$$

$$
= \frac{k^4}{4\pi\varepsilon^2 |E_{inc}|^2}\left| p_x + \frac{\sqrt{\varepsilon_r}\, m_y}{c} - \frac{ik}{6}Q_{xz}\right|
$$

(1.6)

Hence, if Eq. (1.3) is fulfilled, a constructive interference in forward direction is achieved, in addition to the destructive one in backward direction. To illustrate the physical mechanism behind the generalized Kerker condition, the radiation pattern of both an electric dipole p_x and an electric quadrupole moment Q_{xz} in the xz-plane $(\varphi = 0°)$ are shown in Figures 1.8c and d. Here, we consider $m_y = 0$, since the nanoring investigated later exhibits a negligible magnetic response. Then, if Eq. (1.3) holds, the electric field radiated by an electric dipole and electric quadrupole interfere constructively at $\theta = 0°$ [Figure 1.8e], i.e., the radiated fields are in phase in forward direction [cf. Figure 1.8c and d]. On the other hand, there is no backscattering ($\theta = 180°$) because of the destructive interference in this direction.

Kerker conditions are therefore not limited to the notion of Huygens sources and cancellation of either backward or forward can be performed through exploiting higher multipole moments. This concept is crucial to obtain very high efficiency metasurfaces and must be taken into consideration for the design. In particular, leveraging on this concept, several high transmission efficiency components have been reported recently [57–59].

Lately, it has been shown that such conditions are intimately more related to structure symmetries than Kerker found out, and the cylindrical symmetry holds a very important place in this problem [60]. Beside tuning the resonator size to address scattering phase and amplitude control, Prof. Pancharatnam and Sir M. Berry discovered that polarization changes introduce a phase delay, a "geometric phase," on the polarization converted signal.

1.2.6 PANCHARATNAM-BERRY PHASE

Geometric phase, or Pancharatnam-Berry (PB) phase, is a phase shift that is accumulated by a light wave when its polarization is adiabatically changed [61,62]. Polarization conversion is a common process performed by half- or quarter-waveplates that amounts at traveling across the Poincaré sphere. During this "travel" along the sphere, the geometric phase is accumulated, while being highly dependent on the chosen path (see Figure 1.9 from Ref. [63]). Thus, for the same starting and ending polarizations, for example from left to right circular polarization, there exist infinite number of paths allowing continuous control of the added phase. How to choose between different paths? By simply rotating nano-antennas that have polarization conversion properties. Compared to the previous phase control process, it is here a non-resonant control of the phase that is decoupled from resonant properties of the nano-antennas (usually used for polarization conversion) [64–66]. This unexpected phase tuning mechanism has been utilized, mainly by Prof. Hasman group in the early 2000 [67] to realize some of the pioneer Pancharatnam-Berry (PB) metasurfaces.

An illustrative example is given by four v-shaped plasmonic elements designed to produce a controllable phase retardation in cross polarization over the π-phase range. The unit cell is completed by four identical, yet $\pi/2$-rotated, elements such that each one of them scatters light in cross polarization with an additional π retardation. The array of 4 + 4 elements can therefore cover the entire 2π range. The additional phase delay obtained by converting the polarization with a rotated anisotropic element can be understood by considering rigorous Jones matrix analysis [68]. Let us for example consider the reflection from a generic metasurface, representing a two dimensional array of identical but anisotropic plasmonic resonators illuminated by an incident plane wave. The transmission properties of the interface are characterized by the Jones matrix, $T = \begin{pmatrix} t_{uu} & t_{uv} \\ t_{vu} & t_{vv} \end{pmatrix}$ where u and v denote the principal axes of the shape-birefringent nano-structures. Suppose that light is normally incident with a circular polarization with unit vectors defined by $\hat{e}_{\pm}(0) = \frac{\hat{u} \pm i\hat{v}}{\sqrt{2}}$, we can express the transmission matrix in the new basis formed by the Pauli matrices $\{\hat{\sigma}_1, \hat{\sigma}_2, \hat{\sigma}_3\}$ and the identity matrix $\hat{I}$ as:

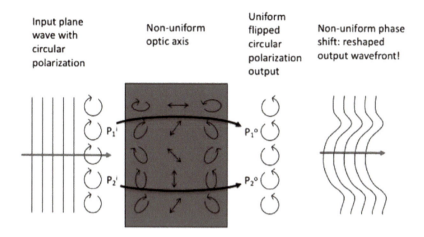

FIGURE 1.9 Working principle of Pancharatnam-Berry (PB) Optical Element. It can be made as a birefringent medium with a uniform half-wave birefringent retardation but an optic axis that is space-variant in the transverse plane. An input circularly polarized plane wave passing through the medium will be transformed into the opposite-handed circular polarization uniformly across the plate. However, the polarization transformations taking place in the medium are different at distinct points in the transverse plane and hence give rise to a space-variant transverse PB phase and a correspondingly reshaped output wavefront. For example, given any two transverse positions P_1 and P_2, the polarization evolution across the medium (black lines) is different and corresponds to two distinct meridians on the Poincaré sphere, sharing the initial and final points (i.e., the poles on the Poincaré sphere, corresponding to opposite circular polarizations). Hence, the two optical rays will acquire a relative PB phase difference given by half the solid angle subtended by these two meridians. (Reprinted with permission from Piccirillo, B. et al., Flat polarization-controlled cylindrical lens based on the Pancharatnam-Berry geometric phase, *Eur. J. Phys.*, 38, 034007, 2017. Copyright 2017, Institute of Physics.)

$$T\left(0\right)=\frac{1}{2}\left[\left(t_{uu}+t_{vv}\right)\hat{I}+\left(t_{uv}-t_{vu}\right)\hat{\sigma}_3+\left(t_{uu}-t_{vv}\right)\hat{\sigma}_1+\left(t_{uv}+t_{vu}\right)\hat{\sigma}_2\right]\qquad(1.7)$$

Now consider the case of a generic resonator rotated in the plane of the metasurface by an angle φ with respect to the reference direction $\left(\varphi=0\right)$. The transmission matrix is then given by $T\left(\varphi\right)=M^{\dagger}\left(\varphi\right)T\left(0\right)M\left(\varphi\right)$ where $M\left(\varphi\right)=e^{i\varphi\hat{\sigma}_3}$ leading to:

$$T\left(0\right)=\frac{1}{2}\left[\left(t_{uu}+t_{vv}\right)\hat{I}+\left(t_{uv}-t_{vu}\right)\hat{\sigma}_3\right]+$$
$$\frac{1}{2}\left[\left(t_{uu}-t_{vv}\right)\left(e^{-i2\varphi}\hat{\sigma}_++e^{i2\varphi}\hat{\sigma}_-\right)+\left(t_{uv}+t_{vu}\right)\left(-e^{-i2\varphi}\hat{\sigma}_++e^{i2\varphi}\hat{\sigma}_-\right)\right]\qquad(1.8)$$

after defining the spin-flip operators $\hat{\sigma}_\pm=\left(\hat{\sigma}_1\pm i\hat{\sigma}_2\right)/2$. This notation is very convenient. It helps understand the origin of the so-called PB phase retardation. Spin-flip comes with an additional phase retardation equal to $\pm2\varphi$ which is related to the non-commutation between $e^{i\varphi\hat{\sigma}_3}$ and $\hat{\sigma}_\pm$. This calculation which can also be performed for reflection, helps understanding the properties of reflect-array metasurfaces.

Various devices of interest such has beam deflectors, holograms and high numerical apertures lenses have been proposed and experimentally demonstrated using PB metasurfaces [69–71].

1.2.7 RESONANCE TUNING

Another way to control phase is to tune the parameters of resonant dielectric elements/pillars, enabling much higher transmission efficiencies than plasmonic antennas. For instance, the building blocks of a recent achromatic meta-lens are composed of solid (inset in Figure 1.10a, also referred to as nano-pillars) and inverse (inset in Figure 1.10b) GaN nano-structures. To satisfy the phase requirement of chromatic meta-lenses, a couple of resonant modes inside the pillars are employed. Although

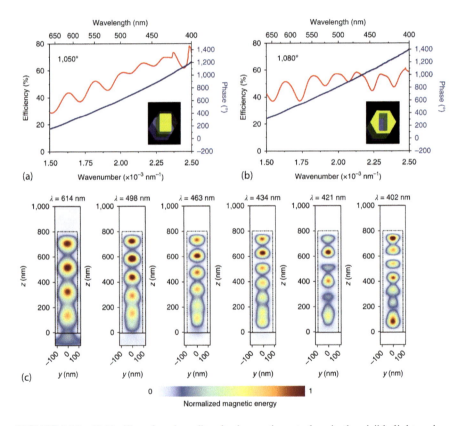

FIGURE 1.10 GaN pillars for a broadband achromatic meta-lens in the visible light region. (a, b), circularly polarized conversion efficiency (red curves) and phase profile (blue curve) for elements with phase compensation of 1,050° (a) and 1,080° (b). Insets illustrate the solid and inverse nano-structures, respectively. (c) Normalized magnetic energy for phase compensation of 1,050° (c) at different incident wavelengths. The black dashed line indicates the boundary of GaN structures. The thickness of all GaN nano-pillars is fixed at 800 nm, standing on an Al$_2$O$_3$ substrate. (Reprinted from Wang, S. et al., *Nat. Nanotechnol.*, 13, 227–232, 2018. With permission from Nature.)

in this example a part of the optical functionality of the metasurface is realized through geometric phase by rotating the elements as explained just before, the phase management offered by the parameters tuning of the nano-pillars geometry helps compensating the chromaticity of the structure [72,73]. Optical coupling between dielectric nano-pillars is weak because of their high refractive index compared with the surrounding environment. They behave as separate waveguides in which light sees an effective index depending on its wavelength and pillar's geometry. Large phase compensation—phase difference between the maximum and the minimum wavelengths within the working bandwidth—is obtained by exciting higher order modes of the waveguide-like cavity resonances, through directly increasing the height of the nano-pillars. Figure 1.10c shows the right-hand circular polarization to left-hand circular polarization conversion efficiency (red curves) and phase profile (blue curves) for phase compensation of 1,050° (Figure 1.10c). The ripples result from the excitation of multi-resonances inside the GaN nano-pillars, which can be verified by checking the near-field distribution, as shown in Figure 1.10c. The chosen wavelengths correspond to the efficiency peaks and dips, indicating that waveguide-like cavity resonances are supported in either the GaN nano-pillar (solid cases) or the surrounding GaN (inverse cases).

These two main ways of managing phase of light, either via geometric phase or resonance tuning, are used separately or together depending on the goal of the metasurface. The fundamental difference between the two methods is the fact that Pancharatnam-Berry phase is solely changed by rotating the same elements, whereas resonance tuning relies on change of geometry of every single nano-element.

1.2.8 INNOVATIVE FABRICATION PROCESSES

For these concepts to be deployed in real-world devices, important efforts in terms of nanoscale fabrication are needed. Metasurfaces is a rather young research topic that progresses hand in hand with advances in nanofabrication. Hence, there is an evident need for innovative processes that could speed up fabrication process, increase quality and decrease the production cost on large-area substrates. This section does not review all possible fabrication techniques but presents a newly developed patterning technique for semiconductor based metasurfaces.

As a matter of fact, our group is currently working on establishing new fabrication method for large-area metasurfaces in view of their utilization as active/amplifying meta-optic components [74]. To this end, we proposed an innovative etching-free fabrication technique that combines nanoimprint lithography and selective area sublimation for high throughput and high resolution, avoiding irreversible damage caused by plasma etching. An example of this method and the light emission performance of an obtained meta-lens are presented in Figure 1.11. This cost-effective large-area fabrication technique maintains the high material qualities after nano-processing and may meet the stringent requirements of industrial manufacturing processes. Gallium Nitride, a promising semiconductor, is selected for the realization, as it offers several advantages: it remains highly transparent in the entire visible range; its high enough refractive index (>2) ensures that GaN nano-structures can possess strong Mie scattering resonances; its considerably high thermal and

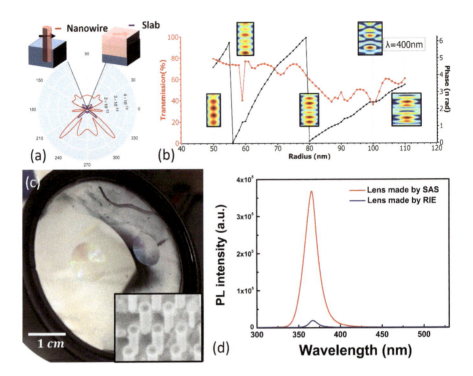

FIGURE 1.11 (a) Numerical study showing the effect of nanostructuring the GaN on the emission cross section of a nano-pillar. Total internal reflection angles, beyond which light is total internally reflected in the case of slab waveguide, are represented by the dashed lines. (b) Transmission (red) and phase (black) values of elements used for a working wavelength of 400 nm. The insets represent the near-field distribution inside the different nano-pillars, showing multi-longitudinal mode operations. (c) Macroscopic images and SEM pictures of the large scale GaN meta-lens fabricated by selective area sublimation. The lenses are fabricated on 2″ sapphire wafer. (d) Room-temperature photoluminescence measurements of light emitting metasurfaces. The strong difference is associated with the absence in selective area sublimation of irreversible damage induced by the dry-etching. (Reprinted from Brière, G. et al., Innovative etching-free approaches towards large scale light emitting metasurfaces, Submitted. With permission.)

chemical stabilities make this material suitable even for extreme application; it is a very mature material, highly compatible with the semiconductor fabrication techniques, with widespread applications in electronics and optoelectronics. With respect to dielectric material, semiconductors have a unique advantage since one can realize active devices by directly embedding amplification functionality at the nano-patterning level.

With new or improved fabrication techniques, the field of meta-optics gains degrees of freedom for structure design. Not to mention the countless efforts that are made to create new metamaterials with electrically tunable optical properties for real-time wavefront control. Understanding the physics, theoretically designing clever meta-optics devices and getting access to efficient fabrication processes can be seen as three pillars that can push further the frontier of meta-surfaces technology.

1.3 ADVANCED METASURFACES

In this paragraph, we have selected several relevant developments in the field of metasurfaces that exploit and develop the concepts discussed above to achieve pecu-liar optical functions. Though not exhaustive, this ensemble may be sufficient to give a concrete idea of what can be realized with metasurfaces in the linear regime.

1.3.1 BEAM SHAPING WITH APERTURE METASURFACES

We start by presenting a different type of metasurface: an apertured holographic one to convert circularly polarized Gaussian wavefront in radially polarized Gaussian beam, which after being designed at a given wavelength, is able to manipulate the three fundamental properties of light (phase, amplitude and polarization) over a broad spectrum [11,74]. The design strategy of these metasurfaces relies on replacing the large openings of conventional apertured holograms by arrays of subwavelength slits, oriented to locally select a particular state of polarization. The resulting optical element can therefore be viewed as the superposition of two independent structures with very different length scales, i.e., a hologram with each of its apertures filled with nanoscale openings to transmit only a desired state of polarization. The studied nano-structured holographic plate has the capability to convert circularly polarized incident light into radially polarized optical beams.

The additional phase imparted to a light beam in the process of polarization con-version from a circularly polarized beam to a radially polarized one using a radial polarizer has a spiral distribution $e^{j\theta}$ [76]. The origin of this spiral phase is the pro-jection of incident light with, for example, right-handed circular polarization onto the state of polarization of the light passing through a circular wire-grid polarizer, i.e., the radial state of polarization, such that the output beam profile can be described by:

$$E_{\text{out}} = e^{j\theta}\mathbf{e}_r. \tag{1.9}$$

The geometric phase term $e^{j\theta}$ is often compensated by additional optical elements, e.g., spiral phase plates [77] that impose a conjugate spiral phase $e^{-j\theta}$, i.e., with the wavefront helicity reversed with respect to the undesired geometric phase. A differ-ent approach to introduce such a spiral phase distribution involves diffracting the light from a binary fork hologram (Figure 1.12b), which achieves phase modulation in the diffracted orders [78–80]. The design of such a fork hologram is performed by calculating the off-axis interference pattern between an object beam (a vortex beam carrying the spiral phase in our case) and a reference beam (usually a Gaussian beam). This interferometric approach has been recently used to design plasmonic interfaces which facilitate the selective detection of light carrying orbital angular momentum [74]. The novelty of this approach relies on replacing the large openings

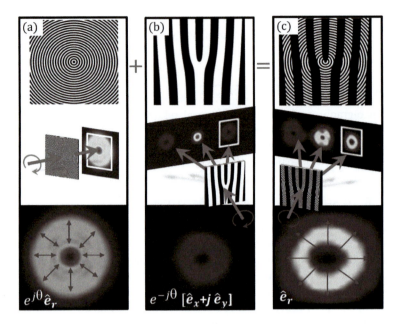

FIGURE 1.12 The technique used to generate radially polarized beams (RPBs) combines subwavelength apertures for polarization control and wavelength-scale diffracting apertures. The structure that generates RPBs (c) is the superposition of a radial polarizer (a) and a fork diffraction hologram (b) to cancel the e^{jq} phase contribution that arises from the projection of the circularly polarized optical field onto the state of radial polarization. The black and white colors in the upper row represent respectively "0" (opaque) and "1" (transparent) in the transmittance function of the device. The periods in (a) and (b) are 200 nm and 1.2 μm, respectively. For each panel, the figure in the middle presents the far-field intensity distribution after traversing the structure with the transmittance function given by the upper part of the panel. The simulated interfaces comprise patterned Au films (150 nm thick) on a glass substrate. The calculated far-field intensity distributions presented in the lower part of each panel are enlarged versions of the diffracted beams located in the white boxes. Each sample is illuminated by a right-handed circularly polarized plane wave incoming at normal incidence. The wavelength of the light in the simulation is 633 nm, but identical results have been obtained across the visible spectrum. All simulations have been performed using commercial finite-difference-time-domain (FDTD) software. (Reprinted with permission from Lin, J. et al., *Nano Lett.*, 13, 4269–4274, 2013. Copyright 2013 by the American Physical Society.)

of holograms by arrays of subwavelength apertures, oriented to locally select a particular state of polarization. To avoid spurious diffraction effects from these polarizing metallic apertures, their size is chosen to be significantly smaller than the wavelength of light. The resulting design can therefore be viewed as a superposition of two independent structures with very different length scales. The larger scale features, as presented in Figure 1.12b, are spaced by roughly $\lambda/\sin(\theta)$, where θ is the first-order diffraction angle for normal incidence; these control the spatial modulation of the phase. The smaller features can be viewed as miniature wire-grid polarizers, which locally control the state of polarization of the transmitted light. To generate a radially polarized beam, the fork hologram creates a spiral phase front, and the subwavelength apertures select the radial state of polarization. For incident light with circular polarization, the overall effect of the structure is to generate a radially polarized beam without the need for an additional PB phase compensation element, Figure 1.12c. For fabrication convenience, we are considering here only the binary version of the hologram, consisting of wavelength-scale openings in an optically opaque gold (Au) film deposited on a SiO_2 substrate. Although these holograms are designed to create a virtual image at a given angle for a specific wavelength, they actually operate over a broad wavelength range, creating virtual images at the corresponding angles. Therefore, any device designed by combining wavelength and subwavelength scale apertures will operate over a broad wavelength range as long as the apertures are non-resonant. The bandwidth comprises wavelengths that are longer than the physical size of the smaller apertures and shorter than the period of the holograms, though the resulting beams appear at different angles, depending on the wavelength, as predicted by the grating equation.

These features extend the functionality of conventional printable wavelength-scale binary holograms, which can manipulate complex scalar fields, to include control of the polarization of light. By incorporating subwavelength features, we have achieved full control of vector optical field over a broad range of wavelengths with a single flat subwavelength binary structure. Phase modulation and polarization modulation are not independent features that can be simply added on each other. As shown for the generation of radially polarized beams, the PB phase induced by the projection of the incident polarization onto the desired state of polarization plays an important role and has to be determined in advance, before designing the holograms. Through capability to simultaneously control amplitude, phase and polarization of light at several wavelengths, metasurface shaping any arbitrary type of wavefront can find widespread applications in photonics, augmented reality devices and various sort of lightweight imaging lightweight. If polarization, phase and amplitude can be relatively well addressed considering the conception methods discussed above, maintaining these performances over a large wavelength range remains challenging. Several approaches for light control over a large bandwidth that have been proposed are discussed in the following paragraph.

1.3.2 DISPERSION MANAGEMENT

Chromatic aberration compensation is a key issue as elements forming a given metasurface are highly dependent on the wavelength of operation, especially the devices relying on resonant tuning. On the one hand, the chosen materials are naturally dispersive. And on the other hand, the geometries are defined at one particular wavelength. Making broadband metasurfaces is a challenging task that has been undertaken through various methods: either by finely designing of the nano-antenna geometries, or by exploiting both resonance tuning and Pancharatnam-Berry phase with birefringent elements.

Recent works have reported metasurfaces with relatively broad reflectivity and transmission spectra [20] by designing resonators with broadband response, in order to achieve the high radiation losses necessary for high scattering efficiency.

Absorption losses give a significantly smaller contribution to the spectral response. Because of this broadband response, the phase function implemented by the meta-surface is relatively constant over a range of wavelengths. However large chromatic aberrations are induced due to the dispersion of the phase accumulated during light propagation, i.e., after the interaction with the metasurface. The idea is to manage the resonator dispersive response, meaning the wavelength-dependent phase shift imparted by the metasurface, to compensate for the dispersion of the propagation phase. To understand the origin of each contribution, let us consider the light interaction with a metasurface. The total accumulated phase after the interface splits into two contributions:

$$\varphi_{tot}(r,\lambda) = \varphi_m(r,\lambda) + \varphi_p(r,\lambda), \tag{1.10}$$

where φ_m is the phase shift imparted at point r by the metasurface and φ_p is the phase accumulated via propagation through free space. The first term is character-ized by a large variation across the resonance as discussed in the previous para-graphs. The second is given by $\varphi_p(r, \lambda) = 2\pi/\lambda l(r)$, where $l(r)$ is the physical distance between the interface at position and the desired wavefront. If one is able to maintain the condition of constructive interference at different wavelengths keep-ing φ_{tot} constant, the metasurface will function as an achromatic device. To do so, the dispersion of light has to be considered and compensated for the wavelengths of interest during the design of the metasurfaces. In recent contribution [71,72], the authors start with the phase profile of the achromatic meta-lens, and stated that it can be described by:

$$\varphi_{AL}(r,\lambda) = -\left[\frac{2\pi}{\lambda}\left(\sqrt{r^2 + f^2} - f\right)\right] + \varphi_{shift}(\lambda) \tag{1.11}$$

where $r^2 = x^2 + y^2$ is the distance between an arbitrary point and the center of the achromatic meta-lens surface (assuming the surface of the component is located in the $z = 0$ plane). λ and f are respectively the working wavelength in free space and the designed focal length. The additional phase shift exhibits an inversely lin-ear relationship with respect to λ, that is $\varphi_{shift}(\lambda) = \frac{a}{\lambda} + b$, with $a = \delta\frac{\lambda_{min}\lambda_{max}}{\lambda_{max} - \lambda_{min}}$ and $b = -\delta\frac{\lambda_{min}}{\lambda_{max} - \lambda_{min}}$. δ denotes the largest additional phase shift while λ_{min} and λ_{max} are the boundaries of the wavelength range of interest. To satisfy the phase requirement as described in equation (1.10), it splits into two parts:

$$\varphi_{AL}(r,\lambda) = \varphi_L(r,\lambda_{max}) + \Delta\varphi'(r,\lambda) \tag{1.12}$$

where $\varphi_L(r,\lambda_{max}) = -\left[\frac{2\pi}{\lambda_{max}}\left(\sqrt{r^2 + f^2} - f\right)\right]$ and $\Delta\varphi'(r,\lambda) = -\left[2\pi\left(\sqrt{r^2 + f^2} - f\right)\right]$ $\left(\frac{1}{\lambda} - \frac{1}{\lambda_{max}}\right) + \varphi_{shift}(\lambda)$.

As already mentioned, the first term in equation (1.11) is dispersionless and the second term is wavelength-dependent. The first term is obtained using the geometric phase through rotation of the elements, while the second term can be realized through resonance tuning of GaN pillars. The phase difference between the maximum and the minimum wavelengths within the working bandwidth is then compensated by the integrated resonances. In more physical terms, dispersionless operation over a certain bandwidth $\Delta\omega$ means that the device should be able to focus a transform limited pulse with bandwidth $\Delta\omega$ and carrier frequency ω_0 to a single spot located at focal length f. Part of the pulse hitting the lens at a distance r from its center needs to experience a pulse delay (i.e., group delay $\partial\varphi(\omega_0)/\partial\omega$) smaller by $\left(\sqrt{r^2+f^2}-f\right)/c$ than part of the pulse hitting the lens at its center. This ensures that parts of the pulse hitting the lens at different locations arrive at the focus at the same time. Also, the carrier delay (i.e., phase delay $\varphi(\omega_0)/\omega_0$) should also be adjusted so that all parts of the pulse interfere constructively at the focus. The zero dispersion case discussed above corresponds to a case where the phase and group delays are equal: $\varphi(\omega_0)/\omega_0 = \partial\varphi(\omega_0)/\partial\omega$ [72]. The elements forming the lens then take the form of subwavelength one-sided resonators, where the group delay is related to the quality factor Q of the resonator, and the phase delay depends on the resonance frequency. Figure 1.13a shows an optical microscopic image of an achromatic meta-lens sample with an NA of 0.106. Owing to the introduction of φ_{shift}, the inverse GaN structure

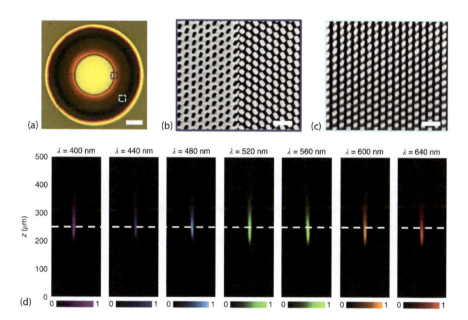

FIGURE 1.13 Experimental verification of achromatic meta-lenses. (a) Optical image of the fabricated element with NA = 0.106. The dashed squares indicate the position of the SEM images shown in (b) and (c). Scale bar: 10 μm. (b and c), Zoomed-in SEM images at the boundary of nano-pillars and Babinet structures (top view, b) and the region of nano-pillars (tilted view, c). Scale bars: 500 nm. (d) Experimental light intensity profiles for the achromatic meta-lens with NA = 0.106 at various incident wavelengths. The white dashed line indicates the position of the focal plane. (Reprinted from Wang, S. et al., *Nat. Nanotechnol.*, 13, 227–232, 2018. With permission from Nature.)

(which is able to offer larger phase compensation) plays the dominant role in the central part of the achromatic meta-lens. Figure 1.13b and c shows the scanning electron microscope images of the fabricated sample. Both the solid and inverse GaN pillars resulting from several hard mask transfer and etching processes are observed.

The advantage of this method is the total absence of coupling between the phase term that produce the optical function and the other term that correct the wavelength dependence of the first one. In principle, any chromatic aberration can be turned to zero in a reasonable wavelength range in circular polarization. Apart from chromaticity and when used at angles for which they have not been designed, flat elements suffer from strong geometrical aberrations such as spherical one, coma and other higher order ones. There are however substantial differences compared to their bulky counterparts as shown in the following section.

1.3.3 Aberrations of Metasurfaces

Here is presented a brief summary of works describing and computing the various aberrations inherited from the planar shape of a metasurface [81]. Optical aberrations arise in optical systems whenever the rays emerging from a point object do not meet all at the same image point. There exist many ways to describe aberrations [22]; the wave aberration function (WAF) is one of the most commonly used. The main advantage of the WAF is that it can be directly accessed via interferometric measurements. Rather than tracing the rays that form an image, WAF represent the difference between the wavefront and an ideal, aberration-free wavefront, i.e., a reference sphere centered in the object point. While for an ideal lens WAF $= 0$, the Maréchal criterion [82], stipulates that when $\langle \mathrm{WAF} \rangle_{\mathrm{rms}} = \langle \mathrm{WAF} \rangle^2 - \langle \mathrm{WAF}^2 \rangle$ is less than $\lambda/14$ where the brackets represent the mean value and λ is the wavelength, the dominant factor limiting the imaging quality is diffraction and therefore, for most applications, aberrations are negligible. The presence of aberrations can be easily visualized using the point spread function (PSF), defined as the image of a point object. If the focusing wavefront differs from the reference sphere, the PSF will deviate from the ideal Airy disk. The Strehl ratio, defined as the ratio between the peak of the PSF and the peak of the PSF of an aberrations-free lens, is used to quantify this deviation.

In a real implementation of a flat lens it is convenient to use a limited set of elements that cover the phase range 0-to-2π by replacing the target phase distribution with an approximated step-function whose accuracy depends on the number of elements (i.e., phase levels) used. For example, in [83], the hyperboloidal distribution was approximated using a set of 8 elements with incremental phase of $\pi/4$. This approximation introduces slight variations to the wavefront compared to an ideal converging wave (Figure 1.14b). In order to quantify the effect of this finite phase resolution, one can calculate the WAF$_{\mathrm{rms}}$ for a flat lens with a hyperboloidal phase distribution for an increasing number of phase levels (Figure 1.14c). The results show that four levels of phase (0, $\pi/2$, π, $3\pi/2$) are sufficient to satisfy the Maréchal criterion. For an increasing number of phase levels the WAF rapidly approaches the aberration-free limit of continuous phase. It is also possible to calculate the Strehl ratio that shows that for a flat lens with more than four discrete elements, the deviation from the Airy

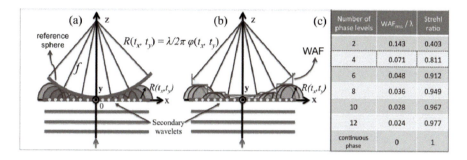

FIGURE 1.14 (a) The wavefront scattered by a flat lens based on a metasurface is given by the envelope of the secondary spherical waves emitted by the antennas with subwavelength separation. For a very dense distribution of antennas, the phase function $\varphi(x, y)$ can be assumed continuous leading to a perfect spherical wavefront. (b) If a flat lens is designed using a limited set of phase elements, the continuous phase function is replaced with a discrete distribution that introduces aberrations (the difference between the spherical wavefront and the discretized wavefront corresponds to the wave aberration function (WAF)). (c) The effect of the discretization of the phase function is evaluated by calculating the root mean square of the wave aberration function (WAF$_{rms}$) and the Strehl ratio for an increasing number of phase levels. (From Aieta, F. et al., *Opt. Express*, 21, 31530–31539, 2013. With permission of Optical Society of America.)

disk becomes negligible (Strehl ratio > 0.8) (Figure 1.14c). Similar considerations regarding the trade-off between the approximation of the phase distribution and diffraction efficiency are known in the context of diffractive Fresnel lenses (i.e., Fresnel lenses with a finite number of thickness levels instead of a continuous profile), where a higher number of phase levels can be obtained at the cost of increasing the number of consecutive fabrication steps. Based on the application requirements, a flat metasurface-based lens can be made with a large number of phase levels in a single fabrication step, i.e., using a single mask.

An aplanatic lens is a lens corrected for both spherical aberrations and coma; this type of lens is widely used in microscope objectives and condenser lenses [23]. Following the approach developed by Murty [84] for Fresnel zone plates, it is possible to design an aplanatic lens based on a metasurface. The Abbe sine condition establishes that an optical system corrected for spherical aberrations will also be free from coma if the ratio between the sine of the angle traced by a ray as it leaves the object and the sine of the angle traced by the same ray as it reaches the image plane is constant for all the rays. It has been shown that it is possible to design an aplanatic lens by patterning the metasurface on a spherical interface. From the generalized law of refraction for spherical interfaces, the phase gradient required for an aplanatic meta-lens reads:

$$\frac{d\varphi}{d\theta} = -n\frac{2\pi}{\lambda}\sin(\theta) \tag{1.13}$$

Performances are compared by tracing the rays through a conventional plano-convex lens, a flat lens, and an aplanatic metasurface with radius $\rho = 1$ mm and NA = 0.5, under the same illumination condition used above ($\alpha = 10°$). For the conventional

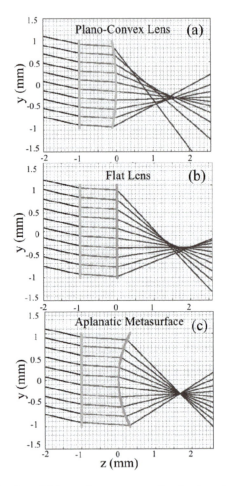

FIGURE 1.15 Ray tracing plot for a plano-convex refractive lens (a), flat lens (b), and an aplanatic metasurface (c). The dieletric supporting both interfaces has a refractive index of (n=3.5). The metasurfaces are positionned on the right side in panel b and c. The parallel illumination forms an angle $\alpha = 10°$ with respect to the optical axis, NA = 0.5. (From Aieta, F. et al., *Opt. Express*, 21, 31530–31539, 2013. With permission of Optical Society of America.)

spherical lens (Figure 1.15a) we assume a refractive index of $n = 3.5$ and a radius of curvature (R_c) as obtained from the Lensmaker equation ($R_c = 4.35$ mm). For the flat lens (Figure 1.15b) and the aplanatic metasurface (Figure 1.15c) the substrate also has a refractive index $n = 3.5$ and the rays are refracted at the first interface. Then, at the metasurfaces (green lines), the angles of refraction are calculated from the generalized laws of refraction, using the distribution of phase discontinuities given by Eq. (1.1) of [83] and Eq. (1.13), respectively for flat lens and aplanatic metasurface. Progressive improvement is obtained for the three designs: the plano-convex lens is affected by both spherical aberrations and coma which produces a degraded focal area for oblique illumination; the flat lens corrects spherical aberrations, but the coma is still present and gives rise to the typical comet-like asymmetrical focal spot; finally the aplanatic meta-surface can compensate for coma as well, producing a good focus even for off-axis illumination. Although it is not simple to fabricate a component on such curved interfaces, it is still possible to use the refracting properties of two compensating metasurfaces, as proposed in [85]. These results promise the realization of compact and extremely light-weight components with relatively good imaging performances, expecting imminent applications of metadevices in various imaging and optical characterization systems.

1.3.4 META-DOUBLET FOR MONOCHROMATIC ABERRATIONS CORRECTION

As explained in the previous section, a metasurface lens can be corrected for coma if patterned on the surface of a sphere, but direct patterning of nano-structures on curved surfaces is challenging. Although conformal metasurfaces discussed in the following paragraph might provide a solution, the resulting device would not be flat. Another approach for correcting monochromatic aberrations of a metasurface lens is through cascading and forming a meta-doublet lens [85,86]. Such a system can be corrected over a wide range of incident angles.

In Ref. [85], the proposed doublet lens is composed of two metasurfaces behaving as polarization insensitive phase plates that are patterned on two sides of a single transparent substrate. In the optimum design, the first metasurface operates as a corrector plate and the second one performs the significant portion of focusing; thus, it is referred to them as correcting and focusing metasurfaces, respectively. The elements are designed for the operation wavelength of 850 nm, and are implemented using the dielectric nano-post platform shown in Figure 1.16a and b. The metasurfaces

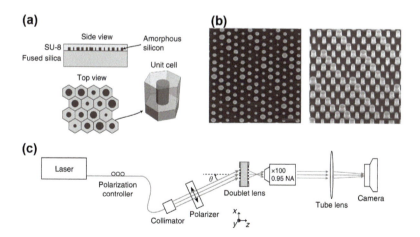

FIGURE 1.16 (a) A schematic illustration of the dielectric metasurface used to implement the meta-doublet lens. It is composed of an array of amorphous silicon nano-posts covered with a layer of SU-8 polymer and arranged in a hexagonal lattice. (b) Scanning electron micrographs showing a top and an oblique view of the amorphous silicon nano-posts composing the metasurfaces. Scale bars, 1 μm. (c) Schematic drawing of the measurement setup.

(Continued)

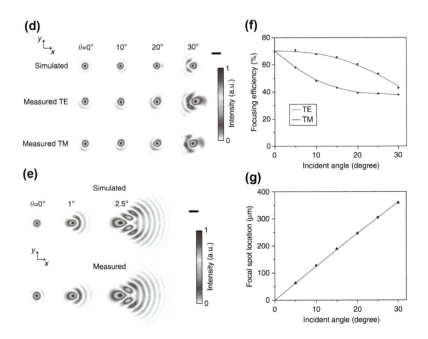

FIGURE 1.16 (Continued) (d) Simulated and measured focal plane intensity profiles of the meta-doublet lens for different incident angles (θ). Simulation results are shown in the top row, and the measurement results for the transverse electric (TE) and the transverse magnetic (TM) polarizations are shown in the second and third rows, respectively. Simulation results are obtained using scalar approximation (that is, ignoring polarization dependence). Scale bar, 2 µm. (e) Simulated and measured focal plane intensity profiles for a metasurface singlet with the same aperture diameter and focal length as the meta-doublet. For the range of angles shown, the measured intensity distributions are polarization insensitive. Scale bar, 2 µm. (f) Measured focusing efficiency of the met-doublet for TE- and TM-polarized incident light as a function of incident angle. The measured data points are shown by the symbols and the solid lines are eye guides. (g) Transverse location of the focal spot for the doublet lens as a function of incident angle. The measured data points are shown by the symbols, and the solid line shows the $f\sin(\theta)$ curve, where $f = 717$ mm is the focal length of the meta-doublet lens. (From Arbabi, A. et al., *Nat. Commun.*, 7, 13682, 2016. With permission from Nature.)

are composed of hexagonal arrays of amorphous silicon nano-posts with different diameters that rest on a fused silica substrate and are covered by the SU-8 polymer. The fabricated doublet is characterized by illuminating it with an 850 nm laser beam at different incident angles (as shown in Figure 1.16c), and by measuring its focal spot and focusing efficiency. For comparison, a spherical-aberration-free singlet metasurface lens with the same aperture diameter and focal length as the doublet lens (phase profile $\varphi(\rho) = -(2\pi/\lambda)\sqrt{f^2 + \rho^2}$, ρ: radial coordinate, $f = 717$ µm: focal length, $D = 800$ µm: aperture diameter) was also fabricated and characterized. The focal spots of the metasurface doublet and singlet lenses were measured with two different polarizations of incident light and are shown along with the corresponding simulation results in Figure 1.16d and e, respectively. The doublet lens has a nearly diffraction limited focal spot for incident angles up to more than $25°$ (with the criterion of Strehl ratio larger than 0.9) while the singlet exhibits significant aberrations even at incident angles of a few degrees. As Figure 1.16d and e shows, simulated and measured spot shapes agree well. In addition, the focusing efficiency (ratio of the focused power to the incident power) for the metasurface doublet lens is shown in Figure 1.16f, and is ~70% for normally incident light.

Based on this work, we can envision realizing any metasurface to correct geometric and chromatic aberrations. The technological progresses are making the design of more complex metasurfaces systems able to both fulfill an arbitrary optical function and to correct all aberrations that are existing. Another step along this direction would be to implement them on curved surfaces to meet the conditions mentioned in the previous section.

1.3.5 Manipulating States of Light

Light waves have different characteristics: amplitude; phase; wavelength; wavefront, and also polarization that is able to carry spin and orbital momentum. All of them can therefore be used to transform waves into data carriers encoding a great deal of information. To this end, two classes of metasurfaces exist: the polarimetric ones that can fully analyze states of light, and the multiplexer ones that are able to make different waves carrying different information travel along the same channel.

Polarimetric metasurfaces mostly exploit asymmetric photonic structures, based on PB phase or gap plasmons, that send light into different directions with respect to its momentums, Stokes parameters, wavelength or even impinging modal wavefront [87–94]. Then, a proper analysis of the different output directions and amplitudes give direct access to the encoded data. Multiplexer metasurfaces play a different role as they can gather different incoming states of light and transform them into orthogonal propagating modes without loss of information [38,95–100]. This way, the waves reach their destination without mutual interaction and information decay. Same types of metasurfaces are eventually used to de-multiplex the wavefronts and get back the initial states of light. Coupling these ideas with fiber optics communication and photonic circuitry could lead to new scenario of data sharing and processing only relying on passive optical elements.

1.3.6 Conformal Boundary Optics

The realization of metasurfaces on flexible and conformable substrates open new design perspectives for the realization of unexpected holographic images from various type of surfaces and geometries, as presented in Figure 1.17 [101]. For user-specified incident, transmitted, and reflected fields and a given surface geometry represented by the conformal coordinate system (u, v, n), one can determine the required surface optical response to project a desired holographic image [102]. Various flexible metasurface components have been reported and in particular Burch et al. have demonstrated that the shape of the interface can determine the properties of the far-field image, including its state of polarization [103–106]. The focusing

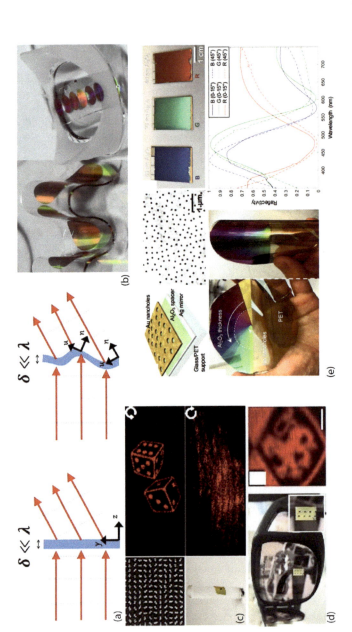

FIGURE 1.17 (a) The realization of planar (left) or conformal (right) metasurfaces requires adapted modeling tools that can account for the effect of the physical distortion. Conformal boundary optics, a theoretical description of conformal boundary conditions, helps investigating reflection and refraction arising at the surface of arbitrarily shaped objects. (Reprinted with permission from Teo, J.Y.H. et al., *Phys. Rev. A*, 94, 023820. Copyright 2016 by the American Physical Society.) (b) Infrared meta-lenses made on flexible substrates. (Reprinted from Kamali, S.M. et al., *Nat. Commun.*, 7, 11618, 2016. With permission from Nature.) (c) Holographic metasurfaces designed for curved substrates can have different response according to both polarization and surface topology. (Reprinted with permission from Burch, J. and Di Falco, A., *ACS Photon.*, 5, 1762–1766, 2018. Copyright 2018 American Chemical Society.) (d) Multiplexed holographic metasurfaces have been conformed to a pair of safety glasses. The inset is a close up of the metasurface and on the right the experimentally obtained image. (From Burch, J. et al., *Sci. Rep.*, 7, 4520, 2017. With permission from Nature.) (e) Nanoholes gold plasmonic metasurfaces made on flexible substrate to control the reflection of RGB visible light. (From Xiong, K. et al., *Adv. Mater.*, 28, 9956–9960, 2016.)

properties of meta-lenses is also affected by the geometry. The possibility of addressing light on arbitrary geometry will have important impact for the design of augmented reality displays. Plasmonic interfaces can also be combined with polymer thin films to realize new flexible and colorful electronic papers, providing fast response time and ultralow power consumption. Understanding the full potential of free-form metasurfaces appears as an elusive task without efficient numerical simulations tools. To this end, a newly developed theoretical framework starts from Generalized Sheet Transmission Conditions (GSTC) to determine the characteristics of an arbitrary conformal metasurface as a function of input and output far-fields [102]. The constructive proof asserts that any combination of input and output wavefronts can be realized thanks to a free-form metasurface with analytically given electric and magnetic susceptibilities (see Figure 1.17a). Shaped components can address the electromagnetic fields along a conformal interface, making it a conformal boundary optical element. A Finite-Difference-Time-Domain (FDTD) algorithm based on this analysis has been described in a recent paper [107] solving the behavior of a free-form metasurface with chosen susceptibilities.

The analytical derivation of the susceptibilities obtained from the conformal boundary optics theory, and the FDTD algorithm that determines the output wavefront created by a free-form metasurface with arbitrary susceptibilities now answer the need for complete simulation tools. They provide various ways to design realistic metasurfaces with chosen optical responses. Once implemented together with an optimization algorithm, it might enable fast and efficient design of compact free-form meta-optics fulfilling arbitrary optical functionalities.

1.4   CONCLUSION

Throughout the last few paragraphs, we have discussed and briefly reviewed some of the important historical steps achieved in the field of linear phased array metasurfaces. This technology has developed at a frantic pace, going from inefficient chromatic devices working at long wavelength – typically at mid-IR and THz frequencies – to high numerical aperture components with subwavelength resolution recording high transmissivity in the visible. Achromatic devices and specific wavelength-dependent optical response can now be designed on demand on a single subwavelength-thick interface. Leveraging on various scattering mechanisms such as Berry-Pancharatnam phase, chirality, Kerker diffusion and zero order diffraction in a subwavelength grating, the concept of planar optics is currently transforming and modernizing photonics, establishing new functional devices for controlling, modulating and steering light at optical frequencies. Today, it is possible to dream about fully addressing the reflection/transmission of light at/across arbitrarily shaped interfaces to realize unexpected effects such as ultrathin cloaking, illusion optics and arbitrarily shaped optical resonators [101,108,109].

Beyond this introductory chapter on linear meta-optics, a great deal of exciting new developments, notably in nonlinear and quantum optics, has to be expected. In order to tease the reader, Figure 1.18 shows perspectives on second-harmonic generation with metasurfaces. The first example of Figure 1.17a relies on the expression

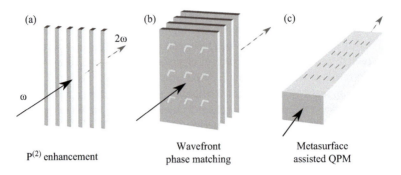

FIGURE 1.18 Scenarii for Second-Harmonic Generation enhancement with metasurfaces: (a) A subwavelength grating of nonlinear nano-rods sustains guided mode and symmetry breaking leads to Bound state In the Continuum, able to concentrate electric field with arbitrary efficiency. This greatly enhances the quantity of nonlinear sources and leads to higher SHG intensity. (b) Nonlinear metasurfaces allow creation of a chosen wavefront at higher harmonics thanks to PB elements. Stacking them in a way that they are phase-matched could lead to linear or even resonant addition of the wavefronts generated by each of the metasurfaces. (c) Quasi-Phase Matching is a key issue in nonlinear waveguides. This can be performed through adding metasurfaces either on top of the nonlinear slab, or directly inside to periodically put back the waves in phase.

of the nonlinear sources term $P^{(2)} = \chi^{(2)} E^2$, showing that an increase of the electric field inside a nonlinear medium greatly enhances nonlinear processes. Such a grating of dielectric nano-rods can behave as an assembly of dipoles able to coherently extinguish incoming light leading to resonant zero reflection if spaced with a subwavelength period. This comes together with electric field enhancement within the nano-rods, which can even become arbitrarily high through a symmetry breaking of the geometry inducing a Bound state In the Continuum [110]. The second idea shown in Figure 1.17b consists in exploiting the capacity of non-centrosymmetric gold nano-antennas to produce higher harmonic holograms thanks to a nonlinear version of PB phase [111]. Starting from one layer, it might be possible to stack many of them to phase match any arbitrarily chosen emitted wavefronts and thus to enhance the intensity of the overall output image. A third application of metasurfaces directly influences Quasi-Phase Matching in nonlinear waveguides. Patterning the top of a LiNbO$_3$ waveguide leads to phase matching free SHG because nonlinear light is directly generated in higher order guided modes that do not interact back with pump light [112]. Another application consists cascading metasurface to achieve interesting phase-matching conditions, including backward second-harmonic generation [113].

The following chapters will bring the readers to new horizons in this field. Leading research group and professors have contributed to this book to offer opportunities for the readers to further explore the frontiers of metasurface technology, helping them to forge new intuitions on this fascinating and emerging field of research in nonlinear photonics.

ACKNOWLEDGMENTS

We acknowledge funding from the European Research Council (ERC) under the European Union's Horizon 2020 research and innovation program (Grant agreement no. 639109).

REFERENCES

1. Pendry, J. B., D. Schurig and D. R. Smith. "Controlling electromagnetic fields." *Science* 312, 5781 (2006): 1780–1782.
2. Yu, N., P. Genevet, M. A. Kats et al. "Light propagation with phase discontinuities: Generalized laws of reflection and refraction." *Science* 334, 6054 (2011): 333–337.
3. Kildishev, A. V., A. Boltasseva, and V. M. Shalaev. "Planar photonics with metasurfaces." *Science* 339, 1232009 (2013). doi:10.1126/science.1232009.
4. Li, Y., B. Liang, Z. Gu, X. Zou, and J. Cheng. "Reflected wavefront manipulation based on ultrathin planar acoustic metasurfaces." *Scientific Reports* 3, 2546 (2013). doi:10.1038/srep02546.
5. Mohammadi Estakhri, N. and A. Alù. "Ultrathin unidirectional carpet cloak and wave-front reconstruction with graded metasurfaces." *IEEE Antennas and Wireless Propagation Letters* 13 (2014): 1775–1778.
6. Sounas, D. L., R. Fleury and A. Alù. "Unidirectional cloaking based on metasurfaces with balanced loss and gain." *Physical Review Applied* 4, 1 (2015): 014005.
7. Selvanayagam, M. and G. V. Eleftheriades. "Experimental demonstration of active electromagnetic cloaking." *Physical Review X* 3, 4 (2013): 041011.
8. Khoo, E. H., E. P. Li and B. Crozier. "Plasmonic wave plate based on subwavelength nanoslits." *Optics Letters* 36, 13 (2011): 2498–2500.
9. Zhao, Y. and A. Alù. "Manipulating light polarization with ultrathin plasmonic metasurfaces." *Physical Review B* 84, 20 (2011): 205428.
10. Aieta, F., P. Genevet, N. Yu, M. A. Kats, Z. Gaburro and F. Capasso. "Out-of-plane reflection and refraction of light by anisotropic optical antenna metasurfaces with phase discontinuities." *Nano Letters* 12, 3 (2012): 1702–1706.
11. Genevet, P., N. Yu, F. Aieta, J. Lin, M. A. Kats, R. Blanchard, M. O. Scully, Z. Gaburro and F. Capasso. "Ultra-thin plasmonic optical vortex plate based on phase discontinuities." *Applied Physics Letters* 100, 1 (2012): 013101.
12. Karimi, E., S. A. Schulz, I. De Leon, H. Qassim, J. Upham and R. W. Boyd. "Generating optical orbital angular momentum at visible wavelengths using a plasmonic metasurface." *Light: Science & Applications* 3, e167 (2014). doi:10.1038/lsa.2014.48.
13. Yang, Y., W. Wang, P. Moitra, I. I. Kravchenko, D. P. Briggs and J. Valentine. "Dielectric meta-reflect array for broadband linear polarization conversion and optical vortex generation." *Nano Letters* 14, 3 (2014): 1394–1399.
14. Pors, A., M. G. Nielsen, R. L. Eriksen and S. I. Bozhevolnyi. "Broadband focusing flat mirrors based on plasmonic gradient metasurfaces." *Nano Letters* 13, 2 (2013): 829–834.
15. Yu, N., P. Genevet, F. Aieta, M. A. Kats, R. Blanchard, G. Aoust, J.-P. Tetienne, Z. Gaburro and F. Capasso. "Flat optics: Controlling wavefronts with optical antenna metasurfaces." *IEEE Journal of Selected Topics in Quantum Electronics* 19, 3 (2013): 4700423.
16. Liu, H., M. Q. Mehmood, K. Huang, et al. "Twisted focusing of optical vortices with broadband flat spiral zone plates." *Advanced Optical Materials* 2, 12 (2014): 1193–1198.
17. Kats, M. A., P. Genevet, G. Aoust, N. Yu, R. Blanchard, F. Aieta, Z. Gaburro and F. Capasso. "Giant birefringence in optical antenna arrays with widely tailorable optical anisotropy." *Proceedings of the National Academy of Science* 109, 31 (2012): 12364–12368.

18. Glybovski, S. B., S. A. Tretyakov, P. A. Belov, Y. S. Kivshar and C. R. Simovski. "Metasurfaces: From microwaves to visible." *Physics Reports* 634, 24 (2016): 1–72.

19. Zhu, A. Y., A. I. Kuznetsov, B. Luk'yanchuk, N. Engheta and P. Genevet. "Traditional and emerging materials for optical metasurfaces." *Nanophotonics* 6, 2 (2016). doi:10.1515/nanoph-2016-0032.

20. Yu, N. and F. Capasso. "Flat optics with designer metasurfaces." *Nature Materials* 13, 2 (2014): 139–150.

21. Genevet, P., F. Capasso, F. Aieta, M. Khorasaninejad and R. Devlin. "Recent advances in planar optics: From plasmonic to dielectric metasurfaces." *Optica* 4, 1 (2017): 139–152.

22. Jackson, D. *Classical Electrodynamics* (3rd ed.), New York: Wiley, 1998.

23. Born, W. and E. Wolf. *Principles of Optics* (7th ed.), Cambridge, UK: Cambridge University Press, 1999.

24. Hecht, E. *Optics* (4th ed.), Reading, MA: Wesley Publishing Company, 2001.

25. Kock, W. E. "Metallic delay lenses." *The Bell System Technical Journal* 27, 1 (1948): 58–82.

26. Stork, W. N. Streibl, H. Haidner and P. Kipfer. "Artificial distributed-index media fabricated by zero-order gratings." *Optics Letters* 16, 24 (1991): 1921–1923.

27. Farn, M. W. "Binary gratings with increased efficiency." *Applied Optics* 31, 22 (1992): 4453–4458.

28. Chang-Hasnain, C. J. and W. Yang. "High-contrast gratings for integrated optoelectronics." *Advances in Optics and Photonics* 4, 3 (2012): 379–440.

29. Arbabi, A., Y. Horie, A. J. Ball, M. Bagheri and A. Faraon. "Subwavelength-thick lenses with high numerical apertures and large efficiency based on high contrast transmitarrays." *Nature Communications* 6 (2015): 1069.

30. Moharam, M. G. and T. K. Gaylord. "Rigorous coupled-wave analysis of planar-grating diffraction." *Journal of the Optical Society of America* 71, 7 (1981): 811–818.

31. Lalanne, P., J.-P. Hugonin and P. Chavel. "Optical properties of deep lamellar gratings: A coupled Bloch-mode insight." *Journal of Lightwave Technology* 24, 6 (2006): 2442–2449.

32. Collin, S. "Nanostructure arrays in free-space: Optical properties and applications." *Reports on Progress in Physics* 77, 12 (2014): 126402.

33. Mateus, C. F. R., M. C. Y. Huang, L. Chen, C. J. Chang-Hasnain and Y. Suzuki. "Broad-band mirror (1.12–1.62 μm) using a subwavelength grating." *IEEE Photonics Technology Letters* 16, 7 (2004): 1676–1678.

34. Fattal, D., J. Li, Z. Peng, M. Fiorentino and R. G. Beausoleil. "Flat dielectric grating reflectors with focusing abilities." *Nature Photonics* 4 (2010): 466–470.

35. Huang, M. C. Y., Y. Zhou and C. J. Chang-Hasnain. "A surface-emitting laser incorporating a high-index-contrast subwavelength grating." *Nature Photonics* 1 (2007): 119–122.

36. Li, J., D. Fattal, M. Fiorentino and R. G. Beausoleil. "Strong optical confinement between nonperiodic flat dielectric gratings." *Physical Review Letters* 106, 19 (2011): 193901.

37. Levy, U., H.-C. Kim, C.-H. Tsai and Y. Fainman. "Near-infrared demonstration of computer-generated holograms implements by using subwavelength gratings with space-variant orientation." *Optics Letters* 30, 16 (2005): 2089–2091.

38. Wen, D., F. Yue, G. Li, et al. "Helicity multiplexed broadband metasurface holograms." *Nature Communications* 6 (2015): 8241.

39. Chen, W. T., K.-Y. Yang, C.-M. Wang, et al. "High-efficiency broadband meta-hologram with polarization-controlled dual images." *Nano Letters* 14, 1 (2014): 225–230.

40. Lalanne, P., S. Astilean, P. Chavel, E. Cambril, and H. Launois. "Blazed binary subwavelength gratings with efficiencies larger than those of conventional échelette gratings." *Optics Letters* 23, 14 (1998): 1081–1083.

41. Lalanne, P. "Waveguiding in blazed-binary diffractive elements." *Journal of the Optical Society of America A* 16, 10 (1999): 2517–2520.

42. Lalanne, P., S. Astilean, P. Chavel, E. Cambril, and H. Launois. "Design and fabrication of blazed binary diffractive elements with sampling periods smaller than the structural cutoff." *Journal of the Optical Society of America A* 16, 5 (1999): 1143–1156.

43. Kats, M. A., N. Yu, P. Genevet, Z. Gaburro and F. Capasso. "Effect of radiation damping on the spectral response of plasmonic components." *Optics Express* 19, 22 (2011): 21748–21753.

44. King, R. W. P. *The Theory of Linear Antennas*, Cambridge, MA: Harvard University Press, 1956.

45. Kuester, E. F., M. A. Mohamed, M. Piket-May, and C. L. Holloway. "Averaged transition conditions for electromagnetic fields at a metafilm." *IEEE Transactions on Antennas and Propagation* 51, 10 (2003): 2641–2651.

46. Kuester, E. F., C. L. Holloway, and M. A. Mohamed. "A generalized sheet transition condition model for a metafilm partially embedded in an interface." *2010 IEEE Antennas and Propagation Society International Symposium* (2010): 11514691. doi:10.1109/APS.2010.5562250.

47. Idemen, M. "Universal boundary relations of the electromagnetic field." *Journal of the Physical Society of Japan* 59, 1 (1990): 71–80.

48. Achouri, K., M. A. Salem, and C. Caloz. "General metasurface synthesis based on susceptibility tensors." *IEEE Transactions on Antennas and Propagation* 63, 7 (2015): 2977–2991.

49. Larouche, S. and D. R. Smith. "Reconciliation of generalized refraction with diffraction theory." *Optics Letters* 37, 12 (2012): 2391–2393.

50. Mie, G., "Beiträge zur Optik trüber Medien, speziell kolloidaler Metallösungen." *Annalen der Physik* 330, 3 (1908): 377–445.

51. Kerker, M., D. S. Wang and C. L. Giles. "Electromagnetic scattering by magnetic spheres." *Journal of the Optical Society of America A* 73, 6 (1983): 765–767.

52. Kerker, M. "Invisible bodies." *Journal of the Optical Society of America A* 65, 4 (1975): 376–379.

53. Nieto-Vesperinas, M., R. Gomez-Medina and J. J. Saenz. "Angle-suppressed scattering and optical forces on submicrometer dielectric particles." *Journal of the Optical Society of America A* 28, 1 (2011): 54–60.

54. Person, S., M. Jain, Z. Lapin, J. J. Saenz, G. Wicks and L. Novotny. "Demonstration of zero optical backscattering from single nanoparticles." *Nano Letters* 13, 4 (2013): 1806–1809.

55. Rolly, B., B. Stout and N. Bonod. "Boosting the directivity of optical antennas with magnetic and electric dipolar resonant particles." *Optics Express* 20, 18 (2012): 20376–20386.

56. Alaee, R., R. Filter, D. Lehr, F. Lederer and C. Rockstuhl. "A generalized Kerker condition for highly directive nanoantennas." *Optics Letters* 40, 11 (2015): 2645–2648.

57. Yu, Y. F., A. Y. Zhu, R. Paniagua-Domìnguez, Y. H. Fu, B. Luk'yanchuk and A. I. Kuznetsov. "High-transmission dielectric metasurface with 2π phase control at visible wavelengths." *Laser & Photonics Review* 9, 4 (2015): 412–418.

58. Liu, S., A. Vaskin, S. Campione, O. Wolf, M. B. Sinclair, J. Reno, G. A. Keeler, I. Staude and I. Brener. "Huygens' metasurfaces enabled by magnetic dipole resonance tuning in split dielectric nanoresonators." *Nano Letters* 17, 7 (2017): 4297–4303.

59. Yang, Y., A. E. Miroshnichenko, S. V. Kostinski, M. Odit, P. Kapitanova, M. Qiu and Y. S. Kivshar. "Multimode directionality in all-dielectric metasurfaces." *Physical Review B* 95, 16 (2017): 165426.

60. Zambrana-Puyalto, X., I. Fernandez-Corbaton, M. L. Juan, X. Vidal and G. Molina-Terriza. "Duality symmetry and Kerker conditions." *Optics Letters* 38, 11 (2013): 1857–1859.

61. Berry, M. V. "The adiabatic phase and Pancharatnam's phase for polarized light." *Journal of Modern Optics* 34, 11 (1987): 1401–1407.
62. Pancharatnam, S., "Generalized theory of interference, and its applications." *Proceedings of the Indian Academy of Sciences – Section A* 44, 5 (1956): 247–262.
63. Piccirillo, B., M. F. Picardi, L. Marrucci and E. Santamato. "Flat polarization-controlled cylindrical lens based on the Pancharatnam-Berry geometric phase." *European Journal of Physics* 38, 3 (2017): 034007.
64. Niv, A., G. Biener, V. Kleiner and E. Hasman. "Propagation-invariant vectorial Bessel beams obtained by use of quantized Pancharatnam-Berry phase optical elements." *Optics Letters* 29, 3 (2004): 238–240.
65. Biener, G., A. Niv, V. Kleiner and E. Hasman. "Formation of helical beams by use of Pancharatnam-Berry phase optical elements." *Optics Letters* 27, 21 (2002): 1875–1877.
66. Bomzon, Z., G. Biener, V. Kleiner and E. Hasman. "Space-variant Pancharatnam-Berry phase optical elements with computer-generated subwavelength gratings." *Optics Letters* 27, 13 (2002): 1141–1143.
67. Bomzon, Z., V. Kleiner, and E. Hasman. "Pancharatnam–Berry phase in space-variant polarization-state manipulations with subwavelength gratings." *Optics Letters* 26, 18 (2001): 1424–1426.
68. Luo, W., S. Xiao, Q. He, S. Sun and L. Zhou. "Photonic spin Hall effect with nearly 100% efficiency." *Advanced Optical Materials* 3, 8 (2015): 1102–1008.
69. Khorasaninejad, M., W. T. Chen, R. C. Devlin, J. Oh, A. Y. Zhu and F. Capasso. "Metalenses at visible wavelengths: Diffraction-limited focusing and subwavelength resolution imaging." *Science* 352, 6290 (2016): 1190–1194.
70. Zheng, G., H. Mühlenbernd, M. Kenney, G. Li, T. Zentgraf and S. Zhang. "Metasurface holograms reaching 80% efficiency." *Nature Nanotechnology* 10 (2015): 308–312.
71. Lin, D., P. Fan, E. Hasman, M. L. Brongersma. "Dielectric gradient metasurface optical elements." *Science* 345, 6194 (2014): 298–302.
72. Wang, S., P. C. Wu, V.-C. Su, et al. "A broadband achromatic metalens in the visible." *Nature Nanotechnology* 13 (2018): 227–232.
73. Chen, W. T., A. Y. Zhu, V. Sanjeev, M. Khorasaninejad, Z. Shi, E. Lee and F. Capasso. "A broadband achromatic metalens for focusing and imaging in the visible." *Nature Nanotechnology* 13 (2018): 220–226.
74. Brière, G., P. Ni, S. Héron, S. Chenot, S. Vézian, V. Brändli, B. Damilano, J.-Y. Duboz, M. Iwanaga and P. Genevet. "An etching-free approach toward large-scale light-emitting metasurfaces." *Advanced Optical Materials* 1801271 (2019).
75. Lin, J., P. Genevet, M. A. Kats, N. Antoniou and F. Capasso. "Nanostructured holograms for broadband manipulation of vector beams." *Nano Letters* 13, 9 (2013): 4269–4274.
76. Zhan Q. and J. R. Leger. "Interferometric measurement of Berry's phase in space-variant polarization manipulations." *Optics Communications* 213, 4–6 (2002): 241–245.
77. Beijersbergen, M. W., R. P. C. Coerwinkel, M. Kristensen and J. P. Woerdman. "Helical-wavefront laser beams produced with a spiral phaseplate." *Optics Communications* 112, 5–6 (1994): 321–327.
78. Bazhenov, V. Yu., M. V. Vasnetsov and M. S. Soskin, "Laser beams with screw dislocations in their wavefronts." *JETP Letters* 52, 8 (1990): 429–431.
79. Heckenberg, N. R., R. McDuff, C. P. Smith, and A. G. White. "Generation of optical phase singularities by computer-generated holograms." *Optics Letters* 17, 3 (1992): 221–223.
79. Genevet, P., J. Lin, M. A. Kats and F. Capasso. "Holographic detection of the orbital angular momentum of light with plasmonic photodiodes." *Nature Communications* 3 (2012): 1278.

81. Aieta, F., P. Genevet, M. A. Kats and F. Capasso. "Aberrations of flat lenses and aplanatic metasurfaces." *Optics Express* 21, 25 (2013): 31530–31539.
82. Maréchal, A. "Mechanical integrator for studying the distribution of light in the optical image." *Journal of the Optical Society of America* 37, 5 (1947): 403.
83. Aieta, F., P. Genevet, M. A. Kats, N. Yu, R. Blanchard, Z. Gaburro and F. Capasso. "Aberration-free ultrathin flat lenses and axicons at telecom wavelengths based on plasmonic metasurfaces." *Nano Letters* 12, 9 (2012) 4932–4936.
84. Murty, M. V. R. K. "Spherical zone-plate diffraction grating." *Journal of the Optical Society of America* 50, 9 (1960): 923.
85. Arbabi, A., E. Arbabi, S. M. Kamali, Y. Horie, S. Han and A. Faraon. "Miniature optical planar camera based on a wide-angle metasurface doublet corrected for monochromatic aberrations." *Nature Communications* 7 (2016): 13682.
86. Groever, B., W. T. Chen and F. Capasso. "Meta-lens doublet in the visible region." *Nano Letters* 17, 8 (2017): 4902–4907.
87. Ding, F., Y. Chen and S. I. Bozhevolnyi. "Metasurface-based polarimeters." *Applied Sciences* 8, 4 (2018): 594.
88. Pors, A., M. G. Nielsen and S. I Bozhevolnyi. "Plasmonic metagratings for simultaneous determination of Stokes parameters." *Optica* 2, 8 (2015): 716–723.
89. Wu, P. C., J.-W. Chen, C.-W. Yin, et al. "Visible metasurfaces for on-chip polarimetry." *ACS Photonics* 5, 7 (2018): 2568–2573.
90. Mueller, J. P. B., K. Leosson and F. Capasso. "Ultracompact metasurface in-line polarimeter." *Optica* 3, 1 (2016): 42–47.
91. Juhl, M., C. Mendoza, J. P. B. Mueller, F. Capasso and K. Leosson. "Performance characteristics of 4-port in-plane and out-of-plane in-line metasurface polarimeters." *Optics Express* 25, 23 (2017): 28697–28709.
92. Wei, S., Z. Yang and M. Zhao. "Design of ultracompact polarimeters based on dielectric metasurfaces." *Optics Letters* 42, 8 (2017): 1580–1583.
93. Chen, W. T., P. Török, M. R. Foreman, C. Y. Liao, W.-Y. Tsai, P. R. Wu and D. P. Tsai. "Integrated plasmonic metasurfaces for spectropolarimetry." *Nanotechnology* 27, 22 (2016): 224002.
94. Ding, F., A. Pors, Y. Chen, V. A. Zenin and S. I. Bozhevolnyi. "Beam-size-invariant spectropolarimeters using gap-plasmon metasurfaces." *ACS Photonics* 4, 4 (2017): 943–949.
95. Kamali, S. M., E. Arbabi, A. Arbabi, Y. Horie, M. Faraji-Dana and A. Faraon. "Angle-multiplexed metasurfaces: encoding independent wavefronts in a single metasurface under different illumination angles." *Physical Review X* 7, 4 (2017): 041056.
96. Arbabi, E, A. Arbabi, M. Kamali, Y. Horie and A. Faraon. "Multiwavelength metasurfaces through spatial multiplexing." *Scientific Reports* 6 (2016): 32803.
97. Zhao, H., B. Quand, X. Wang, C. Gu, J. Li and Y. Zhang. "Demonstration of orbital angular momentum multiplexing and demultiplexing based on a metasurface in the terahertz band." *ACS Photonics* 5, 5 (2018): 1726–1732.
98. Malek, S. C., H.-S. Ee and R. Agarwal. "Strain multiplexed metasurface holograms on a stretchable substrate." *Nano Letters* 17, 6 (2017): 3641–3645.
99. Ye, W., F. Zeuner, X. Li, B. Reineke, S. He, C.-W. Qiu, J. Liu, Y. Wang, S. Zhang and T. Zentgraf. "Spin and wavelength multiplexed nonlinear metasurface holography." *Nature Communications* 7 (2016): 11930.
100. Mehmood, M. Q., S. Mei, S. Hussain, et al. "Visible-frequency metasurface for structuring and spatially multiplexing optical vortices." *Advanced Materials* 28, 13 (2016): 2533–2539.
101. Kim, I., G. Yoon, J. Jang, P. Genevet, K. T. Nam and J. Rho. "Outfitting next generation displays with optical metasurfaces." *ACS Photonics* 5, 10 (2018), 3876–3895.

102. Teo, J. Y. H., L. J. Wong, C. Molardi and P. Genevet. "Controlling electromagnetic fields at boundaries of arbitrary geometries." *Physical Review A* 94, 2 (2016): 023820.

103. Kamali, S. M., A. Arbabi, E. Arbabi, Y. Horie and A. Faraon. "Decoupling optical function and geometrical form using conformal flexible dielectric metasurfaces." *Nature Communications* 7 (2016): 11618.

104. Burch, J. and A. Di Falco. "Surface topology specific metasurfaces holograms." *ACS Photonics* 5, 5 (2018): 1762–1766.

105. Burch, J., D. Wen, X. Chen and A. Di Falco. "Conformable holographic metasurfaces." *Scientific Reports* 7 (2017): 4520.

106. Xiong, K., G. Emilsson, A. Maziz, X. Yang, L. Shao, E. W. H. Jager and A. B. Dahlin. "Plasmonic metasurfaces with conjugated polymers for flexible electronic paper in color." *Advanced Materials* 28, 45 (2016): 9956–9960.

107. Wu, K., P. Coquet and P. Genevet. "Modelling of free-form conformal metasurfaces." *Nature Communications* 9 (2018): 3494.

108. Chen, P.-Y. and A. Alù. "Mantle cloaking using thin patterned metasurfaces." *Physical Review B* 84, 20 (2011): 205110.

109. Liu, Y., J. Xu, S. Xiao, X. Chen and J. Li. "Metasurface approach to external cloak and designer cavities." *ACS Photonics* 5, 5 (2018): 1749–1754.

110. Hsu, C. W., B. Zhen, A. D. Stone, J. D. Joannopoulos and M. Soljačić. "Bound states in the continuum." *Nature Review Materials* 1 (2016): 16048.

111. Almeida, E., O. Bitton and Y. Prior. "Nonlinear metamaterials for holography." *Nature Communications* 7 (2016): 12533.

112. Wang, C., Z. Li, M.-H. Kim, X. Xiong, X.-F. Ren, G.-C. Guo, N. Yu and M. Lončar. "Metasurface-assisted phase-matching-free second harmonic generation in lithium niobate waveguides." *Nature Communications* 8 (2017): 2098.

113. Héron, S., B. Reineke, S. Vézian, B. Damilano, T. Zentgraf and P. Genevet. "Three-dimensional nonlinear plasmonic metamaterials." *SPIE Photonics Europe Paper* 11345–68 (2020).

2 Semiconductor-Loaded Nonlinear Metasurfaces

Mykhailo Tymchenko, Juan Sebastian Gomez-Diaz, Alex Krasnok, Mikhail A. Belkin, and Andrea Alù

CONTENTS

The previous chapter discussed the role of surface plasmon polaritons (SPPs) to increase the efficiency of various linear and nonlinear processes. One important example is the case of second-harmonic generation (SHG) from bare metallic surfaces, for which the nonlinearity is associated with the asymmetry of potentials confining the electron movement at metallic boundaries subjected to a strong orthogonal electric field. Similarly to surface-enhanced Raman spectroscopy (SERS), which greatly benefits from the use of SPPs to obtain 10^{10}–10^{11} enhancement factors, making it an indispensable tool for modern spectroscopy and sensing technology [1,2], surface-enhanced SHG (SE SHG) can be also used for chemical and biological sensing, or as a method to characterize surface roughness, as even small corrugations lead to higher field localization and increased SHG [3].

Besides sensing and characterization applications which only require bringing the power of a nonlinear signal above the noise level of the detector, metallic surfaces, gratings and metasurfaces composed of subwavelength metallic resonators, cavities or holes have experienced a rapidly growing attention as a promising platform to obtain enhanced nonlinear phenomena and high generation efficiencies in a miniaturized footprint. Indeed, thin metallic metasurfaces and gratings, with deliberately introduced sharp profiles and highly localized plasmonic fields, can exhibit nonlinear responses many orders of magnitude larger than those of conventional nonlinear media of similar thicknesses. In addition, if the thickness of a nonlinear medium is sufficiently small (less than the coherence length $L \sim \pi/\Delta k$ with Δk being the momentum mismatch), the phase-matching conditions typically required in nonlinear setups are lifted. The significant relaxation of phase-matching requirements is another attractive feature of nonlinear metasurfaces, which enables more compact and simpler nonlinear setups. In this context, all-metallic metasurfaces have become an important milestone on the route toward miniaturized nonlinear optical devices, because they limit the role of intrinsic nonlinearities in materials and, instead, engineer the geometry of plasmonic resonators to provide large field intensity at the desired frequencies.

For practical frequency conversion and wave-mixing applications, nonlinear metasurfaces have so far have had limited applicability. Despite giant per-volume conversion efficiencies, their small thickness implies that the net power conversion efficiency is still low. For instance, the SHG power conversion efficiency (defined here as $\eta_{SHG} = P_{2\omega}/P_{\omega}$, with P_{ω} and $P_{2\omega}$ being the fundamental and second-harmonic beam powers, respectively) typically obtained in all-metallic metasurfaces is on the order of 10^{-9}, although higher values, up to 10^{-6}, have been also reported after photon-counting corrections [4]. In contrast, bulk nonlinear media can achieve nearly 100% conversion efficiencies with proper phase-matching (at the expense of a much larger size). The efficiency of nonlinear processes relying on $\chi^{(3)}$, such as third-harmonic generation (THG), is more promising, reaching values up to 0.45×10^{-2} [5].

To increase the efficiency of nonlinear generation, we must either apply very large field intensities, or use larger nonlinearities, or preferably both. It is well-known that the efficiency of most nonlinear processes typically increases with an increase in the intensity of a pump beam, until saturation phenomena prevail. In plasmonic structures, however, high field amplitudes are always associated with increased loss rate and heating. Therefore, the maximum efficiency is greatly limited by ablation and damage. In contrast to all-dielectric structures, which can withstand pump intensities in the order of GW/cm^2, ablation of metallic metasurfaces occurs at much lower intensities, about 10^2 MW/cm^2. Thus, it is safe to say that the net conversion efficiency of nonlinear processes, especially $\chi^{(2)}$, in all-metallic metasurfaces has a very low upper threshold. On the other hand, lasers able to generate so much power are physically large and require active cooling, which may not be feasible for practical setups. It is desirable to achieve sufficiently high nonlinear conversion efficiencies at moderate light intensities, on the order of dozens of kW/cm^2, which can be generated by much more compact sources such as quantum-cascade lasers (QCLs).

To this aim, rather than rely only on surface nonlinearity, we can use strong SPP fields to excite bulk nonlinearities of adjacent media [7]. In such geometries, the large field enhancement can be harnessed over a larger volume, rather than just at the interface, leading to significantly improved overall conversion efficiencies. Early experimental and theoretical studies of metallic gratings covered with a few micron thick layer of nonlinear media such as GaAs indeed showed a 2–3 orders of magnitude increase in SHG conversion efficiency in the vicinity of the SPP resonance [5,6]. Later, it has been also envisioned that filling the gaps of subwavelength split-ring resonators (SRRs), where the field intensity is the highest, with a nonlinear medium would lead to dramatically enhanced nonlinear responses using only a small volume of nonlinear media [8]. A similar geometry in which periodically arranged SRRs were filled with a Kerr-type nonlinear material has been also predicted to exhibit an intensity-dependent behavior of its effective electric permittivity and magnetic permeability [9]. All these results highlighted the great potential of hybrid nonlinear metasurfaces for nonlinear optics in deeply subwavelength structures. Most importantly, they offer a promising route to address the issue of efficiency by combining the best of the two worlds: strong local field enhancement provided by plasmonic resonators and large nonlinearities of adjacent bulk media.

In this chapter, we discuss the recent progress in obtaining efficient nonlinear generation and wave-mixing in hybrid nonlinear metasurfaces, mainly in the context of second-harmonic generation. Hybrid nonlinear metasurfaces open new horizons on the way to extremely subwavelength, phase-match-free and highly efficient planarized structures operating at low and moderate light intensities. We also discuss how their flat geometry makes nonlinear metasurfaces a unique platform for robust wave front shaping which is very difficult to realize in setups employing bulk nonlinear media. The ability to efficiently tailor the radiated wavefront allows getting rid of bulky optical elements, such as lenses, polarizers, and beam-splitters, and augments the functionality of nonlinear metasurfaces, enabling effects that so far have been limited to their linear counterparts.

2.1 PLASMONIC METASURFACES COUPLED WITH NONLINEAR MATERIALS

One of the early experimental observations of enhanced nonlinear generation in hybrid nonlinear metasurfaces was performed using gold films with protrusions filled with a bulk nonlinear medium (GaAs), see Figure 2.1a [10,11]. Pumped at mid-IR, such structures being only about 100 nm thick were able to match the SHG conversion efficiency of a ~100 times thicker $LiNbO_3$ crystal (without phase-matching). Subsequent works employed a variety of other geometries, such as core-shell particles consisting of $BaTiO_3$ cores covered by a thin layer of gold [12], dipole, bowtie and SRR antennas placed on top of $\chi^{(2)}$ media (see Figure 2.1b and c) that provide 2–3 orders of magnitude increase in SHG efficiency [13–15], gold nanoparticles covered with a $\chi^{(3)}$ material for intensity-dependent absorption (Figure 2.1d) [16], and dipole antennas with small gaps loaded with $\chi^{(3)}$ media for enhanced THG [17,18] and phase conjugation [19–21] (Figure 2.1e). Nineteen orders of magnitude increase in four-wave-mixing (FWM) efficiency has been predicted in a metasurface consisting of

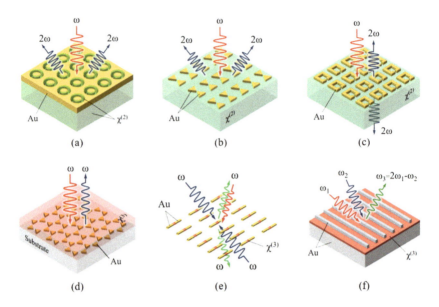

FIGURE 2.1 Schematics of various hybrid metasurfaces employing resonant plasmonic nanostructures and antennas to efficiently excite nonlinear responses of adjacent nonlinear media. (a) Gold film with protrusions filled with a $\chi^{(2)}$ medium for enhanced SHG at oblique angles. (From Fan, W. et al., *Nano Lett.*, 6, 1027–1030, 2006; Fan, W. et al., *Opt. Express*, 14, 9570, 2006.) (b) Bowtie antennas placed on top of a $\chi^{(2)}$ medium for SHG at oblique angles. (From Bar-Lev, D. and Scheuer, J., *Opt. Express*, 21, 29165, 2013.) (c) SRRs placed on top of a $\chi^{(2)}$ medium for SHG in the normal direction. (From Niesler, F.B.P. et al., *Opt. Lett.*, 34, 1997, 2009.) (d) Gold nanoparticles covered with a $\chi^{(3)}$ medium for intensity-dependent absorption. (From Ning, T. et al., *J. Phys. D. Appl. Phys.*, 40, 6705–6708, 2007.) (e) Dipole antennas loaded with a $\chi^{(3)}$ medium for phase conjugation. (From Chen, P. et al., *Nano Lett.*, 11, 5514–5518, 2011.) (f) A thin layer of a $\chi^{(3)}$ medium sandwiched between an array of metallic rods and a ground plane for enhanced four-wave mixing process. (From Jin, B. and Argyropoulos, C., *Sci. Rep.*, 6, 28746, 2016.)

nm-thin resonant gaps formed between silver nanorods and a thick metallic substrate filled with a $\chi^{(3)}$ nonlinear medium [22], as shown in Figure 2.1f. Importantly, all these results were obtained using very small volumes of nonlinear media without any phase-matching in the direction of radiation. At the same time, hybrid metasurfaces are still susceptible to relatively low optical damage thresholds (although they can be larger than those of all-metallic structures), which limit the amplitude of the pump field in the hot spots, and thus, the achievable conversion efficiency.

2.2 PLASMONIC METASURFACES COUPLED TO MULTI-QUANTUM-WELL MEDIA

To further increase the efficiency of nonlinear generation of hybrid metasurfaces without boosting the pump intensity to extreme values we can use nonlinear materials with very large nonlinearity. The largest second-order nonlinear susceptibility of conventional nonlinear media, such as $LiNbO_3$, is typically in the order of

~10^2 pm/V. However, there is one class of artificially engineered media, *multi-quantum-wells* (MQWs), which can exhibit nonlinearities in the range of $\chi^{(2)} \sim 10^2$ to 10^4 nm/V [23–29], 3–4 orders of magnitude larger than any other nonlinear media. In this section, we discuss recent advances associated with MQW-based hybrid non-linear metasurfaces exhibiting record-high efficiencies of nonlinear harmonic generation, more specifically, SHG.

2.2.1 MULTI-QUANTUM-WELLS

MQWs are composed of stacks of thin semiconductor layers with different electron binding energies and bandgap widths. The bandgaps of the stack form a sequence of potential wells and barriers for electrons and holes moving orthogonally to semiconductor layers. This confinement leads to quantization of carriers' energy and emergence of a 1D band structure where free electrons occupy subbands of the conductions band and holes reside in subbands of the valence band, as is schematically depicted Figure 2.2. The bandstructure of natural materials originates from a particular arrangement of atoms and molecules, which means that their linear and nonlinear properties are very hard to change on demand. With the stacks of semiconductor layers, we essentially obtain a quasi-1D atom with a much greater control over its optical response.

Band engineering in semiconductor heterostructures has become a fruitful field of research and has enabled many ubiquitous modern devices such as pn-diodes, field-effect transistors (FETs), sensors, light-emitting diodes (LEDs), compact semiconductor lasers, photodetectors, optical modulators, solar cells, and others [30].

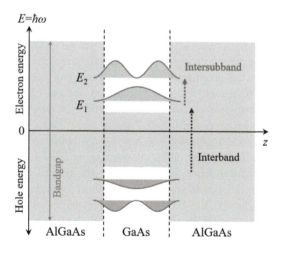

FIGURE 2.2 Schematic of the band diagram and two types of transitions occurring in a single GaAs/AlGaAs quantum well. The shaded area shows the bandgap as a function of the layer growth direction, z. Electron (holes), with probability functions schematically shown in the positive (negative) part of energy spectrum (remember that holes have negative energy), occupy discrete energy levels called *subbands* forming the conduction (valence) band. Intersubband transitions occur between subbands of the same band.

In the context of nonlinear optics, MQWs provide a direct way to engineer giant nonlinearities by controlling intersubband transitions, making them one of the most attractive nonlinear materials. In addition, intersubband nonlinearities in MQWs can be dynamically modified and spectrally tuned using the voltage bias [25,31–38]. Electrical pumping may be used to produce active intersubband structures with full loss-compensation for second- [28,29] and third-order [39] nonlinear processes. At the same time, nonlinear generation in MQWs also possesses a significant challenge: intersubband transitions have mostly z-oriented dipole moments (assuming no extra longitudinal momentum), and therefore, the strong nonlinear response in MQWs can be achieved only for light polarized along the z-direction, i.e., the only relevant nonlinear susceptibility elements are $\chi_{zzz}^{(2)}$, $\chi_{zzzz}^{(3)}$, and so on. The generated strong nonlinear polarization currents are also oriented along the z-axis and cannot radiate orthogonally to MQW layers. Therefore, illuminating MQWs with light coming from the direction perpendicular to their layers (i.e., the z-axis) does not induce strong nonlinear processes. MQWs also exhibit a high permittivity, $\varepsilon_{MQW} \sim 10$, with significant anisotropy and increased absorption around the frequencies of intersubband transitions, which makes light outcoupling at oblique angles very inefficient. Besides, such outcoupling may be forbidden beyond the critical angle due to total internal reflection. All these nuances make efficient nonlinear generation in MQWs a challenging problem. In the following, we show that properly designed plasmonic nanoresonators address these challenges and enable efficient excitation and outcoupling of nonlinear radiation from MQWs to free-space.

2.2.2 SHG FROM MQW-BASED HYBRID NONLINEAR METASURFACES

The possibility of obtaining strong second-harmonic generation in MQWs was originally explored in AlGaAs heterostructures with asymmetric composition gradients, which would provide the required polarization centrosymmetry breaking [40]. More recently, to facilitate the coupling of normally impinging light and convert it into z-polarized electric field within the MQW volume, the use of plasmonic metasurfaces consisting of stripe-, dogbone- or SRR-shaped gold nanoresonators placed on top of MQWs has been proposed [41,42 Todorov et al. PRL 102, 186402 (2009).]. These studies confirmed that, owing to z-polarized resonant electric fields, plasmonic structures indeed enable efficient coupling of light impinging from free-space to intersubband transitions in MQWs. Shortly after, it has been shown that properly designed *doubly*-resonant structures, such as asymmetric nanocrosses (see Figure 2.3a) and SRRs, can also be used to enable efficient second-harmonic radiation outcoupling from an MQW volume into the free-space [43,44]. Such hybrid nonlinear metasurfaces with only 400 nm thick MQWs designed to have a giant $\chi_{zzz}^{(2)} \sim 50$ nm/V at the fundamental frequency of 37 THz were able to reach a very high SHG efficiency of $\eta_{SHG} \sim 10^{-6}$ at pump intensities of only 15 kW/cm^2 (in the continuous pump regime), see Figure 2.3b. This SHG efficiency is 2–3 orders of magnitude higher than that of any previously fabricated structure of comparable volume, and about 8 orders of magnitude larger the efficiency of LiNbO$_3$ crystal of the same thickness. Plasmonic resonators also introduce a strong sensitivity to

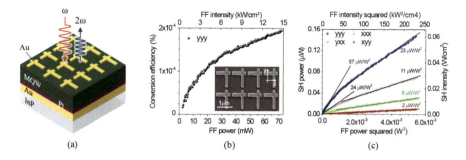

FIGURE 2.3 (a) Design and operation schematic of an MQW-based hybrid nonlinear metasurface aimed at SHG. The metasurface is comprised of a 400 nm-thick MQW layer with large $\chi^{(2)}$, sandwiched between an array of doubly-resonant gold nanocrosses and a thick metallic ground plane. The structure is illuminated at normal incidence by a 34 μm-wide Gaussian beam with a fundamental frequency (FF) of 37 THz. The second-harmonic (SH) beam generated at 74 THz is emitted in the opposite direction orthogonally to the metasurface. (b) Conversion efficiency versus the FF power. The inset shows the FEM image of the fabricated sample. (c) SH power (intensity) versus the FF power (intensity) squared for various input and output polarizations combinations: *yyy*, *xxx*, *yxx* and *xyy*. Here, the first polarization index corresponds to the second-harmonic wave; the other two denote a particular pump field polarization combination. (Panels [b] and [c] were reproduced from Lee, J. et al., *Nature*, 511, 65–69, 2014.)

polarization of the pump and second-harmonic signal, as can be seen in Figure 2.3c, indicating a great potential for efficient polarization control in the nonlinear regime. The largest second-harmonic power yield of about 0.15 μW was achieved for the *yyy* polarization combination.

Figure 2.4a shows another metasurface of a similar design, with gold SRRs placed on top of 700 nm-thick MQW stack. Under pulse illumination and operating in transmission, this metasurface was able to achieve the SHG conversion efficiency about 3×10^{-5} at pump intensities of about 10 kW/cm^2 [45], see Figure 2.4b and the inset therein.

Finally, a record-high SHG conversion efficiency of 7.5×10^{-4} (0.075%) has been experimentally achieved at similar pump intensities in reflection regime using an improved metasurface design with T-shaped plasmonic resonators and 400 nm-thick MQW medium etched around them, see Figure 2.5a and b [46]. The maximum recorded second-harmonic power yield was about 70 μW, as can be seen in Figure 2.5c.

Even higher SHG efficiency, up to 1%, may be achievable using MQWs of higher quality and improved resonator designs [47]. Undoubtedly, one of the main challenges in the quest to further boosting the conversion efficiency is the accurate modeling, design and realization of such hybrid metasurfaces. In the next section, we present a rigorous yet efficient approach to characterize nonlinear metasurfaces aimed at SHG which provides useful insights into the nonlinear generation mechanism and allows to identify optimal design strategies.

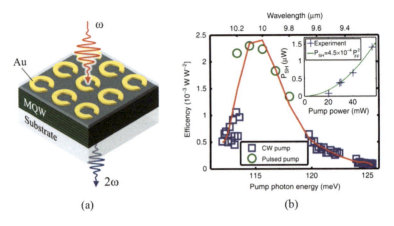

(a)   (b)

FIGURE 2.4 (a) Design and operation schematic of an MQW-based hybrid nonlinear metasurface aimed at SHG. The 700 nm-thick MQW layer with large $\chi^{(2)}$ was grown on top of a transparent substrate. The MQW is capped with an inactive semiconductor medium with a similar permittivity. The array of doubly-resonant gold SRRs is placed on top of MQWs. The metasurface is illuminated in continuous and pulsed pump regimes. (b) SH conversion efficiency, here defined as $\eta_{SHG} = P_\omega / P_\omega^2$, as a function of pump wavelength. The inset shows the SH power dependence on pump power for a pump wavelength of 10.22 μm. The red line is a guide to the eye. (Panel [b] was reproduced from Wolf, O. et al., *Nat. Commun.*, 6, 7667, 2015.)

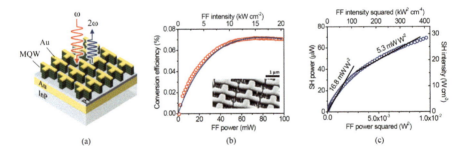

(a)   (b)   (c)

FIGURE 2.5 (a) Improved design and operation schematic of a hybrid metasurface aimed at SHG. The metasurface is composed of a 400 nm-thick MQW layer sandwiched between the array of gold T-shaped doubly-resonant structures and a thick metallic ground plane. The MQW layer is etched around the resonators to boost the *z*-polarized field components at SH and FF frequencies. The metasurface is illuminated by a normally impinging 34 μm-wide beam with an FF of 30 THz. (b) SHG conversion efficiency versus the FF power and intensity. The inset shows the FEM image of the fabricated sample. (c) SH power (intensity) versus the FF power (intensity) squared for various input and output polarizations combinations. (Panels [b] and [c] were reproduced from Lee, J. et al., *Adv. Opt. Mater.*, 4, 664–670, 2016.)

2.3   MODELING AND DESIGN PRINCIPLES OF MQW-BASED HYBRID METASURFACES AIMED AT SHG

In this section, we outline the main principles and design rules of MQW-based metasurfaces aimed at highly efficient SHG. Metasurfaces designed for other non-linear processes usually should follow similar rules adapted to their particular features [48,49]. For efficient SHG in MQW-based hybrid nonlinear metasurfaces, a successful design must address two main challenges: engineering the MQW stack with a large $\chi^{(2)}_{zzz}$ at the desired fundamental frequency and designing a suitable plasmonic nanoresonator or cavity. The former can be accomplished by choosing suitable semiconductor materials and meticulous engineering the intersubband transition frequencies of the MQW stack, including linewidths and carrier doping. The details of semiconductor band structure engineering go beyond the scope of this chapter. However, it is important to mention that there are strong tradeoffs among several parameters, such as the magnitude of the nonlinear susceptibility, transition line-widths, absorption rates, and saturation intensities of these transitions. Saturation, as we will discuss later in detail, plays a critical role in limiting the performance of MQW-based hybrid metasurfaces. Specifically, engineering MQWs with very high $\chi^{(2)}$ may require high doping levels, which inadvertently increases the absorption rate and reduces the saturation intensity. Then, the SHG efficiency of a resulting metasurface might be severely hindered due to significant damping of the induced fields at the resonance and quick saturation of the nonlinear process. Therefore, it is often desirable to design MQWs with lower $\chi^{(2)}$ in exchange for significantly reduced absorption and increased saturation thresholds.

The second part of the metasurface design consists in engineering subwavelength plasmonic resonators, which is also non-trivial, because their role is multifold: (i) they must efficiently couple light impinging from free-space into the metasurface at the fundamental frequency ω; (ii) they must resonantly enhance the z-polarized fields at ω in MQWs; and (iii) they must allow the z-oriented nonlinear polarization currents to efficiently radiate back into free-space, which can be accomplished by making them resonant also at 2ω. In addition, for SHG at normal incidence and radiation, plasmonic nanostructures must be non-centrosymmetric [50]. Figure 2.6a depicts one design of such a structure employing non-centrosymmetric resonators to achieve efficient SHG [46]. The corresponding modal profiles and absorption spectra are shown in Figure 2.6b and c. Alternatively, SHG can be achieved in centrosymmetric structures illuminated and/or observed at oblique angles. In this case, centrosymmetry of the second-harmonic polarization perceived in the observation direction is broken by asymmetric gradients of the field at pump frequency. Designing a structure simultaneously meeting all these requirements can be a tedious and computationally intensive task, further tangled by the nonlinearity of the system.

Several methods are typically employed to model nonlinear optical structures. One of the most commonly used approaches is a finite-difference time-domain (FDTD) method [51] which is widely available in various commercial and open-source electromagnetic simulation packages. It is also the most "natural" way to study all kinds of nonlinearities, since it accurately captures the temporal response of the system to an arbitrary signal at any instant of time and it automatically accounts for frequency

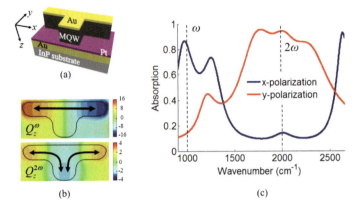

FIGURE 2.6 (a) Example of an MQW-based hybrid nonlinear metasurface designed for highly efficient SHG. The centrosymmetric T-shape of the structure is tailored to support strong plasmonic resonances at the pump frequency ω along the z-axis, and at second-harmonic 2ω along the y-axis. (b) Enhancement factors for the z-polarized field components at ω and 2ω. (c) Absorption spectra for x- and y-polarizations. (Panels [a]–[c] were reproduced from Lee, J. et al., *Adv. Opt. Mater.*, 4, 664–670, 2016.)

conversion and mixing. However, while this method is rigorous, it often requires prolonged simulation times due to vastly different time scales involved in many nonlinear processes. Therefore, this method is often one of the least efficient. Other theoretical and semi-numerical methods for studying nonlinear effects include various coupled-mode theories [52], Green's function approaches [53,54], Bloch mode expansions [55], or nonlinear scattering theories [56]. Fortunately, we can leverage the unique geometry of nonlinear metasurfaces to develop a novel method which simultaneously simplifies the theoretical analysis, can rigorously account for saturation effects and losses, and provides deeper insights into metasurface operation. The method is based on evaluating the effective nonlinear susceptibility tensor of the metasurface using the reciprocity theorem, which helps to significantly speed up the design process, as well as accurately predict the SHG efficiency. Importantly, this method requires a minimum amount of linear full-wave numerical simulations which can be performed in any electromagnetic simulation suite.

2.3.1 Effective Nonlinear Susceptibility Tensor

Theoretical and numerical modeling of linear metasurfaces and metamaterials is often centered around the derivation of homogenized parameters such as effective electric and magnetic susceptibilities which encapsulate all linear scattering phenomena. Here, we follow the same strategy and compute the effective second-order nonlinear susceptibility tensor of a thin metasurface consisting of identical unit-cells arbitrarily filled with a nonlinear medium. For simplicity, we assume that the nonlinear medium is composed of MQWs with the only non-zero susceptibility component $\chi^{(2)}_{zzz}$. Later, we generalize the obtained expressions to the case of an arbitrary nonlinear medium characterized by a susceptibility tensor $\vec{\chi}^{(2)}$. Effective nonlinear susceptibilities of metasurfaces aimed at other nonlinear processes can be derived following similar steps [48,49].

In the following analysis, we make two important assumptions: first, we assume that our nonlinear metasurfaces are very thin and that their unit-cell elements are subwavelength at all involved frequencies. Second, we assume that the nonlinear power conversion efficiency is sufficiently small, which allows us to work in an undepleted pump approximation (i.e., the amplitude of the field at the pump frequency is constant). We start the analysis by considering a plane wave oscillating at the frequency ω and impinging orthogonally onto a nonlinear metasurface consisting of identical unit-cells containing the MQW medium. Without loss of generality, we assume the wave to be x-polarized, $\mathbf{E}_{\text{inc}}^{\omega} = \hat{\mathbf{e}}_x E_{\text{inc},x}^{\omega}$. To account for the internal structure of unit-cells, for instance when MQWs are etched around the resonators as shown in Figure 2.5a, we shall allow the nonlinear susceptibility to be spatially-varying, $\chi_{zzz}^{(2)} = \chi_{zzz}^{(2)}(\mathbf{r})$. Since the intensity of the second-harmonic radiation is negligible compared to the intensity of the pump wave, the nonlinear problem can be split into a homogeneous (source-free) linear scattering problem at the fundamental frequency ω and inhomogeneous linear problem at the frequency 2ω, with a source polarization current that depends on the square of the z-component of the pump field excited in MQWs. Specifically, in each unit-cell, the x-polarized impinging wave induces the local electric field

$$\mathbf{E}_{(x)}^{\omega}(\mathbf{r}) = \hat{\mathbf{e}}_x E_{x(x)}^{\omega}(\mathbf{r}) + \hat{\mathbf{e}}_y E_{y(x)}^{\omega}(\mathbf{r}) + \hat{\mathbf{e}}_z E_{z(x)}^{\omega}(\mathbf{r}), \tag{2.1}$$

which can be easily found from linear full-wave numerical simulations. The local second-order polarization density has only the z-oriented component which can be found as

$$P_{z(xx)}^{2\omega}(\mathbf{r}) = \varepsilon_0 \chi_{zzz}^{(2)}(\mathbf{r})[E_{z(x)}^{\omega}(\mathbf{r})]^2, \tag{2.2}$$

leading to a nonlinear polarization current density within the MQWs of

$$J_{z(xx)}^{2\omega}(\mathbf{r}) = -i\,2\omega P_{z(xx)}^{2\omega}(\mathbf{r}). \tag{2.3}$$

Since the unit-cells are subwavelength at the second-harmonic frequency, this polarization current will emit a single plane wave $\mathbf{E}_{\text{FF}}^{2\omega}$ of a yet unknown amplitude and polarization, see Figure 2.7a. In order to determine the features of the nonlinear field, we can apply the reciprocity theorem [57] which states that for two current sources $\mathbf{J}_1$ and $\mathbf{J}_2$ surrounded by a linear time-invariant medium and radiating the fields $\mathbf{E}_1$ and $\mathbf{E}_2$, respectively, the following relation holds:

$$\int_V \mathbf{J}_1 \cdot \mathbf{E}_2 dV = \int_V \mathbf{J}_2 \cdot \mathbf{E}_1 dV, \tag{2.4}$$

where the integration is performed over the entire domain.

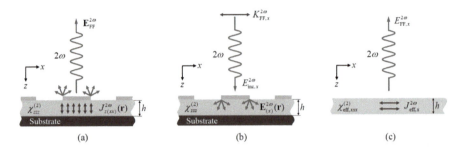

FIGURE 2.7 Evaluation of effective susceptibility tensor using the reciprocity theorem. (a) Nonlinear polarization current $J_{z(xx)}^{2\omega}$ sustains a single second-harmonic plane wave $\mathbf{E}_{FF}^{2\omega}$ in the far-field. The thickness h of the MQW layer is much smaller than the free-space wavelength at 2ω. (b) Fictitious 2D current $K_{FF,x}^{2\omega}$ located in the far field and emitting a plane $E_{inc,x}^{2\omega}$ which induces the field distribution $\mathbf{E}_{(x)}^{2\omega}(\mathbf{r})$ in the MQW volume. (c) The x-polarized component of the second-harmonic plane wave, $E_{FF,x}^{2\omega}$, can be sustained by a thin layer of an effective current $J_{eff,x}^{2\omega}$. The layer is assumed to have $\chi^{(1)} = 0$ at ω and 2ω.

Using Eq. (2.4), we can relate $\mathbf{E}_{FF}^{2\omega}$ radiated by $J_{z(xx)}^{2\omega}$ and the field inside the unit cell $E_{z(x)}^{\omega}(\mathbf{r})$. To do so, let us imagine a fictitious uniform 2D current of an amplitude $\mathbf{K}_{FF}^{2\omega} = \hat{\mathbf{e}}_x K_{FF,x}^{2\omega}$ located far away from the metasurface and emitting an x-polarized plane wave $\mathbf{E}_{inc}^{2\omega} = \hat{\mathbf{e}}_x E_{inc,x}^{2\omega}$ oscillating at frequency 2ω and inducing the field $\mathbf{E}_{(x)}^{2\omega}(\mathbf{r})$ in the unit-cell, as is shown in Figure 2.7b. Applying the reciprocity theorem, we may write:

$$\int_S E_{FF,x}^{2\omega} K_{FF,x}^{2\omega} dS = -i2\omega \int_V E_{z(x)}^{2\omega}(\mathbf{r}) P_{z(xx)}^{2\omega}(\mathbf{r}) dV, \quad \mathbf{r} \in V, \tag{2.5}$$

where S is the area and V is the volume of the unit-cell. From quasi-statics, we may find that

$$K_{FF,x}^{2\omega} = \frac{2}{\eta_0} E_{inc,x}^{2\omega}, \tag{2.6}$$

where $\eta_0 = \sqrt{\mu_0/\varepsilon_0}$ is the free-space impedance. Substituting (2.2) and (2.6) into (2.5) and performing the integration in the l.h.s., we obtain:

$$E_{FF,x}^{2\omega} = \frac{-i\omega}{cS} \int_V \chi_{zzz}^{(2)}(\mathbf{r}) Q_{z(x)}^{2\omega}(\mathbf{r}) \left[E_{z(x)}^{\omega}(\mathbf{r}) \right]^2 dV, \quad \mathbf{r} \in V, \tag{2.7}$$

where $c = 1/\sqrt{\varepsilon_0 \mu_0}$ is the speed of light in free-space, and $Q_{z(x)}^{2\omega}(\mathbf{r})$ denotes the local enhancement factor of a z-polarized field at 2ω excited by an x-polarized second-harmonic impinging wave:

$$Q_{z(x)}^{2\omega}(\mathbf{r}) = \frac{E_{z(x)}^{2\omega}(\mathbf{r})}{E_{inc,x}^{2\omega}}. \tag{2.8}$$

Now we model the radiation sustaining the x-polarized component of the second-harmonic field $E^{2\omega}_{FF,x}$ through an effective nonlinear polarization current $J^{2\omega}_{eff,x} = -i\,2\omega P^{2\omega}_{eff,x}$. We assume that the induced effective current is uniformly distributed across an effective metasurface with the same height as the MQW layer, see Figure 2.7c. For simplicity, we also assume that this effective metasurface is ideally transparent, i.e., $\chi^{(1)}_{eff} = 0$ at ω and 2ω, and therefore there is no need to solve the homogeneous scattering problem. Furthermore, the metasurface is negligibly thin compared to the free-space wavelength at 2ω, which allows us to assume a constant impinging field across the unit-cell volume. Then, in the quasi-static approximation, we obtain

$$J^{2\omega}_{eff,x}(z) = -i2\omega\varepsilon_0\chi^{(2)}_{eff,xxx}(E^\omega_{inc,x})^2, \tag{2.9}$$

where the effective susceptibility tensor relates the nonlinear polarization to the *incident* field, because the metasurface is ideally transparent. Integrating Eq. (2.9) over the metasurface height $h = V/S$ and using the relation (2.6), we obtain

$$E^{2\omega}_{FF,x} = -i\frac{\omega}{c}h\,\chi^{(2)}_{eff,xxx}(E^\omega_{inc,x})^2. \tag{2.10}$$

Finally, substituting (2.10) into (2.7), we obtain the effective second-order susceptibility tensor for the xxx polarization combination:

$$\chi^{(2)}_{eff,xxx} = \frac{1}{V}\int_V \chi^{(2)}_{zzz}(\mathbf{r})Q^{2\omega}_{z(x)}(\mathbf{r})[Q^\omega_{z(x)}(\mathbf{r})]^2\,dV, \quad \mathbf{r}\in V. \tag{2.11}$$

Eq. (2.11) can be also generalized for any combination of polarizations as

$$\chi^{(2)}_{eff,ijk} = \frac{1}{V}\int_V \chi^{(2)}_{zzz}(\mathbf{r})Q^{2\omega}_{z(i)}(\mathbf{r})Q^\omega_{z(j)}(\mathbf{r})Q^\omega_{z(k)}(\mathbf{r})dV, \quad \mathbf{r}\in V, \tag{2.12}$$

where i, j, k can be x or y. Here, the z-component is not included because waves at both ω and 2ω propagate along the z-axis. From this expression, we can draw several important conclusions. First, the magnitude of the second-harmonic field (when it is small compared to the field at ω) is proportional to a spatial overlap integral between the modal profiles $Q^\omega_z(\mathbf{r})$ and $Q^{2\omega}_z(\mathbf{r})$. It also means that by tailoring these profiles we can engineer virtually any component of the effective susceptibility tensor $\chi^{(2)}_{eff,ijk}$ from intrinsic $\chi^{(2)}_{zzz}$. Second, this modal overlap automatically requires breaking of second-harmonic polarization centrosymmetry to obtain strong SHG. If Q^ω_z is even and $Q^{2\omega}_z$ is odd across the unit-cell, then the integral in Eq. (2.12) is strictly zero. It also explains why some metasurfaces composed of symmetric structures being illuminated by a normally incident wave can radiate only obliquely. In these cases, Q^ω is even, and the overlap is non-zero only when $Q^{2\omega}$ is strongly asymmetric which is achieved away from the normal direction. For the case when the intrinsic nonlinear susceptibility tensor has more than one element, Eq. (2.12) takes the form

$$\chi^{(2)}_{\text{eff},ijk} = \frac{1}{V} \int_V \mathbf{Q}^{2\omega}_{(i)}(\mathbf{r}) \cdot \vec{\chi}^{(2)}(\mathbf{r}) : \mathbf{Q}^{\omega}_{(j)}(\mathbf{r})\mathbf{Q}^{\omega}_{(k)}(\mathbf{r})dV, \qquad \mathbf{r} \in V, \qquad (2.13)$$

where $\mathbf{Q}_{(i)}(\mathbf{r})$ is now a three-dimensional field enhancement factor, and the double dot denotes the dyadic tensor product. We may further generalize (2.13) to the case of an arbitrary second-order process

$$\chi^{(2)}_{\text{eff},ijk}(\omega_3,\omega_1,\omega_2) = \frac{1}{V} \int_V \mathbf{Q}^{\omega_3}_{(i)}(\mathbf{r}) \cdot \vec{\chi}^{(2)}(\omega_3,\omega_1,\omega_2;\mathbf{r}) : \mathbf{Q}^{\omega_1}_{(j)}(\mathbf{r})\mathbf{Q}^{\omega_2}_{(k)}(\mathbf{r})dV, \mathbf{r} \in V, \quad (2.14)$$

where $\omega_3 = \omega_1 \pm \omega_2$. Finally, we note that, in principle, the metasurface can radiate on both sides with different efficiencies, which we can account for by writing

$$\chi^{(2)\pm}_{\text{eff},ijk}(\omega_3,\omega_1,\omega_2) = \frac{1}{V} \int_V \mathbf{Q}^{\omega_3}_{(i,\mp)}(\mathbf{r}) \cdot \vec{\chi}^{(2)}(\omega_3,\omega_1,\omega_2;\mathbf{r}) : \mathbf{Q}^{\omega_1}_{(j)}(\mathbf{r})\mathbf{Q}^{\omega_2}_{(k)}(\mathbf{r})dV, \mathbf{r} \in V, \quad (2.15)$$

where $\mathbf{Q}^{\omega_3}_{(i,\mp)}(\mathbf{r})$ denotes the second-harmonic enhancement factor for illumination on either side of the metasurface.

Having derived the effective nonlinear susceptibility tensor, it is straightforward to evaluate the SHG efficiency. The intensity of the radiated second-harmonic signal can be found using Eq. (2.10) as [50]

$$I_{2\omega} = 2c\varepsilon_0 \left| \mathbf{E}^{2\omega}_{\text{FF}} \right|^2 = \frac{2\omega^2 h^2 d^2_{\text{eff}}}{\varepsilon_0 c^3} I^2_\omega, \qquad (2.16)$$

with d_{eff} given by

$$d_{\text{eff}} = \frac{1}{2} \left| \hat{\mathbf{e}}^{2\omega} \cdot \vec{\chi}^{(2)}_{\text{eff}} : \hat{\mathbf{e}}^{\omega} \hat{\mathbf{e}}^{\omega} \right|, \qquad (2.17)$$

where $\hat{\mathbf{e}}^{\omega}$ and $\hat{\mathbf{e}}^{2\omega}$ are unit-vectors corresponding to polarizations of the pump and second-harmonic wave. Taking the radius of the illuminated spot to be w and the intensity to have a simple Gaussian distribution, $I_\omega(r) = I^0_\omega \exp(-2r^2/w^2)$, we obtain

$$\eta_{\text{SHG}} = \frac{P_{2\omega}}{P_\omega} = \frac{4\omega^2 h^2 d^2_{\text{eff}}}{\varepsilon_0 c^3 \pi w^2} I^0_\omega \int_0^{2\pi} d\theta \int_0^\infty \exp\left(-\frac{4r^2}{w^2}\right) r dr =$$

$$= \frac{\omega^2 h^2 d^2_{\text{eff}}}{\varepsilon_0 c^3} I^0_\omega = \frac{2\omega^2 h^2 d^2_{\text{eff}}}{\varepsilon_0 c^3 \pi w^2} P_\omega, \qquad (2.18)$$

with $P_\omega = \pi w^2 I^0_\omega/2$. To verify the accuracy of this expression, in Figure 2.8 we show the second-harmonic power (intensity) for the metasurface shown in Figure 2.3 as a function of pump power (intensity) squared, evaluated using Eqs. (2.18), (2.17), and

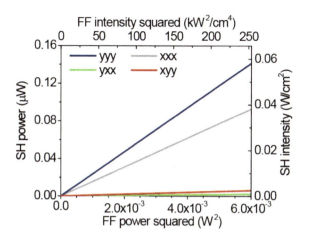

FIGURE 2.8 Theoretically evaluated second-harmonic power (intensity) generated by the metasurface shown in Figure 2.3 versus the FF power (intensity) squared for various input and output polarizations combinations. The results agree well with experimental data shown in Figure 2.3c. (Reproduced from Lee, J. et al., *Adv. Opt. Mater.*, 4, 664–670, 2016.)

(2.12) for four different polarization combinations. It is seen that despite a peculiar resonator geometry and rather complex spatial modal profiles at ω and 2ω, the analytical results correctly predicted that *yyy* polarization combination provides the highest conversion efficiency, followed by *xxx*. Theoretically estimated second-harmonic power is also close to experimentally obtained results. These facts prove the robustness and high accuracy of the proposed method. However, one can also see that the theoretically computed efficiencies of the SHG process scale linearly with the applied pump power, while the experimental curves in Figure 2.3c significantly deviate from the linear trend. The origin of this discrepancy, as we show below, is the saturation of MQWs stemming from high local field intensity in the vicinity of plasmonic resonators. Even at relatively low light intensities, on the order of a few kW/cm², saturation effects negatively affect the metasurface performance. The impact of saturation is especially evident in the experimentally measured SHG conversion efficiency from the metasurface with T-shaped structures, see Figure 2.5b, where it essentially stops growing beyond the intensity of ~15 kW/cm². Although applying more power still results in more second-harmonic power, the power conversion efficiency does not increase. Evidently, saturation phenomena cannot be dismissed and must be included in our theoretical model. In the next paragraph, we improve our theoretical framework by explicitly accounting for saturation effects and losses in MQWs, which leads to a much better agreement between theoretical and experimental results.

2.3.2 SATURATION OF MQW-BASED NONLINEAR METASURFACES AIMED AT SHG

Experimental measurements of second-harmonic power generated by MQW-loaded plasmonic metasurfaces shown in Figures 2.3 and 2.5 indicated that saturation effects in MQWs strongly affect the overall conversion efficiency even at relatively low pump intensities [43,46]. This is consistent with earlier works modeling the nonlinear response of bulk MQW intersubband transitions, which revealed a strong impact of saturation and losses on the efficiency of nonlinear processes [58,59]. In case of MQW-loaded nonlinear plasmonic metasurfaces, saturation is primarily caused by very large field enhancement near metallic resonators, which depopulates the ground subband. If the population difference between the two subbands is significantly reduced, the corresponding transition saturates [50], quenching nonlinear generation from the saturated domain. Owing to very strong local fields in hybrid metasurfaces, even low pump intensities lead to ground level depopulation which negatively impacts the efficiency of the SHG process. Intuitively, this occurs because, as the pump intensity increases, more and more volume of MQWs becomes saturated. Beyond a certain critical intensity which is far below the optical damage threshold, most of the MQW volume becomes saturated and does not generate many second-harmonic photons, which effectively entails a drop in the SHG conversion efficiency. Rigorous modeling of saturation effects and losses in MQW-loaded plasmonic metasurfaces is therefore critical for accurately predicting their performance as well as for finding optimal designs before actual experimental fabrication.

Consider an MQW semiconductor stack with three subbands tuned for the SHG process. The corresponding transition energies are $E_{ij} = \hbar\omega_{ij}$ where i and j can be 1, 2, or 3. For the SHG process, the transitions of interest are $1 \to 2$, $2 \to 3$, and $3 \to 1$. For SHG, the MQWs bands have been engineered to have $\omega_{12} \approx \omega_{23}$ and $\omega_{31} \approx \omega_{12} + \omega_{23}$. In the vicinity of these transitions, the zzz-component of the intersubband nonlinear susceptibility is approximately given by [60]

$$\chi_{zzz}^{(2)}(I_z^\omega) \approx \frac{NS(I_z^\omega)}{\varepsilon_0 \hbar^2} \frac{e^3 z_{12} z_{23} z_{31}}{\left[(\omega_{12} - \omega) - i\gamma_{12}\right]\left[(\omega_{31} - 2\omega) - i\gamma_{31}\right]}, \tag{2.19}$$

where $-e$ is the electron charge, N is the average electron density, z_{ij} is the corresponding dipole matrix element, and γ_{ij} is the transition linewidth (Figure 2.9).

The magnitude of the second-order susceptibility depends on the population density in each of the three subbands, which we account for using a dimensionless saturation factor [60],

$$S(I_z^\omega) \approx \frac{N_1(I_z^\omega) - 2N_2(I_z^\omega) + N_3(I_z^\omega)}{N}, \tag{2.20}$$

where N_i is the electron population density of the i-th subband which depends on the "acting" local intensity of the z-polarized electric field inside MQWs at the pump frequency,

$$I_z^\omega = 2n_{zz}^{\text{MQW}}(\omega)\varepsilon_0 c \left|E_z^\omega\right|^2, \tag{2.21}$$

with n_{zz}^{MQW} denoting the real part of the refractive index of MQWs in the z-direction [61] (see Appendix A). When the input intensity is low, the electron population predominantly resides in the ground state ($N_1 \approx N$, $N_2, N_3 \approx 0$ even at $T = 300$ K) and

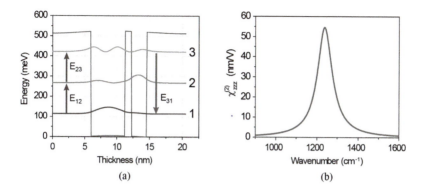

FIGURE 2.9 (a) Conduction band diagram of one period of an InGaAs/AlInAs coupled quantum well structure designed for giant nonlinear response for SHG (Ref. [43]). The moduli squared of the electron wavefunctions for subbands 1, 2, and 3 are shown and labeled accordingly. Arrows indicate the corresponding transitions between pairs of electron subbands, E_{12}, E_{23}, and E_{31}. (b) Intersubband nonlinear susceptibility $\chi_{zzz}^{(2)}$ as a function of wavenumber. (Panels [a] and [b] were reproduced from Lee, J. et al., *Nature*, 511, 65–69, 2014.)

therefore, the saturation effects are negligible, i.e., $S(I_z^\omega) \approx 1$. As the input intensity increases, the populations in the subbands vary, modifying the saturation and loss rate. In such a three-level system, the population of the first two subbands can be found from the coupled-rate equations in the steady-state as [62]

$$\frac{\partial N_1}{\partial t} = -\frac{\alpha_{12}(\omega)I_z^\omega}{\hbar\omega} - \frac{\alpha_{13}(2\omega)I_z^{2\omega}}{\hbar 2\omega} + \frac{N_2 - N_2^0}{\tau_{12}} + \frac{N_3 - N_3^0}{\tau_{13}} = 0, \qquad (2.22)$$

$$\frac{\partial N_2}{\partial t} = \frac{\alpha_{12}(\omega)I_z^\omega}{\hbar\omega} - \frac{\alpha_{23}(\omega)I_z^\omega}{\hbar\omega} - \frac{N_2 - N_2^0}{\tau_{12}} + \frac{N_3 - N_3^0}{\tau_{23}} = 0, \qquad (2.23)$$

where $\tau_{ij} = 1/\gamma_{ij}$ is the intersubband relaxation time, $I_z^{2\omega}$ is the acting intensity of the second-harmonic field in the MQWs, and N_i^0 is the carrier concentration in the i-th subband in thermal equilibrium. In addition, the average carrier density does not change with the pump intensity, and therefore $N_3 = N - N_1 - N_2$ and $N_3^0 = N - N_1^0 - N_2^0$ hold. The absorption coefficient α_{ij} also depends on the carrier concentration as [63]:

$$\alpha_{ij}(\omega') = \frac{N_i - N_j}{N}\alpha_{ij}^0(\omega'), \qquad (2.24)$$

where α_{ij}^0 is the absorption rate in the low-intensity limit

$$\alpha_{ij}^0(\omega') = \frac{N\omega'e^2 z_{ij}^2}{n_{zz}^{MQW}(\omega')\varepsilon_0 c\hbar^2\left[(\omega' - \omega_{ij})^2 + (\gamma_{ij})^2\right]}. \qquad (2.25)$$

Furthermore, the field intensity at 2ω is negligible compared to the intensity at ω, and so we can take $I_z^{2\omega} \approx 0$. As we discussed above, we can also assume $N_2^0 \approx N_3^0 \approx 0$. Neglecting the effects of optical heating and the energy relaxation into lattice phonons [60], we can rewrite the coupled-rate equations (2.22) and (2.23) as:

$$-N_1\left[\frac{I_z^\omega}{2I_{s12}^\omega}+\frac{\tau_{12}}{\tau_{13}}\right]+N_2\left[\frac{I_z^\omega}{2_{s12}^\omega}-\frac{\tau_{12}}{\tau_{13}}+1\right]+N\frac{\tau_{12}}{\tau_{13}}=0, \qquad (2.26)$$

$$N_1\left[\frac{I_z^\omega}{2I_{s12}^\omega}\frac{\tau_{23}}{2\tau_{12}}-\frac{I_z^\omega}{2I_{s23}^\omega}-1\right]-N_2\left[\left(\frac{I_z^\omega}{2I_{s12}^\omega}+1\right)\frac{\tau_{23}}{\tau_{12}}+\frac{I_z^\omega}{I_{s23}^\omega}+1\right]+N\left[\frac{I_z^\omega}{2I_{s23}^\omega}+1\right]=0, \quad (2.27)$$

where $I_{s,ij}^{\omega'}$ denotes the saturation intensity at the frequency ω' defined as

$$I_{sij}^{\omega'} = \frac{N\hbar\omega'}{2\alpha_{ij}^0(\omega')\tau_{ij}}. \qquad (2.28)$$

Solving for N_1, N_2, and N_3 (see Appendix B for the full expressions), and inserting the results into (2.20), we obtain the following expression for the saturation factor [47]:

$$S(I_z^\omega) = \frac{2(B-C)+4D}{3+2(B+2C)+4D}, \qquad (2.29)$$

where

$$B = \frac{I_{s12}}{I_z^\omega}\left(1+\frac{\tau_{12}}{\tau_{13}}\right), \quad C = \frac{I_{s23}}{I_z^\omega}\left(1+\frac{\tau_{23}}{\tau_{13}}\right), \quad D = \frac{I_{s12}I_{s23}}{(I_z^\omega)^2}\left(1+\frac{\tau_{23}}{\tau_{13}}\right). \qquad (2.30)$$

Figure 2.10 depicts an example of evolution of the population in each of the three subbands and plots the saturation factor in the MQWs versus the acting field intensity, I_z^ω (during the SHG process). It is evident that saturation is associated predominantly with the depopulation of the first subband and the increased population of the second subband.

Combining (2.12) and (2.19) with (2.29) and (2.30) allows us to obtain the intensity-dependent effective nonlinear susceptibility of the metasurface as

$$\chi_{\text{eff},ijk}^{(2)}(I_\omega) = \frac{1}{V}\int_V \chi_{zzz}^{(2)}(\mathbf{r},I_z^\omega(\mathbf{r}))Q_{z(i)}^{2\omega}(\mathbf{r})Q_{z(j)}^\omega(\mathbf{r})Q_{z(k)}^\omega(\mathbf{r})dV, \qquad (2.31)$$

where $\chi_{zzz}^{(2)}(\mathbf{r},I_z^\omega(\mathbf{r})) \equiv \chi_{zzz}^{(2)}(\mathbf{r})S(I_z^\omega(\mathbf{r}))$. The local acting intensity can be computed from the pump beam intensity as

$$I_z^\omega(\mathbf{r}) = n_{zz}^{\text{MQW}}(\omega)\left[\left|Q_{z(j)}^\omega(\mathbf{r})\right|^2+\left|Q_{z(k)}^\omega(\mathbf{r})\right|^2\right]I_\omega. \qquad (2.32)$$

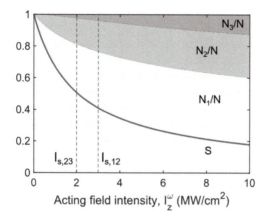

FIGURE 2.10   Evolution of the subband populations N_i and the saturation factor S versus the acting field intensity, I_z^ω. The two saturation intensities are set to be $I_{s12} = 3 \text{ MW/cm}^2$ and $I_{s23} = 2 \text{ MW/cm}^2$. The intersubband relaxation times are taken to be $\tau_{12} = 2.2$ ps, $\tau_{13} = 1$ ps, $\tau_{23} = 2$ ps.

Finally, the efficiency of the SHG process accounting for saturation can be evaluated as

$$\eta_{\text{SHG}} = \frac{P_{2\omega}}{P_\omega} = \frac{2\omega^2 h^2}{\varepsilon_0 c^3 w^2} I_\omega^0 \int_0^\infty |d_{\text{eff}} (I_\omega(r))|^2 \exp\left(-\frac{4r^2}{w^2}\right) r dr, \tag{2.33}$$

where

$$d_{\text{eff}} (I_\omega) = \frac{1}{2} \left| \hat{e}^{2\omega} \cdot \ddot{\chi}_{\text{eff}}^{(2)}(I_\omega) : \hat{e}^\omega \hat{e}^\omega \right|. \tag{2.34}$$

In Figure 2.11, we compare the results for $P_{2\omega} = \eta_{\text{SHG}} P_\omega$ computed using Eqs. (2.31)–(2.34) and plotted versus P_ω^2 for the metasurface considered in Ref. [43] for two different polarization combinations, xxx and yyy. One can appreciate the excellent agreement between the theory and experimental data. This fact also proves that it is saturation, not the optical damage threshold, which determines the upper bound for the conversion efficiency in MQW-loaded plasmonic nonlinear metasurfaces. For comparison, we also plot the second-harmonic power without accounting for saturation effects.

It is also instructive to show that, with a further increase in pump power, the second-harmonic efficiency ceases to grow and starts to drop, as shown in Figure 2.12. It means that a significant fraction of the MQW volume is now saturated and cannot generate nonlinear waves. It is also seen that intensities at which the SHG conversion efficiency and second-harmonic power reach their respective maxima are very different. In fact, the maximum of the theoretically predicted second-harmonic power is achieved when a significant fraction of the MQW volume is saturated. At the same time, the presence of this peak clearly shows that MQW-based hybrid nonlinear metasurfaces are best suited for low- and medium-power nonlinear setups.

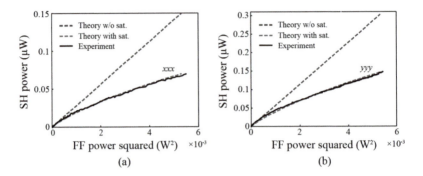

FIGURE 2.11 Generation of second-harmonic beams using the nonlinear plasmonic meta-surface reported in [43]. Results are computed with (red dashed line) and without (blue dashed line) considering saturation effects. Experimental results are shown with black solid lines. (a) *xxx* polarization combination. (b) *yyy* polarization. The impinging Gaussian beam has a radius of 17 μm. (Panels [a] and [b] were reproduced from Gomez-Diaz, J.S. et al., *Phys. Rev. B*, 92, 125429, 2015.)

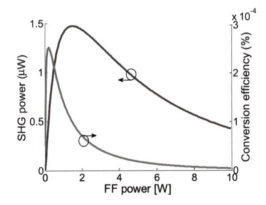

FIGURE 2.12 Simulated SHG power and conversion efficiency of a nonlinear metasurface from shown in Figure 2.3 (see Ref. [43]) for the *yyy* polarization combination versus the power of an incident 34 μm-wide Gaussian beam. (Reproduced from Gomez-Diaz, J.S. et al., *Phys. Rev. B*, 92, 125429, 2015.)

2.3.3 Saturation of MQW-Based Nonlinear Metasurfaces Aimed at DFG

Based on the results of the previous section, we can develop a similar saturation model for an MQW-based nonlinear metasurface aimed at difference-frequency generation (DFG), where two pump waves with distinct frequencies ω_1 and ω_2 produce a third wave $\omega_3 = \omega_1 - \omega_2$. For the DFG process, the intrinsic susceptibility in the vicinity of transitions is approximately given as

$$\chi_{zzz}^{(2)}(I_z^{\omega_1}, I_z^{\omega_2}) \approx \frac{Ne^3 z_{12} z_{13} z_{23}}{\varepsilon_0 \hbar^2 [(\omega_3 - \omega_{23}) - i\gamma_{23}]} \left[\frac{S_1(I_z^{\omega_1}, I_z^{\omega_2})}{(\omega_1 - \omega_{31}) - i\gamma_{31}} + \frac{S_2(I_z^{\omega_1}, I_z^{\omega_2})}{(\omega_{21} - \omega_2) - i\gamma_{21}} \right], \quad (2.35)$$

with the saturation factors $S_1(I_z^{\omega_1}, I_z^{\omega_2}) \approx (N_1 - N_3)/N$ and $S_2(I_z^{\omega_1}, I_z^{\omega_2}) \approx (N_1 - N_2)/N$. Again, solving the coupled-rate equation system (2.22) and (2.23), we arrive at the following expression for the saturation factors

$$S_1(I_z^{\omega_1}, I_z^{\omega_2}) \approx \frac{2D_{13}}{B_{13}(2+C_{12})+2D_{13}(1+C_{12})}, \tag{2.36}$$

$$S_2(I_z^{\omega_1}, I_z^{\omega_2}) \approx \frac{(B_{13}+2D_{13})(1-C_{12})}{B_{13}(2+C_{12})+2D_{13}(1+C_{12})}. \tag{2.37}$$

Here, the dimensionless parameters B_{ij}, C_{12}, and D_{ij} are

$$B_{ij} = \frac{I_{sij}^{\omega_1}}{I_z^{\omega_1}} + \frac{I_{sij}^{\omega_2}}{I_z^{\omega_2}}, \tag{2.38}$$

$$C_{12} = \frac{B_{12}+B_{13}D_{12}[1+2(1+\tau_{13}/\tau_{23})]^{-1}}{B_{12}+D_{12}}, \tag{2.39}$$

$$D_{ij} = \frac{I_{sij}^{\omega_1}}{I_z^{\omega_1}} \frac{I_{sij}^{\omega_2}}{I_z^{\omega_2}} \left(1+\frac{\tau_{ij}}{\tau_{12}}\right), \tag{2.40}$$

where $I_z^{\omega'}$ and $I_{sij}^{\omega'}$ are acting and saturation intensities for each wave defined by (2.21) and (2.28), respectively.

In Figure 2.13a, we show an example of an MQW-based hybrid nonlinear metasurface aimed at highly efficient DFG. The design is similar to the one aimed at SHG, with plasmonic resonances tailored to match the corresponding transition frequencies of MQWs. The DFG power conversion efficiency

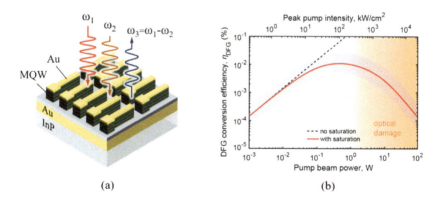

(a)	(b)

FIGURE 2.13 (a) Design and operation schematic of the MQW-based hybrid nonlinear metasurface aimed at DFG. (b) Theoretically computed power conversion efficiency versus the pump beam power and peak intensity. The area shaded in light orange indicates the region of potential optical damage. (Panel [b] was reproduced from Tymchenko, M. et al., *J. Opt.*, 19, 104001, 2017.)

computed by means of the overlap integral (2.14) and the above saturation model is shown in Figure 2.13a. Similar to metasurfaces aimed at SHG, this metasurface is susceptible to strong saturation. The peak theoretically computed power conversion efficiency is found to be on the order of 10^{-4} achieved at intensities of about 10^2 kW/cm^2 [49].

2.4 ADVANCED PHASE CONTROL USING HYBRID NONLINEAR METASURFACES

The interest in flat nonlinear optics and nonlinear metasurfaces is in part associated by the absence of phase-matching requirements. This property opens exciting possibilities to tailor nonlinear fields at subwavelength scales. Similar to linear metasurfaces, which can efficiently manipulate the phase and amplitude of scattered fields by tailoring effective electric and/or magnetic currents across their surface, in this section we discuss how nonlinear metasurfaces can tailor the profile of *nonlinear* polarization currents to obtain advanced radiation control functionalities. For this purpose, we can apply reverse engineering to obtain the complex impedance profiles, i.e., the necessary electric (and magnetic) currents that can sustain the desired radiated fields, required to realize nonlinear beam-steering, focusing, helical beam-shaping, holography, etc. It should be emphasized that standard nonlinear setups employing bulk nonlinear media cannot implement such functionalities and rely on external optical elements based on propagation phase delay, such as lenses, polarizers and beam-splitters, to manipulate the nonlinear field.

If the radiation is directed toward broadside, rather than engineering an exact impedance profile it is sufficient to realize an accurate spatial *phase* distribution of effective polarization currents. Polarization phase gradients induce a local transverse momentum along the metasurface. This approach, which in linear optics is known as generalized Snell's law, has been extensively used to achieve efficient field tailoring [64]. In the realm of linear metasurfaces, various strategies have been devised to obtain a desired spatially varying phase distribution of the field, such as changing the size, orientation, aspect ratio, shape, and arrangement of resonant elements [65]. Nonlinear fields, however, are much more sensitive to small variations of the local phase and amplitude of fields at pump frequency (or frequencies). Even small errors may lead to large phase discrepancies and amplitude variations. Consequently, the number of available phase tailoring techniques is limited. For instance, it has been shown that phase control functionalities can be achieved in thin layers of nonlinear media with specific poling patterns [66–72]. Another approach, conceptually similar to poling, leverages the non-centrosymmetric shapes of metallic inclusions such as SRRs to periodically change the sign of second-harmonic currents by flipping the SRRs' orientations [73,74]. In Figure 2.14, we show an example of an MQW-based metasurface employing this "poling" technique to steer the second-harmonic radiation away from the normal direction [45]. Unfortunately, the poling approach lacks spatial resolution, because the modulation period must be larger than the free-space wavelength at the frequency of radiation. As a result, undesirable nonlinear radiation associated with non-zero diffraction orders will be generated.

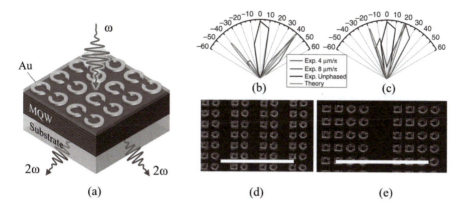

FIGURE 2.14 (a) Design and operation schematic of a highly efficient MQW-based hybrid nonlinear metasurface employing the periodic "poling" approach to steer the second-harmonic radiation away from normal. (b) and (c) Experimental and theoretical far-field patterns of the second-harmonic field for the two fabricated samples with different periodicity. Corresponding SEM images of the samples are shown in (d) and (e). The scale bars are 10 μm. (Panels [b]—[e] were reproduced from Wolf, O. et al., *Nat. Commun.*, 6, 7667, 2015.)

Recently, it has been shown that instead of using simple binary phase patterns obtained through "poling," the nonlinear phase can be controlled with subwavelength precision by gradually rotating polarization-selective elements such as SRRs [48,75,76]. This approach employs a so-called Pancharatnam-Berry (PB) geometric phase to induce unequal phase shifts between the two circular-polarized components of the radiated field. Although the concept of PB geometric phase has been known to the optical community for several decades, it currently sees a new wave of interest due to numerous exciting advances associated with linear metasurfaces. Compared to other phase-front control techniques employing inclusions of varying sizes and shapes, the PB phase approach provides a key advantage: it allows 360° phase manipulation while keeping the amplitude nearly constant. In Figure 2.15a, we show an MQW-based hybrid nonlinear metasurface which employs the PB geometric phase concept with linear orientation gradients to deflect and steer the second-harmonic radiation toward specific directions. The design of the metasurface is conceptually similar to the one shown in Figure 2.5a, but instead of the T-shaped structures, more compact SRRs are used to allow rotation and minimize cross-coupling among the unit-cells. As is clearly seen in Figure 2.15b and c, a linear gradient of SRRs' orientation leads to right-hand and left-hand circular polarization (RCP and LCP) components of the second-harmonic signal that is almost fully radiated in two distinct directions away from normal. This is in contrast to other nonlinear structures radiating obliquely at normal illumination which usually have relatively symmetric radiation far-field patterns with respect to the *z*-axis. Small additional lobes in Figure 2.15b and c correspond to other diffraction orders existing as a result of the gradient metasurface periodicity being larger than the free-space wavelength at 2ω. Importantly, this metasurface employing a PB phase approach successfully

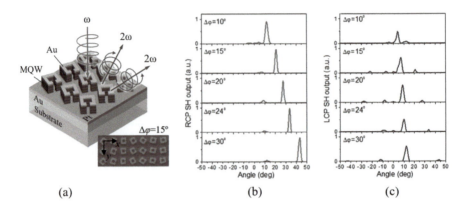

FIGURE 2.15 (a) Highly efficient MQW-based hybrid nonlinear metasurface employing the linear orientation gradient to impart the additional tangential momentum to steer the RCP and LCP second-harmonic radiation away from the z-axis. The metasurface is illuminated by a normally incident RCP beam. The inset shows the FEM image of a fabricated sample with orientation step of $\Delta\varphi = 15°$ along the x axis. (b), (c) The measured second-harmonic signal for RCP and LCP polarizations, respectively, for samples with various $\Delta\varphi$ ranging from 10° to 30°. (Panels [b] and [c] were reproduced from Nookala, N. et al., *Optica*, 3, 283, 2016.)

combines precise nonlinear phase control with high SHG efficiency [76,77]. Other all-metallic structures employing the PB concept have recently demonstrated third-harmonic radiation steering [78] and nonlinear holographic image encoding using second-harmonic radiation [79].

2.4.1 Pancharatnam-Berry Geometric Phase in Nonlinear Metasurfaces

In view of all these exciting advances, it is necessary to develop a theoretical method to accurately model nonlinear metasurfaces with very large periodicity or even without periodicity at all. It would be ideal to avoid full-scale numerical simulations which would obviously require massive computational resources. Fortunately, the theoretical formalism based on effective susceptibility tensor introduced in Sections 2.2 and 2.3 can be expanded to include this scenario. To this purpose, the metasurface should meet several additional restrictions: (i) the spatial variation of elements' orientation should be adiabatic (i.e., sufficiently smooth changes between adjacent cells); (ii) the incidence angle—measured with respect to the direction perpendicular to the structure—should be small; and (iii) the generated SHG signal should be directed not too far away from $\pm z$ direction. If these conditions are met, the nonlinear response of each unit-cell can be still accurately approximated using the effective nonlinear susceptibility (2.12) evaluated in the assumption of a perfectly periodic metasurface, thus minimizing the required amount of full-wave simulations.

Let us begin our analysis by considering a nonlinear metasurface aimed at SHG and consisting of unit-cells located at $\mathbf{r}_n = (x_n, y_n, 0)$, where n is the unit-cell index. Under the assumptions given above, we can describe their second-order nonlinear response with an effective nonlinear susceptibility tensor $\vec{\chi}_{\text{eff},n}^{(2)} \equiv \vec{\chi}_{\text{eff}}^{(2)}(\mathbf{r}_n)$. In what follows, for the sake of brevity, we drop the subscript "eff" throughout the analysis. The effective second-harmonic nonlinear polarization density induced in the n-th cell is related to the *incident* pump field as

$$\mathbf{P}_n^{2\omega} = \varepsilon_0 \, \vec{\chi}_n^{(2)} : \mathbf{E}_n^{\omega} \, \mathbf{E}_n^{\omega}. \tag{2.41}$$

Here, we assume that all quantities are written in Cartesian coordinates. By analogy with linear optics, we anticipate that element rotation will locally induce a phase difference between circularly polarized states. Hence, it is convenient to transform (2.41) to a circular polarization (CP) basis $\{\hat{\mathbf{e}}_R, \hat{\mathbf{e}}_L\}$ corresponding to the right-hand and left-hand circular polarization states, respectively, defined as

$$\hat{\mathbf{e}}_R = \frac{1}{\sqrt{2}}(\hat{\mathbf{e}}_x - i\hat{\mathbf{e}}_y), \qquad \hat{\mathbf{e}}_L = \frac{1}{\sqrt{2}}(\hat{\mathbf{e}}_x + i\hat{\mathbf{e}}_y). \tag{2.42}$$

Note that we choose $\hat{\mathbf{e}}_z$ to point inside the metasurface. The relation between the CP and Cartesian bases can be also written in the matrix form $\hat{\mathbf{e}}_u = \sum_i C_{ui} \hat{\mathbf{e}}_i$, where polarization indexes u and i run through R, L and x, y, respectively, and the matrix $\mathbf{C}$ is

$$\mathbf{C} = \frac{1}{\sqrt{2}}\begin{pmatrix} 1 & -i \\ 1 & i \end{pmatrix}. \tag{2.43}$$

In the CP basis, the amplitude of the pump wave and the nonlinear polarization is related to those in Cartesian basis as

$$\mathbf{E}^{\omega} = \mathbf{D} \cdot \mathbf{E}_{\odot}^{\omega}, \tag{2.44}$$

$$\mathbf{P}^{2\omega} = \mathbf{D} \cdot \mathbf{P}_{\odot}^{2\omega}, \tag{2.45}$$

where $\mathbf{D} = \mathbf{C}^T$ with T denoting the matrix transpose, and the symbol "$\odot$" means that the vector or matrix is written in the CP basis. Substituting (2.44) and (2.45) into (2.41), we obtain:

$$\mathbf{P}_{\odot,n}^{2\omega} = \varepsilon_0 \, \vec{\chi}_{\odot,n}^{(2)} : \mathbf{E}_{\odot,n}^{\omega}\mathbf{E}_{\odot,n}^{\omega}, \tag{2.46}$$

with

$$\vec{\chi}_{\odot}^{(2)} = \mathbf{D}^{-1} \cdot \vec{\chi}^{(2)} : \mathbf{D}\mathbf{D}. \tag{2.47}$$

Explicit transformation relations for the nonlinear susceptibility tensor elements from Cartesian to the CP base are given in Appendix C. Now, we can proceed to calculate the nonlinear susceptibility tensor of an element rotated by an angle φ. The rotation matrix relating the field in the rotated coordinate system to the initial field, $\mathbf{E}(\varphi) = \mathbf{R}_{\varphi} \cdot \mathbf{E}$, is given as

$$\mathbf{R}_\varphi = \begin{pmatrix} \cos\varphi & \sin\varphi \\ -\sin\varphi & \cos\varphi \end{pmatrix}. \tag{2.48}$$

By analogy with (2.47), the rotation matrix in the CP basis can be computed as

$$\mathbf{R}_{\circlearrowleft,\varphi} = \mathbf{D}^{-1} \cdot \mathbf{R}_\varphi \cdot \mathbf{D} = \begin{pmatrix} e^{-i\varphi} & 0 \\ 0 & e^{i\varphi} \end{pmatrix}. \tag{2.49}$$

Neglecting the cross-coupling between the cells, the susceptibility tensor of the structure rotated by an angle φ is given by

$$\vec{\chi}_{\circlearrowleft}^{(2)}(\varphi) = \mathbf{R}_{\circlearrowleft,\varphi}^{-1} \cdot \vec{\chi}_{\circlearrowleft}^{(2)} : \mathbf{R}_{\circlearrowleft,\varphi} \mathbf{R}_{\circlearrowleft,\varphi}. \tag{2.50}$$

Since the matrix $\mathbf{R}_{\circlearrowleft,\varphi}$ is diagonal, all tensor elements in Eq. (2.50) can be easily computed as

$$\chi_{uvw}^{(2)}(\varphi) = \chi_{uvw}^{(2)} e^{i\varphi(-u+v+w)}, \tag{2.51}$$

where each of the indexes u, v, and w in the exponential factor should be replaced according to the following rule: $R = -1, L = 1$. We stress that he phase factor $e^{i\varphi(-u+v+w)}$ is only due to the local element orientation and, thus, is of a purely geometrical nature. The origin of this geometrical phase can be understood at an intuitive level: by rotating a polarization-selective element by an angle φ we introduce unequal phase factors $e^{\mp i\varphi}$ to RCP and LCP components of radiating currents and additional $e^{\pm i\varphi}$ phase factors to RCP and LCP components of the pump fields sensed by SRRs. Despite its apparent simplicity, the emergence of this phase is a closely related to the Pancharatnam-Berry geometric phase, which plays a profound role in optics and quantum mechanics. In metasurfaces, geometric phase gradients break the inversion symmetry $\mathbf{r} \to -\mathbf{r}$ by imprinting the transverse momentum $\nabla\varphi(x,y)$, leading to a splitting of the dispersion relation of states with opposite optical helicity [80].

Geometric phase also imposes additional requirements on the symmetry of the metasurface elements [81]. According to the Neumann principle, any coordinate transformations that lead to the same structure geometry must also preserve all quantities characterizing its physical properties such as susceptibility tensors. For a structure with an N-fold rotation symmetry around the z-axis, we must have

$$\chi_{uvw}^{(2)} = e^{i\frac{2\pi}{N}(-u+v+w)} \chi_{uvw}^{(2)}, \tag{2.52}$$

which can be satisfied only when the following polarization selection rule is met:

$$-u + v + w = nN, \tag{2.53}$$

where n is an integer. For any combination of polarization indexes $-u + v + w$ can be either ± 1 or ± 3. Thus, the relation (2.53) can only be satisfied for $N = 1$ and $N = 3$.

For $N = 1$, such as in the case of SRRs, any polarization combination is possible, and so we end up with the following simple expressions:

$$\chi_{RRR}^{(2)}(\varphi) = \chi_{RRR}^{(2)} e^{i\varphi}, \qquad\qquad \chi_{LLL}^{(2)}(\varphi) = \chi_{LLL}^{(2)} e^{-i\varphi},$$

$$\chi_{RRL}^{(2)}(\varphi) = \chi_{RLR}^{(2)}(\varphi) = \chi_{RRL}^{(2)} e^{-i\varphi}, \qquad \chi_{LLR}^{(2)}(\varphi) = \chi_{LRL}^{(2)}(\varphi) = \chi_{LLR}^{(2)} e^{i\varphi}, \qquad (2.54)$$

$$\chi_{RLL}^{(2)}(\varphi) = \chi_{RLL}^{(2)} e^{-i3\varphi}, \qquad\qquad \chi_{LRR}^{(2)}(\varphi) = \chi_{LRR}^{(2)} e^{i3\varphi}.$$

For $N = 3$, the only allowed polarization combinations are $\chi_{RLL}^{(2)}(\varphi) = \chi_{RLL}^{(2)} e^{-i3\varphi}$ and $\chi_{LRR}^{(2)}(\varphi) = \chi_{LRR}^{(2)} e^{i3\varphi}$, and all other nonlinear susceptibility components are strictly zero. Interestingly, for structures with odd-fold rotation symmetry higher than 3, the SHG is forbidden even despite their non-centrosymmetric geometry. Similar polarization selection rules can be obtained for higher harmonic generation processes [82].

Finally, the relations (2.51)–(2.54) were obtained in the assumption that the second-harmonic signal is radiated along the $+z$-direction (in transmission). If the SHG occurs in the $-z$-direction (in reflection), the RCP (LCP) nonlinear polarization currents become sources for LCP (RCP) second-harmonic waves. This fact can be easily taken into account if we write

$$\chi_{uvw}^{(2)\pm}(\varphi) = \chi_{uvw}^{(2)} e^{i\varphi(\mp u + v + w)}, \qquad (2.55)$$

where "$\pm$" denotes the direction of SHG along the z-axis.

2.4.2 FREE-SPACE SECOND-HARMONIC RADIATION OF GRADIENT NONLINEAR METASURFACES

Having derived the effective nonlinear susceptibility of each rotated element, we are now ready to compute the far-field second-harmonic radiation. The amplitude of the second-harmonic electric field $\mathbf{E}^{2\omega}(\mathbf{r})$ can be found as a sum of radiation from independent effective dipole moments of each cell, $\mathbf{d}_{\ominus,n}^{2\omega} = V\mathbf{P}_{\ominus,n}^{2\omega}$, with V denoting the unit-cell volume, as

$$\mathbf{E}_{\ominus}^{2\omega}(\mathbf{r}) = \frac{k_{2\omega}^2}{\varepsilon_0} V \sum_n \ddot{\mathbf{G}}(\mathbf{r}, \mathbf{r}_n) \cdot \mathbf{P}_{\ominus,n}^{2\omega}, \qquad (2.56)$$

where $k_{2\omega} = 2\omega/c$, and $\ddot{\mathbf{G}}(\mathbf{r}, \mathbf{r}_n)$ is a dyadic Green's function. We also neglected the effect of cross-coupling at 2ω due to fact that the second-harmonic fields are much weaker than the pump field. Since all linear scattering phenomena are encapsulated by the effective susceptibility tensor, we shall assume that the metasurface is suspended in free-space, i.e., we should employ the far-field free-space Green's function [83].

$$\ddot{\mathbf{G}}_{FF}(\mathbf{r}, \mathbf{r}_n) = [\ddot{\mathbf{I}} - \hat{\mathbf{e}}_r \hat{\mathbf{e}}_r] \frac{e^{ik_{2\omega}r}}{4\pi r} e^{-ik_{2\omega}\hat{\mathbf{e}}_r \cdot \mathbf{r}_n}, \qquad (2.57)$$

with $\mathbf{r} = r\hat{\mathbf{e}}_r$ being the location of the observer.

Let us consider now an arbitrary polarized plane wave $\mathbf{E}^{\omega}_{\text{in},\ominus} e^{i\mathbf{k}_{\omega}\cdot\mathbf{r}}$ with a wave-number $\mathbf{k}_{\omega} = k_{\omega}\hat{\mathbf{e}}^{\omega}_r$, where $\hat{\mathbf{e}}^{\omega}_r$ is a unit-vector defined by a pair of polar angles (θ_1, ϕ_1), as shown in Figure 2.16. The incidence angle θ_1 is assumed to be small to justify the use of the effective nonlinear susceptibility computed at normal incidence. The amplitude of the impinging wave can be presented in its own CP basis $\{\hat{\mathbf{e}}^{\omega}_R, \hat{\mathbf{e}}^{\omega}_L\}$ as

$$\hat{\mathbf{e}}^{\omega}_R = \frac{1}{\sqrt{2}}(\hat{\mathbf{e}}^{\omega}_\theta - i\hat{\mathbf{e}}^{\omega}_\phi), \qquad \hat{\mathbf{e}}^{\omega}_L = \frac{1}{\sqrt{2}}(\hat{\mathbf{e}}^{\omega}_\theta + i\hat{\mathbf{e}}^{\omega}_\phi), \qquad (2.58)$$

where $\hat{\mathbf{e}}^{\omega}_\theta$ and $\hat{\mathbf{e}}^{\omega}_\phi$ are unit-vectors of a spherical coordinate system, defined uniquely by (θ_1, ϕ_1), see Figure 2.16. The tangential components of the pump field in the CP basis associated with the metasurface, $\mathbf{E}^{\omega}_\ominus$, can be approximately found as

$$\mathbf{E}^{\omega}_\ominus \approx \mathbf{U}_\omega \cdot \mathbf{E}^{\omega}_{\text{in},\ominus}, \qquad (2.59)$$

where

$$\mathbf{U}_\omega = \frac{1}{2}\begin{bmatrix} (\cos\theta_1 + 1)e^{-i\phi_1} & (\cos\theta_1 - 1)e^{i\phi_1} \\ (\cos\theta_1 - 1)e^{-i\phi_1} & (\cos\theta_1 + 1)e^{i\phi_1} \end{bmatrix}. \qquad (2.60)$$

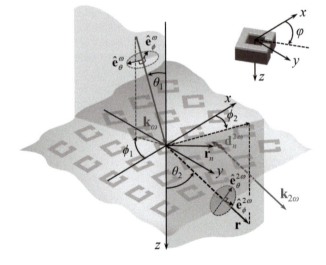

FIGURE 2.16 Nonlinear metasurface consisting of elements with spatially varying $\chi^{(2)}$, and $\chi^{(1)} = 0$ at all frequencies. The metasurface is suspended in free-space and illuminated by an obliquely incident plane wave of arbitrary polarization. Each unit-cell located at $\mathbf{r}_n$ independently induces a nonlinear dipole moment $\mathbf{d}^{2\omega}_n$. The inset in the top right depicts a realistic unit-cell with an SRR rotated by an angle φ.

Here, we neglected the z-polarized component of the pump field in the free-space. Combining (2.41), (2.56), (2.57), and (2.59), after simple manipulations we obtain

$$\mathbf{E}_{\mathrm{FF},\circlearrowleft}^{2\omega}(\mathbf{r}) = \frac{e^{ik_{2\omega}r}}{4\pi r} k_{2\omega}^2 V \mathbf{U}_{2\omega}^{-1} \cdot (\ddot{\chi}_{\circlearrowleft}^{(2)} \circ \mathrm{AF}_{\circlearrowleft}^{2\omega}) : (\mathbf{U}_\omega \cdot \mathbf{E}_{\mathrm{in},\circlearrowleft}^\omega)(\mathbf{U}_\omega \cdot \mathbf{E}_{\mathrm{in},\circlearrowleft}^\omega), \qquad (2.61)$$

where the matrix $\mathbf{U}_{2\omega}$ projects the second-harmonic field in the observation direction $\hat{\mathbf{e}}_r = \hat{\mathbf{e}}_r(\theta_2, \phi_2)$ (again, see Figure 2.16) onto the CP basis $\{\hat{\mathbf{e}}_R, \hat{\mathbf{e}}_L\}$:

$$\mathbf{U}_{2\omega} = \frac{1}{2}\begin{bmatrix} (\cos\theta_2 + 1)e^{i\phi_2} & (\cos\theta_2 - 1)e^{i\phi_2} \\ (\cos\theta_2 - 1)e^{-i\phi_2} & (\cos\theta_2 + 1)e^{-i\phi_2} \end{bmatrix}. \qquad (2.62)$$

In Eq. (2.61), the symbol "$\circ$" denotes the Hadamard entry-wise product of tensors, and the tensor $\mathrm{AF}_{\circlearrowleft}^{2\omega}$ is the linear array factor [84] computed independently for each element of the nonlinear susceptibility tensor $\chi_{uvw}^{(2)}$:

$$\mathrm{AF}_{uvw}^{2\omega} = \sum_n p_{n,vw} e^{i(2\mathbf{k}_\omega - k_{2\omega}\hat{\mathbf{e}}_r)\cdot\mathbf{r}_n - i\varphi_n(u-v-w)}, \qquad (2.63)$$

where the factor $0 < p_{n,vw} < 1$ accounts for variations of pump field amplitudes in the n-th cell when the nonlinear metasurface is illuminated by a finite-size beam. In case that the pump signal is a plane wave, $p_{n,vw} = 1$. This formulation enables fast computation of the field radiated by nonlinear metasurfaces when θ_1 is small and θ_2 is not very far from 0 or π corresponding to SHG in transmission and reflection, respectively. Most importantly, numerical simulations are required only to compute the modal overlap integral in an ideally periodic environment at normal incidence. One should also remember that $\ddot{\chi}_{\circlearrowleft}^{(2)} \equiv \ddot{\chi}_{\circlearrowleft}^{(2)}$ for $0 < \theta_2 < \pi/2$, and $\ddot{\chi}_{\circlearrowleft}^{(2)} \equiv \ddot{\chi}_{\circlearrowleft}^{(2)-}$ for $\pi/2 < \theta_2 < \pi$.

To benchmark the performance of our nonlinear phased array theory, we consider the simplest case of a metasurface consisting of $N = 24$ unit-cells along x and infinitely periodic along y. The metasurface periodicity is d in both directions, and it has a linear orientation gradient along x with the orientation step is $\Delta\varphi$, i.e., $\varphi_n = n\Delta\varphi$ where $n = 1...N$. We assume the metasurface is illuminated by a plane pump with a wavenumber $\mathbf{k}_\omega = k_\omega(\sin\theta_1, 0, \cos\theta_1)$. In this case, the array factor (2.63) becomes

$$\mathrm{AF}_{uvw}^{2\omega} = \sum_{n=1}^N e^{inX_{u-v-w}} = e^{iX_{u-v-w}(N-1)/2}\left[\frac{\sin\left(\dfrac{N}{2}X_{u-v-w}\right)}{\sin\left(\dfrac{1}{2}X_{u-v-w}\right)}\right], \qquad (2.64)$$

with

$$X_l = (2k_\omega \sin\theta_1 - 2k_{2\omega}\sin\theta_2\cos\varphi_2)d - l\Delta\varphi. \qquad (2.65)$$

Note that in contrast to linear phased arrays, the pre-factor $e^{iX_{u-v-w}(N-1)/2}$ generally cannot be dropped when the susceptibility tensor has more than one element. Similar expressions can be obtained for 2D linear orientation gradients [48].

In Figure 2.17, we show far-field second-harmonic directivity patterns for RCP and LCP radiation components for a metasurface with $\Delta\varphi = 15°$ difference of orientation of adjacent elements along the x-direction, computed by means of Eq. (2.60)–(2.65) for normal and slightly oblique incidence. Theoretical results are plotted against full-scale numerical simulations. It is seen that at normal incidence the agreement is nearly perfect. At oblique incidence, theoretical results overall agree very well, and, as expected, deviate only close to the end-fire direction.

Using this simple yet efficient framework, a more complicated second-harmonic beam-shaping functionalities, such as 2D pencil and vortex beam-shaping can be also easily realized, as is shown in Figure 2.18. Analogous theoretical frameworks can be developed for metasurfaces aimed at other nonlinear processes [48].

Finally, we would like to stress that this approach can be also successfully used for analysis of metasurfaces employing other phase-front tailoring techniques. If the geometry of neighboring unit-cells does not differ substantially, it is still possible to evaluate the effective susceptibility tensor for each particular unit-cell in an ideally periodic environment, and then apply the Green's function formalism to obtain the radiated far-field.

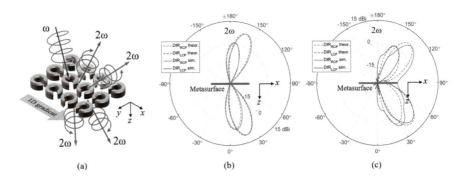

(a)   (b)   (c)

FIGURE 2.17 (a) Schematic of a hybrid nonlinear metasurface aimed at SHG, consisting of identical SRR elements with 1D orientation gradient along the x-direction. The orientation step is $\Delta\varphi = 15°$. (b) Far-field directivity patterns (in dBi) in the x-z plane for RCP and LCP components of second-harmonic radiation when the metasurface is illuminated by a normally impinging RCP Gaussian beam. Theoretical and numerical results are shown with dashed and solid lines, respectively. (c) Same as in (b), but for oblique incidence with $\theta_1 = 20°$. (Panels [b] and [c] were reproduced from Tymchenko, M. et al., *Phys. Rev. B*, 94, 214303, 2016.)

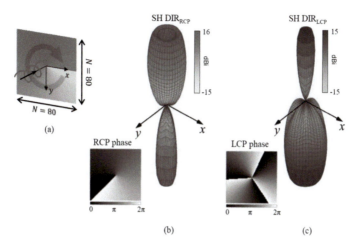

FIGURE 2.18 (a) Phase profile of nonlinear metasurface consisting of 80×80 unit-cells, with a helical variation of elements' orientation, $\varphi(x, y) = \phi(x, y)$, where $\phi(x, y)$ is a polar angle. The metasurface is illuminated by a normally incident RCP wave. Second-harmonic directivity patterns (in dBi) for RCP and LCP polarizations are shown in (b) and (c), respectively. The insets show the spatial phase profile of helical beams radiated "in transmission." (Panels [a]—[c] were reproduced from Tymchenko, M. et al., *Phys. Rev. B*, 94, 214303, 2016.)

2.5 CONCLUSIONS

Hybrid nonlinear metasurfaces comprised of sub-micrometer-thick MQW layer coupled with an array of plasmonic resonators became a giant leap toward efficient harmonics generation in extremely subwavelength structures. Current state-of-the-art structures are able to reach SHG conversion efficiencies in the order of 0.1%, which is a gigantic number considering their vanishingly small thicknesses. Their ultrathin geometries lift phase-matching requirements commonly encountered in nonlinear optical setups and enable robust wave front tailoring functionalities. The latter is of paramount practical importance, since it allows one to get rid of many optical elements and significantly miniaturize and simplify the whole setup. At the same time, hybrid nonlinear metasurfaces have been shown to suffer from saturation effects reducing their performance at low and moderate light intensities, making such structures not practical for high-power setups.

APPENDICES

APPENDIX A

The anisotropic linear permittivity near the intersubband transitions of an MQW stack with 3-subbands tailor for SHG can be modeled as

$$\varepsilon_{xx}(\omega) = \varepsilon_{yy}(\omega) = \varepsilon_{\infty} + \frac{N}{\varepsilon_0 \hbar} \left[\frac{(ez_{12})^2}{(\omega_{12} - \omega) - i\gamma_{12}} + \frac{(ez_{31})^2}{(\omega_{31} - \omega) - i\gamma_{31}} \right], \quad (2.66)$$

$$\varepsilon_{zz}(\omega) = \varepsilon_\infty + i\,\frac{Ne^2\tau}{\varepsilon_0\omega m^*(1-i\omega\tau_D)}, \tag{2.67}$$

where $\varepsilon_\infty \approx 10$ [43].

APPENDIX B

The population densities in each of the three subbands are functions of the acting intensity of the z-polarized field at the pump frequency, I_z^ω, and given as

$$N_1 = N\,\frac{1+2B+2C+4D}{3+2B+4C+4D}, \tag{2.68}$$

$$N_2 = N\,\frac{1+2C}{3+2B+4C+4D}, \tag{2.69}$$

$$N_3 = N\,\frac{1}{3+2B+4C+4D}, \tag{2.70}$$

where

$$B = \frac{I_{s12}}{I_z^\omega}\left(1+\frac{\tau_{12}}{\tau_{13}}\right),\quad C = \frac{I_{s23}}{I_z^\omega}\left(1+\frac{\tau_{23}}{\tau_{13}}\right),\quad D = \frac{I_{s12}I_{s23}}{(I_z^\omega)^2}\left(1+\frac{\tau_{23}}{\tau_{13}}\right). \tag{2.71}$$

APPENDIX C

Relations between the effective second-order susceptibility tensor elements in Cartesian and circular polarization bases computed using Eq. (2.47) can be found to be as follows

$$\chi_{RRR}^{(2)} = \frac{1}{2\sqrt{2}}\left[\chi_{xxx}^{(2)} - \chi_{xyy}^{(2)} + 2\chi_{yxy}^{(2)} + i(\chi_{yxx}^{(2)} - \chi_{yyy}^{(2)} - 2\chi_{xxy}^{(2)})\right],$$

$$\chi_{RRL}^{(2)} = \chi_{RLR}^{(2)} = \frac{1}{2\sqrt{2}}\left[\chi_{xxx}^{(2)} + \chi_{xyy}^{(2)} + i(\chi_{yxx}^{(2)} + \chi_{yyy}^{(2)})\right],$$

$$\chi_{RLL}^{(2)} = \frac{1}{2\sqrt{2}}\left[\chi_{xxx}^{(2)} - \chi_{xyy}^{(2)} - 2\chi_{yxy}^{(2)} + i(\chi_{yxx}^{(2)} - \chi_{yyy}^{(2)} + 2\chi_{xxy}^{(2)})\right], \tag{2.72}$$

$$\chi_{LLL}^{(2)} = \frac{1}{2\sqrt{2}}\left[\chi_{xxx}^{(2)} - \chi_{xyy}^{(2)} + 2\chi_{yxy}^{(2)} - i(\chi_{yxx}^{(2)} - \chi_{yyy}^{(2)} - 2\chi_{xxy}^{(2)})\right],$$

$$\chi_{LRL}^{(2)} = \chi_{LLR}^{(2)} = \frac{1}{2\sqrt{2}}\left[\chi_{xxx}^{(2)} + \chi_{xyy}^{(2)} - i(\chi_{yxx}^{(2)} + \chi_{yyy}^{(2)})\right],$$

$$\chi_{LRR}^{(2)} = \frac{1}{2\sqrt{2}}\left[\chi_{xxx}^{(2)} - \chi_{xyy}^{(2)} - 2\chi_{yxy}^{(2)} - i(\chi_{xxy}^{(2)} - \chi_{yyy}^{(2)} + 2\chi_{xxy}^{(2)})\right].$$

REFERENCES

1. E. C. Le Ru, E. Blackie, M. Meyer, and P. G. Etchegoin, "Surface enhanced Raman scattering enhancement factors: A comprehensive study," *J. Phys. Chem. C*, vol. 111, no. 37, pp. 13794–13803, 2007.
2. B. Sharma, R. R. Frontiera, A.-I. Henry, E. Ringe, and R. P. Van Duyne, "SERS: Materials, applications, and the future," *Mater. Today*, vol. 15, no. 1–2, pp. 16–25, 2012.
3. C. K. Chen, A. R. B. de Castro, and Y. R. Shen, "Surface-enhanced second-Harmonic generation," *Phys. Rev. Lett.*, vol. 46, no. 2, pp. 145–148, 1981.
4. A. Krasnok, M. Tymchenko, and A. Alù, "Nonlinear metasurfaces: A paradigm shift in nonlinear optics," *Mater. Today*, 2017.
5. M. S. Nezami, D. Yoo, G. Hajisalem, S.-H. Oh, and R. Gordon, "Gap plasmon enhanced metasurface third-harmonic generation in transmission geometry," *ACS Photonics*, vol. 3, no. 8, pp. 1461–1467, 2016.
6. C. Dafu, C. Zhenghao, Z. Yueliang, L. Huibin, Y. Guozhen, and G. Shijie, "Surface-enhanced second harmonic generation at GaAs-AI interface," *Phys. Scr.*, vol. 37, no. 5, pp. 746–749, 1988.
7. A. A. Kovalev, P. S. Kondratenko, and B. N. Levinskiĭ, "Second harmonic generation during excitation of surface waves at a periodic metal–nonlinear crystal interface," *Sov. J. Quantum Electron.*, vol. 19, no. 5, pp. 663–665, 1989.
8. J. B. B. Pendry, A. J. J. Holden, D. J. J. Robbins, and W. J. J. Stewart, "Magnetism from conductors and enhanced nonlinear phenomena," *IEEE Trans. Microw. Theory Tech.*, vol. 47, no. 11, pp. 2075–2084, 1999.
9. A. A. Zharov, I. V. Shadrivov, and Y. S. Kivshar, "Nonlinear properties of left-handed metamaterials," *Phys. Rev. Lett.*, vol. 91, no. 3, p. 037401, 2003.
10. W. Fan *et al.*, "Second harmonic generation from a nanopatterned isotropic nonlinear material," *Nano Lett.*, vol. 6, no. 5, pp. 1027–1030, 2006.
11. W. Fan, S. Zhang, K. J. Malloy, S. R. J. Brueck, N. C. Panoiu, and R. M. Osgood, "Second harmonic generation from patterned GaAs inside a subwavelength metallic hole array," *Opt. Express*, vol. 14, no. 21, p. 9570, 2006.
12. Y. Pu, R. Grange, C.-L. Hsieh, and D. Psaltis, "Nonlinear optical properties of core-shell nanocavities for enhanced second-harmonic generation," *Phys. Rev. Lett.*, vol. 104, no. 20, p. 207402, 2010.
13. B. Metzger, L. Gui, J. Fuchs, D. Floess, M. Hentschel, and H. Giessen, "Strong enhancement of second harmonic emission by plasmonic resonances at the second harmonic wavelength," *Nano Lett.*, vol. 15, no. 6, pp. 3917–3922, 2015.
14. F. B. P. Niesler *et al.*, "Second-harmonic generation from split-ring resonators on a GaAs substrate," *Opt. Lett.*, vol. 34, no. 13, p. 1997, 2009.
15. D. Bar-Lev and J. Scheuer, "Efficient second harmonic generation using nonlinear substrates patterned by nano-antenna arrays," *Opt. Express*, vol. 21, no. 24, p. 29165, 2013.
16. T. Ning *et al.*, "Large third-order optical nonlinearity of periodic gold nanoparticle arrays coated with ZnO," *J. Phys. D. Appl. Phys.*, vol. 40, no. 21, pp. 6705–6708, 2007.
17. H. Aouani, M. Rahmani, M. Navarro-Cía, and S. A. Maier, "Third-harmonic-upconversion enhancement from a single semiconductor nanoparticle coupled to a plasmonic antenna," *Nat. Nanotechnol.*, vol. 9, no. 4, pp. 290–294, 2014.
18. B. Metzger *et al.*, "Doubling the efficiency of third harmonic generation by positioning ITO nanocrystals into the hot-spot of plasmonic gap-antennas," *Nano Lett.*, vol. 14, no. 5, pp. 2867–2872, 2014.
19. P.-Y. Chen and A. Alù, "Optical nanoantenna arrays loaded with nonlinear materials," *Phys. Rev. B*, vol. 82, no. 23, p. 235405, 2010.
20. P. Chen, A. Alù, A. Alu, and A. Alù, "Subwavelength imaging using phase-conjugating nonlinear nanoantenna arrays," *Nano Lett.*, vol. 11, no. 12, pp. 5514–5518, 2011.

21. H. Aouani, M. Navarro-Cía, M. Rahmani, and S. A. Maier, "Unveiling the origin of third harmonic generation in hybrid ITO-plasmonic crystals," *Adv. Opt. Mater.*, vol. 3, no. 8, pp. 1059–1065, 2015.

22. B. Jin and C. Argyropoulos, "Enhanced four-wave mixing with nonlinear plasmonic metasurfaces," *Sci. Rep.*, vol. 6, no. 1, p. 28746, 2016.

23. M. M. Fejer, S. J. B. Yoo, R. L. Byer, A. Harwit, and J. S. Harris Jr., "Observation of extremely large quadratic susceptibility at 9.6–10.8 μm in electric-field-biased AlGaAs quantum wells," *Phys. Rev. Lett.*, vol. 62, no. 9, pp. 1041–1044, 1989.

24. E. Rosencher, P. Bois, J. Nagle, and S. Delattre, "Second harmonic generation by inter-sub-band transitions in compositionally asymmetrical MQWs," *Electron. Lett.*, vol. 25, no. 16, p. 1063, 1989.

25. F. Capasso, C. Sirtori, and A. Y. Cho, "Coupled quantum well semiconductors with giant electric field tunable nonlinear optical properties in the infrared," *IEEE J. Quantum Electron.*, vol. 30, no. 5, pp. 1313–1326, 1994.

26. E. Rosencher, A. Fiore, B. Vinter, V. Berger, P. Bois, and J. Nagle, "Quantum engineering of optical nonlinearities," *Science (80-.).*, vol. 271, no. 5246, pp. 168–173, 1996.

27. C. Gmachl *et al.*, "Optimized second-harmonic generation in quantum cascade lasers," *IEEE J. Quantum Electron.*, vol. 39, no. 11, pp. 1345–1355, 2003.

28. M. A. Belkin *et al.*, "Terahertz quantum-cascade-laser source based on intracavity difference-frequency generation," *Nat. Photonics*, vol. 1, no. 5, pp. 288–292, 2007.

29. K. Vijayraghavan *et al.*, "Broadly tunable terahertz generation in mid-infrared quantum cascade lasers," *Nat. Commun.*, vol. 4, no. 1, p. 2021, 2013.

30. J. H. Davies, *The Physics of Low-dimensional Semiconductors: An Introduction*, 1st ed. Cambridge University Press, New York, 1997.

31. M. Whitehead and G. Parry, "High-contrast reflection modulation at normal incidence in asymmetric multiple quantum well Fabry-Perot structure," *Electron. Lett.*, vol. 25, no. 9, p. 566, 1989.

32. R.-H. Yan, R. J. Simes, and L. A. Coldren, "Surface-normal electroabsorption reflection modulators using asymmetric Fabry-Perot structures," *IEEE J. Quantum Electron.*, vol. 27, no. 7, pp. 1922–1931, 1991.

33. C. C. Barron, C. J. Mahon, B. J. Thibeault, and L. A. Coldren, "Design, fabrication and characterization of high-speed asymmetric Fabry-Perot modulators for optical interconnect applications," *Opt. Quantum Electron.*, vol. 25, no. 12, pp. S885–S898, 1993.

34. Y.-H. Kuo *et al.*, "Quantum-confined stark effect in Ge/SiGe quantum wells on Si for optical modulators," *IEEE J. Sel. Top. Quantum Electron.*, vol. 12, no. 6, pp. 1503–1513, 2006.

35. L. Lever, Z. Ikonić, A. Valavanis, J. D. Cooper, and R. W. Kelsall, "Design of Ge–SiGe quantum-confined stark effect electroabsorption heterostructures for CMOS compatible photonics," *J. Light. Technol.*, vol. 28, no. 22, pp. 3273–3281, 2010.

36. R. K. Schaevitz *et al.*, "Simple electroabsorption calculator for designing 1310 nm and 1550 nm modulators using germanium quantum wells," *IEEE J. Quantum Electron.*, vol. 48, no. 2, pp. 187–197, 2012.

37. E. H. Edwards *et al.*, "Ge/SiGe asymmetric Fabry-Perot quantum well electroabsorption modulators," *Opt. Express*, vol. 20, no. 28, p. 29164, 2012.

38. J. Lee *et al.*, "Ultrafast electrically tunable polaritonic metasurfaces," *Adv. Opt. Mater.*, vol. 2, no. 11, pp. 1057–1063, 2014.

39. A. Hugi, G. Villares, S. Blaser, H. C. Liu, and J. Faist, "Mid-infrared frequency comb based on a quantum cascade laser," *Nature*, vol. 492, pp. 229–233, 2012.

40. M. Gurnick and T. DeTemple, "Synthetic nonlinear semiconductors," *IEEE J. Quantum Electron.*, vol. 19, no. 5, p. 791, 1983.

41. A. Gabbay *et al.*, "Interaction between metamaterial resonators and intersubband transitions in semiconductor quantum wells," *Appl. Phys. Lett.*, vol. 98, no. 20, p. 203103, 2011.

42. A. Benz *et al.*, "Optical strong coupling between near-infrared metamaterials and intersubband transitions in III-nitride heterostructures," *ACS Photonics*, vol. 1, no. 10, pp. 906–911, 2014.

43. J. Lee *et al.*, "Giant nonlinear response from plasmonic metasurfaces coupled to intersubband transitions," *Nature*, vol. 511, no. 7507, pp. 65–69, 2014.

44. S. Campione, A. Benz, M. B. Sinclair, F. Capolino, and I. Brener, "Second harmonic generation from metamaterials strongly coupled to intersubband transitions in quantum wells," *Appl. Phys. Lett.*, vol. 104, no. 13, p. 131104, 2014.

45. O. Wolf *et al.*, "Phased-array sources based on nonlinear metamaterial nanocavities," *Nat. Commun.*, vol. 6, no. May, p. 7667, 2015.

46. J. Lee *et al.*, "Ultrathin second-harmonic metasurfaces with record-high nonlinear optical response," *Adv. Opt. Mater.*, vol. 4, no. 5, pp. 664–670, 2016.

47. J. S. Gomez-Diaz, M. Tymchenko, J. Lee, M. A. A. Belkin, and A. Alù, "Nonlinear processes in multi-quantum-well plasmonic metasurfaces: Electromagnetic response, saturation effects, limits, and potentials," *Phys. Rev. B*, vol. 92, no. 12, p. 125429, 2015.

48. M. Tymchenko, J. S. S. Gomez-Diaz, J. Lee, N. Nookala, M. A. A. Belkin, and A. Alù, "Advanced control of nonlinear beams with Pancharatnam-Berry metasurfaces," *Phys. Rev. B*, vol. 94, no. 21, p. 214303, 2016.

49. M. Tymchenko, J. S. Gomez-Diaz, J. Lee, M. A. Belkin, and A. Alù, "Highly-efficient THz generation using nonlinear plasmonic metasurfaces," *J. Opt.*, vol. 19, no. 10, p. 104001, 2017.

50. R. W. Boyd, *Nonlinear Optics*, 3rd ed. Academic Press, Burlington, MA/San Diego, CA, 2008.

51. R. M. Joseph and A. Taflove, "FDTD Maxwell's equations models for nonlinear electrodynamics and optics," *IEEE Trans. Antennas Propag.*, vol. 45, no. 3, pp. 364–374, 1997.

52. G. D'Aguanno *et al.*, "Generalized coupled-mode theory for $\chi^{(2)}$ interactions in finite multilayered structures," *J. Opt. Soc. Am. B*, vol. 19, no. 9, p. 2111, 2002.

53. M. J. Steel and C. M. de Sterke, "Second-harmonic generation in second-harmonic fiber Bragg gratings," *Appl. Opt.*, vol. 35, no. 18, p. 3211, 1996.

54. G. D'Aguanno, N. Mattiucci, M. Scalora, M. J. Bloemer, and A. M. Zheltikov, "Density of modes and tunneling times in finite one-dimensional photonic crystals: A comprehensive analysis," *Phys. Rev. E - Stat. Physics, Plasmas, Fluids, Relat. Interdiscip. Top.*, vol. 70, no. 1, p. 12, 2004.

55. C. M. de Sterke and J. E. Sipe, "Envelope-function approach for the electrodynamics of nonlinear periodic structures," *Phys. Rev. A*, vol. 38, no. 10, pp. 5149–5165, 1988.

56. K. O'Brien *et al.*, "Predicting nonlinear properties of metamaterials from the linear response," *Nat. Mater.*, vol. 14, no. 4, pp. 379–383, 2015.

57. D. M. Pozar, *Microwave Engineering*, 4th ed. John Wiley & Sons, Hoboken, NJ, 2011.

58. J. Khurgin, "Second-order intersubband nonlinear-optical susceptibilities of asymmetric quantum-well structures," *J. Opt. Soc. Am. B*, vol. 6, no. 9, p. 1673, 1989.

59. G. Almogy, M. Segev, and A. Yariv, "Adiabatic nonperturbative derivation of electric-field-induced optical nonlinearities in quantum wells," *Phys. Rev. B*, vol. 48, no. 15, pp. 10950–10954, 1993.

60. J. R. Meyer, C. A. Hoffman, F. J. Bartoli, E. R. Youngdale, and L. R. Ram-Mohan, "Momentum-space reservoir for enhancement of intersubband second-harmonic generation," *IEEE J. Quantum Electron.*, vol. 31, no. 4, pp. 706–714, 1995.

61. K. L. Vodopyanov, V. Chazapis, C. C. Phillips, B. Sung, and J. S. Harris, "Intersubband absorption saturation study of narrow III–V multiple quantum wells in the spectral range," *Semicond. Sci. Technol.*, vol. 12, no. 6, pp. 708–714, 1997.

62. I. Vurgaftman, J. R. Meyer, and L. R. Ram-Mohan, "Optimized second-harmonic generation in asymmetric double quantum wells," *IEEE J. Quantum Electron.*, vol. 32, no. 8, pp. 1334–1346, 1996.

63. L. C. West and S. J. Eglash, "First observation of an extremely large-dipole infrared transition within the conduction band of a GaAs quantum well," *Appl. Phys. Lett.*, vol. 46, no. 12, pp. 1156–1158, 1985.

64. N. Yu *et al.*, "Light propagation with phase discontinuities: Generalized laws of reflection and refraction," *Science (80-.).*, vol. 334, pp. 333–337, 2011.

65. N. Yu and F. Capasso, "Flat optics with designer metasurfaces," *Nat. Mater.*, vol. 13, no. 2, pp. 139–150, 2014.

66. T. Ellenbogen, A. Ganany-Padowicz, and A. Arie, "Nonlinear photonic structures for all-optical deflection," *Opt. Express*, vol. 16, no. 5, p. 3077, 2008.

67. N. Voloch, T. Ellenbogen, and A. Arie, "Radially symmetric nonlinear photonic crystals," *J. Opt. Soc. Am. B*, vol. 26, no. 1, p. 42, 2009.

68. T. Ellenbogen, N. Voloch-Bloch, A. Ganany-Padowicz, and A. Arie, "Nonlinear generation and manipulation of Airy beams," *Nat. Photonics*, vol. 3, no. 7, pp. 395–398, 2009.

69. N. Voloch-Bloch, T. Davidovich, T. Ellenbogen, A. Ganany-Padowicz, and A. Arie, "Omnidirectional phase matching of arbitrary processes by radial quasi-periodic nonlinear photonic crystal," *Opt. Lett.*, vol. 35, no. 14, p. 2499, 2010.

70. L. L. Lu, P. Xu, M. L. Zhong, Y. F. Bai, and S. N. Zhu, "Orbital angular momentum entanglement via fork-poling nonlinear photonic crystals," *Opt. Express*, vol. 23, no. 2, p. 1203, 2015.

71. S. Trajtenebrg-Mills and A. Arie, "Shaping light beams in nonlinear processes using structured light and patterned crystals," *Opt. Mater. Express*, vol. 7, no. 8, p. 2928, 2017.

72. N. V. Bloch, K. Shemer, A. Shapira, R. Shiloh, I. Juwiler, and A. Arie, "Twisting light by nonlinear photonic crystals," *Phys. Rev. Lett.*, vol. 108, no. 23, pp. 1–5, 2012.

73. N. Segal, S. Keren-Zur, N. Hendler, and T. Ellenbogen, "Controlling light with metamaterial-based nonlinear photonic crystals," *Nat. Photonics*, vol. 9, no. 3, pp. 180–184, 2015.

74. S. Keren-Zur, O. Avayu, L. Michaeli, and T. Ellenbogen, "Nonlinear beam shaping with plasmonic metasurfaces," *ACS Photonics*, vol. 3, no. 1, pp. 117–123, 2016.

75. M. Tymchenko, J. S. Gomez-Diaz, J. Lee, N. Nookala, M. A. Belkin, and A. Alù, "Gradient nonlinear Pancharatnam-Berry metasurfaces," *Phys. Rev. Lett.*, vol. 115, no. 20, p. 207403, 2015.

76. N. Nookala *et al.*, "Ultrathin gradient nonlinear metasurface with a giant nonlinear response," *Optica*, vol. 3, no. 3, p. 283, 2016.

77. M. Tymchenko and A. Alu, "Pancharatnam-Berry metasurfaces with giant nonlinear response," *IEEE Antennas Propag. Soc. AP-S Int. Symp.*, vol. 2015, pp. 246–247, 2015.

78. G. Li *et al.*, "Continuous control of the nonlinearity phase for harmonic generations," *Nat. Mater.*, vol. 14, no. 6, pp. 607–612, 2015.

79. W. Ye *et al.*, "Spin and wavelength multiplexed nonlinear metasurface holography," *Nat. Commun.*, vol. 7, p. 11930, 2016.

80. N. Shitrit, S. Maayani, D. Veksler, V. Kleiner, and E. Hasman, "Rashba-type plasmonic metasurface," *Opt. Lett.*, vol. 38, no. 21, pp. 4358–61, 2013.

81. K. Konishi, T. Higuchi, J. Li, J. Larsson, S. Ishii, and M. Kuwata-Gonokami, "Polarization-controlled circular second-harmonic generation from metal hole arrays with threefold rotational symmetry," *Phys. Rev. Lett.*, vol. 112, no. 13, pp. 1–5, 2014.

82. G. Li, S. Zhang, and T. Zentgraf, "Nonlinear photonic metasurfaces," *Nat. Rev. Mater.*, vol. 2, no. 5, p. 17010, 2017.

83. L. Novotny, *Principles of Nano-Optics*. Cambridge University Press, New York, 2006.

84. C. A. Balanis, *Advanced Engineering Electromagnetics*, 2nd ed. John Wiley & Sons Inc., 111 River Street, Hoboken, NJ 07030-5774, USA.

3 Technological Challenges and Material Platforms

V. F. Gili, M. Rahmani, C. Jagadish, and Giuseppe Leo

CONTENTS

The topic of light-matter interaction with nanoscale particles has a long history, and it is of a fundamental importance for research in different areas of optical physics. By taking advantage of the rapid progress in nanoscience and nanotechnology, many types of nanoparticles, e.g., metallic, dielectric, and semiconductor have been fabricated being widely used for applications including sensing, solar cells, and optical communications.

Together with the improvement of numerical simulation via super-computers, advancements in nanofabrication and characterization technologies and techniques have increased the quest to employ light-matter interaction at nanoscale in the last decade.

While the field of nanophotonics started more than two decades ago with suspensions of nanocrystals and metal colloids (see Figure 3.1), our focus is on 2D solid-state photonic nanostructures on-chip (see Figure 3.2). Within this scope, the last years have witnessed the progressive development of nanostructures and metasurfaces of different types: amorphous dielectric (see Chapter 1), metallic (see Chapter 4), silicon or III-V (see Chapter 9), and colloidal (see Chapter 10). Finally, hybrid devices have also been developed, combining plasmonic confinement and enhanced light-matter interaction in either dielectric or semiconductor nanoparticles (see Chapter 9). For obvious topological reasons, at variance with photonic crystals (see Chapter 7),

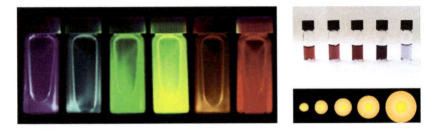

FIGURE 3.1 Left: Emission from ZnS-coated CdSe nanocrystals of different sizes, dispersed in hexane (reproduced with permission from nanocluster.mit.edu). Right: Suspensions of gold nanoparticles of various sizes. Size difference causes the color difference (Wikipedia).

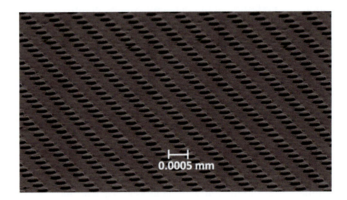

FIGURE 3.2 Silicon nanostructures on a sapphire substrate.

2D arrangements of nanoparticles cannot take the form of suspended membrane. Therefore, this chapter will deal with nanostructures that either lie on their original monolithic wafer or undergo a transfer step on another substrate.

In both these cases, the common technology for nanopatterning is nanolithography. To fabricate all nanostructures, first an ultraclean substrate/wafer is covered with a thin layer of electro- or photo-sensitive polymer, a so-called resist. Then the wafer is placed in a tool that exposes the sample via a writing electron beam or light source through a mask. Subsequently, the patterned resist is placed in a developer chemical which washes away the exposed (or unexposed) resist, based on the type of resist. After these lithographic steps, the fabrication procedures can follow different specific approaches, which will be illustrated hereafter.

To optimize the readability of this chapter, we decided to extract the descriptions of standard techniques from its main stream and we propose them in the form of concise technical sheets. Due to their generality, Electron-Beam Lithography, deposition, and etching are the subject of the first three sheets.

TECHNICAL SHEET: E-BEAM LITHOGRAPHY

Electron-Beam Lithography (EBL) is a technique that allows transferring a pattern on an electron-sensitive material by selective exposure to an electron beam, the so-called resist. When the latter is exposed to electrons it experiences profound structural changes, such that when immersed in a specific basic solution, called developer, the pattern is permanently transferred. It is possible to distinguish two types of resist: positive and negative. The former are such that the exposed areas increase their solubility and are removed after development, while for the latter the opposite occurs. The choice of electron-beam exposure rather than UV light, as in photolithography, is crucial in nanofabrication as it allows sub-10 nm resolution. Photolithography pattern instead hardly exhibit ~300 nm features (diffraction limit). Despite the greater accuracy, the patterning rate in EBL is no greater than a few mm/h, much slower than in photolithography.

Depending on the positive or negative tone of the e-beam resist, after development it is either the exposed or the unexposed resist area respectively that is dissolved in the developer solution (Figure 3.3). The choice between these two possibilities strictly depends on the desired structure and the other steps of fabrication that might damage the resist, e.g., deposition at high temperature. For example, in order to fabricate plasmonic nanostructures, a lift-off approach is normally preferred, and thus the metals are directly evaporated on a low-index substrate on which a positive-tone lithographic mask is patterned at room temperature (see Section 3.1). In contrast, for semiconductor fabrication, an etch-down approach is preferable, and thus negative-tone e-beam resists are employed on semiconductors pre-grown at high temperatures. In this case, especially for nanoscale structures, the patterns are instead transferred to the semiconductor with etching process (see the technical sheet on etching).

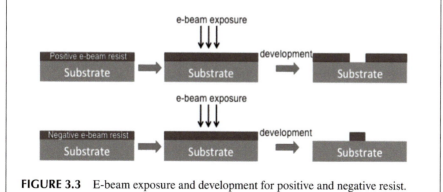

FIGURE 3.3   E-beam exposure and development for positive and negative resist.

TECHNICAL SHEET: DEPOSITION

We will describe six broadly used deposition techniques: electron-beam evaporation, thermal evaporation, sputtering, plasma-enhanced chemical vapor deposition, pulsed laser deposition, and epitaxy.

Electron-Beam Evaporation

The electron-beam (e-beam) evaporation of thin metal films is carried out as follows: a tungsten filament in a high-vacuum chamber ($\approx 10^{-8}$) generates an electron beam by thermionic emission that is accelerated to a plateau voltage (see Figure 3.4). The electrons trajectory is controlled with a magnetic induction field and is directed toward a crucible. The latter contains the metal species of interest, with a purity as high as 99.99% (4N in industrial language), and it is shielded from the filament to avoid chemical contaminations. Once the electron beam hits the target metal, the local temperature increases up to a point in which the material starts to melt and then evaporate, spreading itself in all directions. The access to a second chamber, kept at a pressure of $\approx 10^{-7}$ Torr and containing the sample, is controlled by means of a shutter. Once the crucible temperature is high enough to generate the metal vapor, this first shutter opens. In the meantime, in the sample chamber, a constant potential is applied to a quartz crystal, causing it to oscillate with a precise frequency. As the metal vapor is injected into the sample chamber, its flow will cause a variation in the oscillation frequency. Measuring this difference allows to precisely control the deposition rate.

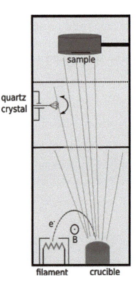

FIGURE 3.4   Schematic of an e-beam evaporation machine.

(Continued)

Once the latter reaches the desired value, a second shutter opens and the metal vapor finally reaches the sample, on which it condenses and forms the thin metal layer. Importantly, this technique does not heat the sample up. Therefore, the electron-resist does not get damaged during this deposition process.

Thermal Evaporation

In thermal evaporation, the target material is heated in a vacuum chamber until its surface atoms have sufficient energy to leave the surface. Therefore, the evaporated atoms will traverse the vacuum chamber, at a thermal energy that is generally lower than 1 eV. Accumulation of evaporated atoms coats a substrate positioned above the evaporating material. Similar to e-beam evaporation, the pressure in the chamber must be below the average distance that an atom or molecule can travel in a vacuum chamber before it collides with another particle. This is to avoid disturbing its direction to some degree. The suitable pressure is typically 3.0×10^{-4} Torr or lower. The main reason to run at the high end of the pressure range is to allow an ion beam source to be employed simultaneously for film densification or other property modification. Like e-beam evaporation, this technique does not heat either the sample or electron resist up, insofar heating is only used to evaporate the target materials.

Sputtering

The sputtering method employs a controlled gas, which usually is the chemically inert argon, into a vacuum chamber. In this technique the target is electrically energized to establish a self-sustaining plasma. The exposed surface of the target, which acts as cathode, is a slab of the material to be coated onto the substrates.

The atoms gas lose electrons inside the plasma to become positively charged ions. Subsequently, the plasma is accelerated into the target and strike with sufficient kinetic energy to dislodge atoms or molecules of the cathode (target material). The sputtered materials then constitute a vapor stream, which traverses the chamber and hits the substrate, sticking to it as a thin film. Sputtering can be performed at various temperatures depending on the fabrication requirements.

Plasma-Enhanced Chemical Vapor Deposition

Plasma-Enhanced Chemical Vapor Deposition (PECVD) is a method for depositing thin films through chemical reactions after creation of various gases by means of generated plasma in an ultra-high-vacuum chamber. Unlike sputtering, in which a solid target is energized to form the plasma, in PECVD, the plasma is formed by atoms or molecules of various gases.

In PECVD, generally the plasma is generated via a direct current discharge between two electrodes, the space between which is filled with the reacting gases, however in some cases the plasma is generally created by radio frequency. PECVD is often used in dielectric and semiconductor manufacturing. This is one of the fastest deposition techniques, as compared with sputter deposition and thermal/

(Continued)

electron-beam evaporation. Amorphous or polycrystalline silicon, silicon oxide, silicon nitride are among most popular films that can be deposited by PECVD. The usually employed precursor gases are silane (SiH_4) and ammonia (NH_3). The main drawback of this technique is that high-temperature environments are not avoidable, making it unfavorable to protect photo- and electron-resists.

Pulsed Laser Deposition

Pulsed laser deposition is a physical vapor deposition technique where a high peak-power pulsed laser impinges on a target made of the material to be deposited (See, Figure 3.5). The latter is placed inside a chamber under ultra-high vacuum or in controlled oxygen atmosphere. The material is vaporized from the target and forms a plasma plume that forms a thin film on the sample. A heater (normally a hot plate) is placed above the sample to decrease nucleation of the material to be deposited and favor a homogeneous deposition. The advantage of this technique is that it can be used to deposit almost any kind of materials. Its drawback is that this technique generally causes sample heating, which is detrimental for e-beam resist protection. Laser can be also used as a versatile tool for other fabrication purposes, such as reducing oxides [1].

Epitaxy

Epitaxy (from the ancient Greek ἐπί = above, and τάξις = ordered manner) is the deposition of a crystalline layer on a crystalline substrate. When the two crystals are identical, it is called homoepitaxy, otherwise it is referred

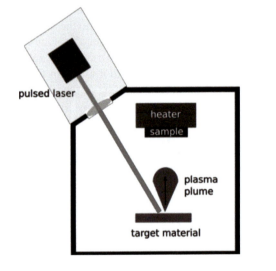

FIGURE 3.5   Schematic of a pulsed laser evaporation machine.

(Continued)

to as heteroepitaxy. Among different types of epitaxy, the most common are molecular beam epitaxy and metal-organic chemical vapor deposition.

1. Molecular Beam Epitaxy (MBE)

   MBE is an epitaxial growth technique based on the interaction of different elements adsorbed from molecular beams on a heated crystalline substrate. Molecular beams are provided by the so-called "effusion cells" by evaporating or sublimating high-purity materials contained in radiatively heated crucibles [2].

   A typical MBE reactor (see Figure 3.6) consists of a stainless steel growth chamber, kept under pressures of the order of 10^{-11} Torr (ultrahigh vacuum) so that the mean free path of molecules in the beams is much greater than the distance between the cells and the substrate. Samples are transferred in and out of the chamber through a vacuum-preserving load-lock system. Molecular beams are individually selected by means of shutters placed in front of the effusion cells. Reflection high-energy electron diffraction (RHEED) is often used for in-situ monitoring of the crystal growth. Deposition rates are generally on the order of 10^3 nm/hour. Among the advantages of MBE it should be mentioned the ability to grow structures in which carriers are quantum confined in regions smaller than their de Broglie wavelength.

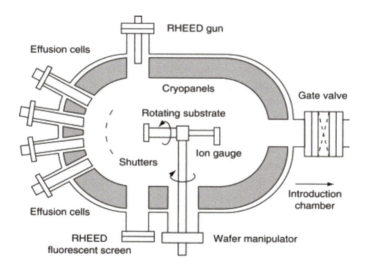

FIGURE 3.6 Schematics of a typical MBE system. (From Frigeri, P. et al., 3.12: Molecular beam epitaxy: An overview, In P. Bhattacharya, R. Fornari, and H. Kamimura, *Comprehensive Semiconductor Science and Technology*, Vol. 3, Elsevier, Amsterdam, the Netherlands, pp. 480–522, 2011.)

(Continued)

Depending if the confinement is on 1 or 3 dimensions, these structures are denominated quantum wells or quantum dots, respectively.

2. Metal-Organic Chemical Vapor Deposition (MOCVD)

Unlike MBE, which is a physical vapor deposition technique, MOCVD relies on a chemical vapor deposition. Vapor precursors in this case are usually in solid or liquid state. As such, they are converted into vapor into an evaporator mixer with the help of a carrier gas (usually hydrogen in arsenide and phosphide growth or nitrogen for nitride). The MOCVD process can occur at atmospheric pressure, thus it is not crucial to maintain the reactor chamber at ultra-high-vacuum conditions and hence the reactor is cost effective. Films deposited by MOCVD start growing in island mode followed by subsequent coalescence of such formed islands. Importantly, the growth rates can be as high as 5 μm per hour, which makes MOCVD eligible for producing thin films at mass production. The GaN technology on large area Si/SiC/sapphire wafers heavily employs MOCVD.

TECHNICAL SHEET: ETCHING

Once a negative-tone lithographic mask is defined with EBL and development, the patterns can be transferred on the material with different etching processes (see scheme of Figure 3.7).

Inductively Coupled Plasma—Reactive Ion Etching

Reactive Ion Etching (RIE) is the most frequently adopted technique for III-V semiconductor nanofabrication. This being a selective etching method, i.e., based on the employed gases, only certain materials can be etched. The operating principle of this machine is illustrated in Figure 3.8. While placing the sample in a vacuum chamber with a load-lock system, a high-density plasma is generated in a separated chamber under high-vacuum conditions by sending a radio frequency (RF) signal coming from the so-called etching source into a coil. The plasma is then injected into the sample chamber, where it is

FIGURE 3.7 Example of a fabrication protocol involving an etching step.

(Continued)

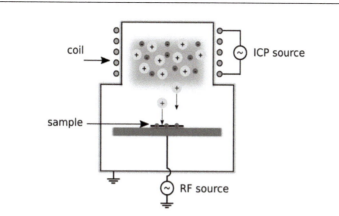

FIGURE 3.8 Schematics of RIE machine.

submitted to a second RF signal coming from a different source connected to the sample holder. The second RF accelerates electrons and ions of the plasma toward the sample. As electrons have a higher mobility than ions, they strike the sample first, charging it negatively to a value called RF bias. Ions strike the sample in a second moment, performing mechanical sputtering [3].

In reality however, the etching mechanism is much more complicated than the sole sputtering. In many cases, collisions within the plasma not only generate ions, but also neutral free radicals which can chemically react with the sample surface. This happens even if the temperature is kept constant as the ions locally change the temperature of the sample surface through collisions. In many systems the etching depth can be monitored via a Fabry-Perot (FP) system consisting of a continuous wave (CW) laser and a high reflectivity mirror forming an optical cavity with the sample surface.

In principle, protocols can be optimized to boost chemical reactions on the surface and obtain partially-isotropic etchings (see for example [4]).

Ion Beam Milling

This method is mostly used for very high precision metallic nanostructuring, e.g., in data storage industry. Unlike RIE, ion beam milling is a mechanical etching technique that is not selective, i.e., etches all the materials (though, with different rates). In this method, submicron ion particles are accelerated and bombard the surface of the target while it is mounted on a rotating stage inside a vacuum chamber. While the resist protects the underlying material during the etching process, everything else exposed to the ion beam gets etched during the process cycle, even the resist itself. The key point is that the resist's etch rate is lower than that of the material that is being etched. Therefore, a precise optimization is required to determine the exact process time.

(Continued)

Different resist thicknesses can be used depending on the thickness, type, and amount of material to be removed.

To ensure uniform removal of waste materials in straight side walls for all features with zero undercutting, target rotation during argon ions striking is required. It leads to a perfectly repeatable circuit time after time. This precision repeatability is the key strength of the wide collimated ion beam milling process. Other methods, such as the chemical process or laser simply do not deliver the same level of precision as compared to an ion beam. Figure 3.9 shows a simplified operating scheme of the ion beam miller. Argon ions contained within plasma are formed via an electrical discharge which is accelerated by a pair of optically aligned grids. The highly collimated beam is focused on a tilted work plate that rotates during the milling operation. A neutralization filament prevents the build-up of positive charges on the work plate. Importantly, the work plate should be cooled and rotate during the process so as to ensure uniformity of the ion beam bombardments. Meanwhile the work plate can be angled to address specific requirements.

Wet Etching

Wet etching is a material removal process that uses liquid chemicals for etching. In this technique, material parts that are not protected by lithographic masks are etched away by liquid chemicals, usually highly concentrated acids or bases. In most cases, a wet etching process involves multiple chemical reactions that consume the original reactants and produce new reactants. The liquid etchant diffuses into the structure to be removed and the reaction between the liquid etchant and the material takes place, resulting into the material getting dissolved.

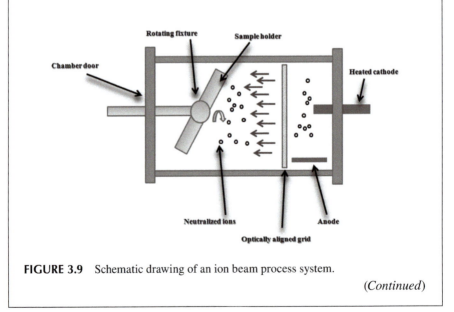

FIGURE 3.9 Schematic drawing of an ion beam process system.

(Continued)

Wet etching can be isotropic or anisotropic. For isotropic wet etching, the material is removed laterally at a rate similar to the downward etching speed. This lateral and downward etching process takes places even with the above-described isotropic dry etching procedures. If directionality is crucial for high-resolution pattern transfer, wet chemical etching is normally not used. In the anisotropic case, the etchants dissolve crystalline materials at different rates depending on which crystal face is exposed to the etchant. It is important to stress that in all the above-mentioned cases the resist or mask thickness is very important for the etching process, as during the etching procedure the mask itself gets always etched as well, though with much lower rate. Therefore, the mask must not give up until the etching process is over.

3.1 FABRICATION OF METALLIC NANOSTRUCTURES

The first nanoparticles used for bridging the gap between conventional optics and highly integrated nanophotonics were metallic. They rely on the collective oscillation of free electrons on the surface, so-called surface plasmons, which in turn induce electric dipoles and higher-order modes [5]. Plasmonics found many applications ranging from super-scattering and cloaking to sensing and artificial antiferromagnetism, etc. Most of the related nanostructures are fabricated through lift-off, including electron-beam or thermal evaporation deposition techniques. Since both techniques can be used at room temperature, the thin metal film deposition process does not damage the electron-beam resist. Therefore, high-quality nanoantennas, patterned by an electron beam on the resist, can be transferred to the metal films by a lift-off technique, leading to minimal variations [6]. It is worth noting that as the metal is deposited on the whole sample a subsequent step, called lift-off, is needed to keep the lithographic patterns only. In this case the sample undergoes a solvent bath that dissolves the resist. This causes the unnecessary metal to be lifted-off together with the residual resist. In this way the metallic lithographic patterns only survive this process. This process is only applicable as far as the resist thickness is larger than the metal layer thickness (Figure 3.10).

An important capability of plasmonics is that by reversibly converting propagating electromagnetic waves into volumes that are orders of magnitude smaller than the diffraction limit of localized light, so-called hot spots, nanostructures can act as actual antennas: this is the key to strong optical nonlinearity. The most

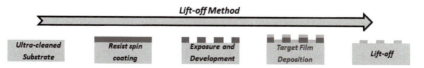

FIGURE 3.10 Example of a fabrication protocol involving a lift-off procedure.

studied metallic nanoantennas are dimers, where the antenna consists of a pair of bars, disks, bowties etc., with a nanoscale gap in between. The usefulness of such designs is the possibility to create electromagnetic hotspots in these gaps, with field enhancement up to a factor 1000. This has enabled for instance the demonstration of surface-enhanced Raman scattering [7] and enhanced nonlinear optical effects like second-harmonic generation (SHG) [8] and third-harmonic generation (THG) [9].

Despite these interesting features, metals are not transparent in the visible range. Because of strong light absorption and ohmic losses, causing important sample heating, metals tend to have high-energy dissipation and low heat resistance, which appears to be a key limiting factor to nonlinear optics. Another consequence of light absorption issue is that metallic nanostructures cannot be easily integrated on-chip. Although several plasmonic applications have faced the above-mentioned limitations, metallic nanoparticles stay a valuable means for developing new strategies for light-matter interactions at nanoscale [10,11]. While for more discussions on this point the reader can see Chapter 4, hereafter let us focus on high-refractive index dielectric and semiconductor nanostructures, which have been recently exploited as an alternative to plasmonics.

Dielectrics and semiconductors benefit from negligible resistive losses at photon energies lower than the energy gap. This advantage allows to excite them with high light intensities with significant electromagnetic field concentration at the nanoscale, which is not limited to interfaces. However, fabrication of nanostructures of dielectrics and semiconductors are not as straightforward as plasmonics. In the following sections, after briefly recalling the benefits of employing various kinds of dielectrics and semiconductors, we will illustrate the main techniques used by several groups in the last few years to fabricate dielectric and semiconductor metasurfaces.

3.2   FABRICATION OF AMORPHOUS DIELECTRIC AND AMORPHOUS NANOSTRUCTURES

After the pioneer work of Lalanne's group two decades ago [12], amorphous nanostructures have been greatly developed by Capasso [13] and several other groups [14–16]. Their physical behavior differs profoundly from metallic nanostructures because the electromagnetic field penetrates in the bulk of their constituent material. In this case, a high-refractive index contrast is needed to grant total internal reflection and high scattering efficiency with respect to a low-index environment. Both the above groups used titanium dioxide (TiO_2) on silica for linear-optics applications spanning from blazed binary diffraction lenses to versatile metalenses operating at either a fixed wavelength (633 nm) or RGB systems (R 660 nm, G 532 nm, B 405 nm), and from polarization-insensitive to circular-polarization formats, with spacing-over-λ ratios around 0.5 and efficiencies around 80% (Figure 3.11).

From the technological point of view, the fabrication of amorphous nanostructures can be similar to that of metallic nanostructures. In case of low-temperature deposition techniques like e-beam evaporation and sputtering, a lift-off protocol will be preferred. This applies mostly to materials like Germanium [17], TiO_2 [18], Indium Tin Oxide (ITO) [19], and ZnO [20]. These are fabricated via

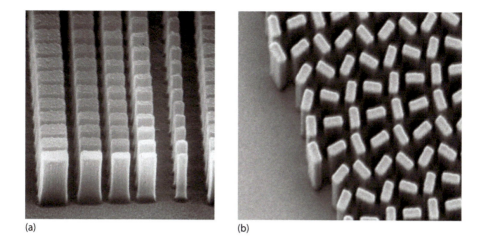

(a) (b)

FIGURE 3.11 SEM images of: (a) blazed binary lens. (From Lalanne, P. et al., *Opt. Lett.*, 23, 14, 1998.) and (b) Metalens. (From Khorasaninejad, M. et al., *Science*, 352, 6290, 2016.)

EBL, followed by Atomic Layer Deposition and etching through ICP and final spin coating of protective layers of polymethylmethacrylate (PMMA).

TiO$_2$ has also been studied for its interesting $\chi^{(3)}$ properties, both in its rutile ($n \approx 2.3$, $n_2 \approx -2.17 \times 10^{-17}$ m^2/V) and anatase ($n \approx 2.2$, $n_2 \approx -6.32 \times 10^{-17}$ m^2/V) phases [18]. However, Miller's rule and its generalizations [21] imply that higher nonlinear coefficients can only be achieved with higher-index materials. Moreover, amorphous materials being centrosymmetric, they do not lend themselves to $\chi^{(2)}$ optics. For both these reasons, and taking also advantage from mature micro- and nano-electronics fabrication [22], the last years have witnessed a huge interest for nonlinear nanophotonics in both silicon and III-V semiconductor platforms.

More recently, Grinblat and coworkers [17] developed a technique to fabricate germanium antennas on borosilicate glass. Arrays of germanium nanodisks were fabricated on borosilicate glass using Electron-Beam Lithography. The substrate was first coated with positive-tone PMMA resist. Then the nanostructures were defined by an electron-beam exposure, followed by a development procedure. Germanium thermal evaporation at 1.5 Å/s, and lift-off processes were the final steps. Figure 3.12 shows the SEM image of final array for germanium disks. Consequently, they obtained the efficient THG from germanium nanoantennas two orders of magnitude larger than unstructured germanium film.

The other, and recently heavily studied, high-refractive index material with large third-order susceptibility is silicon [23]. Unlike germanium, silicon film cannot be deposited through evaporation or sputtering. Therefore, amorphous silicon films should be deposited through high-temperature deposition techniques (200°C–300°C), e.g., PECVD, thus the lift-off method cannot be used anymore. This happens because most of resists (generally organic ones) cannot stand high temperature. On the other hand, PECVD allows growing silicon films on any substrate that can be followed by ICP-RIE etching for nanofabrication. Recently, Xu and coworkers have demonstrated dielectric resonators on a metallic mirror that can significantly enhance the

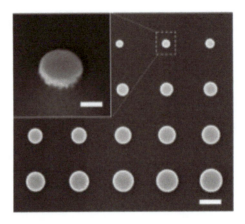

FIGURE 3.12 SEM micrograph of germanium nanodisks fabricated on borosilicate glass. (From Grinblat, G. et al., *Nano Lett.*, 16, 4635, 2016.)

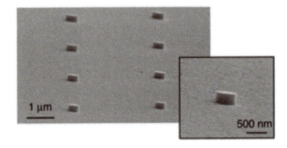

FIGURE 3.13 SEM image of the fabricated amorphous silicon resonators on an Au film. (From Xu, L. et al., *Light Sci. Appl.*, 7, 44, 2018.)

third-harmonic emission, compared to a typical resonator on an insulator substrate. This leads to a significant near-field enhancement that facilitates the nonlinear process [24] (Figure 3.13).

3.3 FABRICATION OF CRYSTALLINE SEMICONDUCTOR NANOSTRUCTURES

Since Chapter 9 of this book is entirely devoted to the recent progress in silicon and III-V platforms for $\chi^{(3)}$ and $\chi^{(2)}$ nano-optics, respectively, here we will just briefly recall that crystalline silicon (Si), gallium phosphide (GaP), gallium arsenide (GaAs) and aluminum gallium arsenide (AlGaAs) resonant nanoparticles are emerging as a promising alternative to metallic nanoparticles for a wide range of nanophotonic applications that utilize localized resonant modes. Such nanoparticles offer unique opportunities for the study of nonlinear effects due to the combination of very low optical losses and multipolar characteristics of both electric and magnetic resonant optical modes. Importantly, the nonlinear optical effects of a magnetic origin can have fundamentally different properties when compared with those of an electric origin. When nonlinearities of both electric and magnetic origin are present, the nonlinear response can be modified substantially, being accompanied by nonlinear mode mixing and magneto-electric coupling studied so far only at microwave frequencies.

On one hand, silicon has high-refractive index and third-order nonlinearity; hence, one can expect strong enhancement in the nonlinear optical response of Si nanoparticles on a low-index substrate. Indeed, Scherbakov et al. have recently demonstrated that by engineering the resonant modes of such nanoparticles, one can control the locally enhanced electromagnetic fields, giving rise to two orders of magnitude enhancement of THG with respect to bulk silicon [25,26].

On the other hand, SHG, SFG, and SPDC have been demonstrated in GaAs and AlGaAs nanoantennas on insulator, taking advantage of a few key properties of AlGaAs: a huge non-resonant $\chi^{(2)}$ nonlinearity, a large direct bandwidth that can be varied with aluminum concentration becoming two-photon-absorption free in the C-band of optical communications, and the mature technology of heterostructure laser diodes [27–29].

The most straightforward way to employ Mie resonances with III-V semiconductors is to etch down a pillar of the material directly on the semiconductor substrate. This technique has recently been demonstrated through reported SHG enhancement in gallium phosphide (GaP) nanostructures [30]. The resonators are fabricated via reactive ion etching on the sample by exploiting nanoscale masks to create the pillars (Figure 3.14).

Despite promising results, this kind of resonator are not fabricated on a low-index substrate. In principle a monolithic GaP-on-AlOx platform is feasible, as thermal wet oxidation of AlGaP has been previously reported [31]. Such monolithic platform would allow to better confine the electromagnetic field in the nanostructures, with the possibility to excite Mie type resonances.

The all-GaP nanoantennas reported in [30] can be directly patterned on a GaP wafer. Therefore, hereafter we will not dedicate a specific attention on its fabrication technique. In the following sections, we will review in greater detail the recent monolithic and nonmonolithic approaches, that are employed to fabricate nanoresonators on a low-index substrate.

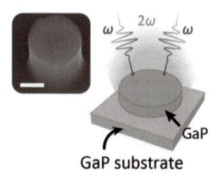

FIGURE 3.14 Scanning electron micrograph of a 200 nm radius GaP disk (scale bar = 200 nm), and schematic view of the experimental setup for a disk emitting green SH light. (From Cambiasso, J. et al., *Nano Lett.*, 17, 1219, 2017.)

3.3.1 CRYSTALLINE SILICON-ON-INSULATOR

When one refers to integrated photonics materials, the first thing that comes to mind is silicon-on-insulator (SOI). The almost ideal SiO_2-Si interface has made the integration of over a million silicon transistors on a single chip possible. Beyond electronics, SOI wafers have been significantly employed in optics as well. SOI building block is a planar shallow waveguide that strongly confines light thanks to the high-refractive index contrast between silicon core and a SiO_2 substrate. This platform has been to date the most successful and the most employed one even at an industrial level in both linear and nonlinear photonic integrated circuits (PICs) [32]. The reason of its success is mainly due to the high field confinement in Si structures, provided by a significant index contrast: $n_{Si} \approx 3.5$, $n_{SiO2} \approx 1.4$ in the Near Infrared (NIR). For silicon nanophotonics, the starting point is provided by the SOI wafers that are typically produced by SOITEC (https://www.soitec.com/en). Then, following the lithographic and etching techniques illustrated in Sections 3.1 through 3.3, state-of-the-art nanofabrication is possible. The general nanofabrication method with SOI substrates is to generate a set of masks on the top of the wafers by EBL, followed by the transfer of the mask geometries into the silicon layer via ICP-RIE. The last step is removing the mask via wet or dry etching.

There is another form of commercially available crystalline silicon-on-insulator, which is silicon-on-sapphire (SOS). In this case the crystalline silicon is epitaxially grown on the sapphire substrate (see Figure 3.2), where the fabrication method is similar to SOI substrates.

3.3.2 AlGaAs-ON-INSULATOR

Nonlinear photonics in silicon is limited by TPA at telecommunications wavelengths (1.55 μm), as no means are currently known to engineer the bandgap of the material (1.1 eV). Moreover, as silicon is a centrosymmetric material, it does not allow for second-order nonlinear optical mixing.

Several TPA-free non-centrosymmetric materials are available, among which lithium niobate ($LiNbO_3$) is the most advanced in terms of integrated microstructures. However, its non-stoichiometry complicates the material processing at the nanoscale, although high-contrast $LiNbO_3$-on-insulator (LNOI) wafers have become available commercially [33] and impressive results have been recently obtained with them [34]. In addition, its relatively low quadratic nonlinearity ($d_{33} \approx 27$ pm/V [35]) limits the miniaturization possibilities of nonlinear integrated chips.

In this context, III-V zinc-blende semiconductors GaAs and $Al_xGa_{1-x}As$, (with $0 < x < 1$) are promising nonlinear materials for the following reasons:

- GaAs has a huge second-order nonlinear coefficient ($d_{14} \approx 110$ pm/V at 1.55 μm), and its third-order nonlinearity is comparable to that of silicon ($n_2 \simeq 1.55 \; 10^{-13}$ cm^2/W).
- Not only GaAs bandgap is larger than in Si, but it can be engineered by modulating the Al molar fraction x, so as to avoid TPA at 1.55 μm for $x \geq 0.18$ ($E_g = 1.424 + 1.266x + 0.26 \, x^2$ eV, for $x < 0.45$). Moreover the gap

of AlGaAs is direct for $0 < x < 0.42$, enabling electrical injection of light-emitting devices.

- GaAs transparency window spans from the visible to the mid-IR ($0.9 \div 16$ μm) and is even broader for AlGaAs.
- AlGaAs laser technology is at a mature stage.

AlGaAs integrated nonlinear optics has attracted the interest of researchers worldwide for more than two decades. The related fabrication techniques follow two distinct approaches.

On the one hand there are the monolithic platforms, implemented either by the selective oxidation of an Al-rich AlGaAs layer to achieve a low-index substrate, or by the exploitation of the refractive index change in $Al_xGa_{1-x}As$ as a function of x, or by suspended structures in air with selective chemical etching. In all these cases the aim is to obtain a high enough index contrast, and therefore a good field confinement by total internal reflection in the bulk of a highly-nonlinear region.

On the other hand there are the nonmonolithic techniques, among which we can distinguish epitaxial lift-off procedures and wafer-bonding methods to report the AlGaAs structures on a low-index substrate (usually glass).

Hereafter we will provide a quick review of what has been achieved with all these methods, and highlight the approach developed within our laboratories in Canberra and Paris.

3.3.2.1 Monolithic Approaches

The simplest monolithic approach developed so far takes advantage of the refractive index change in $Al_xGa_{1-x}As$ as a function of the Al molar fraction x. This variation happens because the refractive index is related to the bandgap via the Kramers-Kronig relations [21]. This implies as well that refractive index variation is monotonic with x. The extremal values are thus exhibited by GaAs ($x = 0$) and AlAs ($x = 1$), and for $\lambda = 900$ nm they correspond to $n_{GaAs} \approx 3.6$ and $n_{AlAs} \approx 3.0$, respectively [36]. Following this principle, in 2011 two groups independently demonstrated AlGaAs monolithic waveguides consisting of stacked AlGaAs layers with different molar concentration for efficient, low-power driven FWM [37], and SHG [38].

In both cases, the stacking was engineered so as to have high field confinement in an Al-poor layer ($x = 0.18$ in [37], $x = 0.2$ in [38]), with TPA-free operation at 1550 nm, by sandwiching it between two higher-x, i.e., lower refractive index, layers. The main limitation of this approach is that the vertical index contrast did not exceed $\Delta n \approx 0.2$, which is considerably lower than the $\Delta n \approx 1.6$ achieved in nonmonolithic platforms. The consequence is that in order to properly confine the electromagnetic field inside the active region, etchings deeper than 1 μm are needed, limiting the miniaturization of the structures.

Until recently, the full development of an AlGaAs monolithic platform was thus hindered by the difficulty of fabricating shallow waveguides and cavities as in the silicon-on-insulator system, and in particular by the shortcomings of wet selective oxidation of AlGaAs epitaxial layers. The latter, discovered in 1990, results in non-stoichiometric alumina (AlOx) with optical and electrical properties similar to SiO_2 [40]. The use of AlOx layers thinner than 100 nm is common in VCSEL technology [41] and also resulted in the demonstration of an AlGaAs guided-wave optical parametric oscillator [42].

However, fabricating high-quality µm-thick AlOx optical substrates is critical because the selective oxidation of AlGaAs layers induces a strong contraction of the oxide. This typically results in high optical losses in integrated photonic devices, due to defects at the interface between AlOx and the adjacent crystal [43]. In 2016 the fabrication was reported of an AlGaAs heterostructure over an AlOx layer, the latter with a sufficient thickness to confine light in the semiconductor heterostructure by total internal reflection and thus behave as an optical substrate [27].

According to the procedure reported in [27,39,44] and sketched in Figure 3.15 and 3.16, a GaAs buffer layer of 500 nm is first grown on [100] non-intentionally doped GaAs wafer, followed by 1000 nm of $Al_{0.98}Ga_{0.02}As$ to be later oxidized, sandwiched between two 90 nm transition layers where the Al molar fraction is linearly varied. Finally, the growth ends with a 100 to 400 nm $Al_{0.18}Ga_{0.82}As$ layer, on which the nanostructures are patterned. In the unit cell of the grown GaAs crystals, gallium, and arsenic atoms exist in equal proportions. AlGaAs crystals have the same structure, with a fraction of gallium atoms replaced by aluminum ones. Since the lattice constant does not vary much with Al molar fraction x ($a = 5.6533 + 0.0078x$ Å), AlGaAs can be easily grown on GaAs with the epitaxial techniques described in the technical sheet "Deposition," with minimal stress and atomic sharp interfaces. The nanostructures are obtained through Electron-Beam Lithography (EBL) followed by Inductively Coupled Plasma – Reactive Ion Etching (ICP). The etching depth, controlled by a laser interferometer, defines the nanoantennas and reveals the aluminum-rich substrate. Such etched sample is then oxidized on the guidelines of the technical sheet "Selective oxidation of AlGaAs." After the oxidation, sub-wavelength optical confinement can occur in each $Al_{0.18}Ga_{0.82}As$ nanoantenna thanks to the low refractive index of the AlOx substrate. A similar structure was fabricated by Sandia Labs [29,45] based on a multilayer GaAs/AlO$_x$ nanoantenna with no transition regions between these two materials, according to the protocol sketched in Figure 3.17.

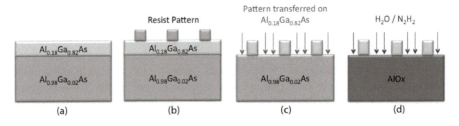

FIGURE 3.15 Fabrication steps of AlGaAs-on-AlOx nanostructures: (a) Epitaxial growth; (b) Electron-beam lithography; (c) Reactive Ion Etching; (d) Selective oxidation. (From Carletti, L. et al., *Nanotechnology*, 28, 114005, 2017.)

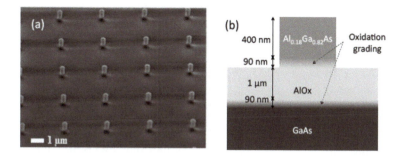

FIGURE 3.16 (a) SEM image of AlGaAs-on-AlOx nanocylinders array. (b) Nanoantenna vertical section. (From Gili, V.F. et al., *Opt. Express*, 24, 14, 2016.)

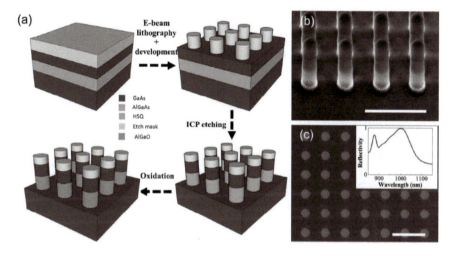

FIGURE 3.17 Fabrication protocol for multilayer GaAs/AlOx nanopillars. (From Liu, S. et al. *Nano Lett.*, 16, 9, 2016.)

TECHNICAL SHEET: SELECTIVE OXIDATION OF AlGaAs

Qualitatively speaking, the thermal wet oxidation of AlGaAs enables to eject As atoms from AlGaAs crystal, which combine with water or hydrogen atoms from water, and obtain low-index amorphous AlO_x. A quantitative model in case of AlAs has been developed in the last two decades, based on Raman spectroscopy [40,46,47], secondary-ion mass spectrometry [48] and X-ray absorption fine-structure spectroscopy [49] measurements. All these efforts allowed the community to come up with the following set of equations:

$$2AlAs_{(s)} + 6H_2O_{(g)} \rightarrow Al_2O_{3(s)} + As_2O_{3(l)} + 6H_{2(g)} \qquad \Delta G = -473 \text{ kJ/mol}$$

(Continued)

$$2\text{AlAs}_{(s)} + 8\text{H}_2\text{O}_{(g)} \rightarrow \text{Al}_2\text{O}_{3(s)} + \text{As}_2\text{O}_{5(l)} + 8\text{H}_{2(g)} \qquad \Delta G = -298 \text{ kJ/mol}$$

$$2\text{AlAs}_{(s)} + 3\text{H}_2\text{O}_{(g)} \rightarrow \text{Al}_2\text{O}_{3(s)} + 2\text{AsH}_{3(l)} + 8\text{H}_{2(g)} \qquad \Delta G = -451 \text{ kJ/mol}$$

$$\text{AlAs}_{(s)} + 2\text{H}_2\text{O}_{(g)} \rightarrow \text{AlO(OH)}_{(s)} + 2\text{AsH}_{3(l)}$$

where the subscripts (s, g, l) refer to the species state, ΔG is the Gibbs free-energy variation during the process calculated at a temperature of ~700 K (~400°C), and its negativity ensures the thermodynamic favorability of the reaction. These first stage reactions happen in the so-called "oxidation front" region and have as a by-product the formation of AsH_3, As_2O_3 and As_2O_5 molecules. These are unstable at 700 K and almost completely degrade into As atoms, which in turn are highly volatile and evacuate through diffusion channels open in the generated amorphous oxide.

$$2\text{As}_2\text{H}_{3(l)} + 3\text{H}_2\text{O}_{(g)} \rightarrow \text{As}_2\text{O}_{3(s)} + 6\text{H}_{2(g)} \qquad \Delta G = -22 \text{ kJ/mol}$$

$$\text{As}_2\text{O}_{3(l)} + 3\text{H}_{2(g)} \rightarrow 2\text{As}_{(g)} + 3\text{H}_2\text{O}_{(g)} \qquad \Delta G = -131 \text{ kJ/mol}$$

$$\text{As}_2\text{O}_{3(l)} + 6\text{H}_{(g)} \rightarrow 2\text{As}_{(g)} + 3\text{H}_2\text{O}_{(g)} \qquad \Delta G = -1226 \text{ kJ/mol}$$

Experimental analysis of $\text{Al}_x\text{Ga}_{1-x}\text{As}$ in the range $0.9 < x < 1$ has led to the following modification [50]:

$$2\text{Al}_x\text{Ga}_{1-x}\text{As}_{(s)} + 6\text{H}_2\text{O}_{(g)} \rightarrow x\text{Al}_2\text{O}_{3(s)} + (1-x)\,\text{Ga}_2\text{O}_{3(s)} + \text{As}_2\text{O}_{3(l)} + 6\text{H}_{2(g)}$$

It was also shown that in this case, especially near the interface with the GaAs substrate, the As atoms can form amorphous precipitates, according to [50];

$$\text{As}_2\text{O}_{3(l)} + 2\text{GaAs}_{(s)} \rightarrow 4\text{As}_{(s)} + \text{Ga}_2\text{O}_{3(s)} \qquad \Delta G = -267 \text{ kJ/mol}$$

The presence of such precipitates has been shown to alter the AlOx optical properties, including a two-order of magnitude increase in the leakage current and up to 30% increase in the material relative permittivity for $x = 0.9$. The use of $\text{Al}_x\text{Ga}_{1-x}\text{As}$ with $x = 0.98$ proved instead to mitigate such effects, with in particular a relative permittivity increase of 7%. In order to formalize the evolution of the oxidation process, a phenomenological model was developed, following a previous work on silicon oxidation [51]. The model assumes that the time $t_{ox}(y)$ necessary to obtain an oxide layer of thickness y is given by the sum of the time necessary to diffuse the oxidant species at

(Continued)

the semiconductor/oxide interface situated at the position y, and the typical duration of the chemical reaction:

$$t_{ox}(y) = t_{diff}(y) + t_{reac}(y)$$

The two terms on the right hand side of the equation depend on the thickness y itself and two parameters A, B in the following way:

$$t_{ox}(y) = \frac{y^2}{A} + y\left(\frac{B}{A}\right)^{-1}$$

where A is proportional to the diffusion coefficient D, and B/A is a function of the reaction constant and the transfer coefficient of the oxidant. As a consequence, the dependence of these two parameters on the temperature will follow the Arrhenius law, as expected from a diffusive process:

$$A = A_0 \exp\left(-E_a / k_B T\right)$$

$$\frac{B}{A} = \frac{B_0}{A_0} \exp\left(-E_a / k_B T\right)$$

where k_B is the Boltzmann constant, and E_a is the activation energy, which for this process is usually ~1.5 eV [52]. The oxidation rate also depends on various other factors. Most importantly, an exponential drop in the oxidation rate with decreasing Al molar content, which is almost 0 for Al molar fractions <0.8 [52] can be appreciated (Figure 3.18). As it will be clear in Chapter 4, this feature has been widely exploited to fabricate AlGaAs-on-AlOx nanoantennas and metasurfaces. A Transmission Electron Microscope (TEM) analysis of our $Al_{0.18}Ga_{0.82}As$-on-AlOx structures confirmed that, while the 1 μm $Al_{0.98}Ga_{0.02}As$ layer is entirely oxidized, only a few-nm oxide layer appears at the nanoantenna surface [53].

The oxidation rate was also shown to weakly depend on the layer thickness [52], however when the latter is larger than ~60 nm the rate is basically constant (Figure 3.18). A less important aspect is represented by the carrier gas, which has been shown to weakly affect the oxidation rate as long as it is not O_2, which completely suppresses AlGaAs oxidation [52]. The temperature of the humid gas mixture consisting of water vapor and carrier gas has instead a considerable impact, and experiments suggest the adoption of temperatures larger than 90°C, while its flow has no significant impact above a certain threshold [52] (Figure 3.19).

It is worth stressing that oxidation must be performed immediately after exposing the Al-rich layer to air with ICP. As a matter of fact, AlGaAs with Al molar fractions >0.8 is very reactive in air, and native oxidation is much more violent than the controlled one, causing cracking, delamination and expansion. Interestingly, in controlled oxidation processes, the volume shrinks instead

(Continued)

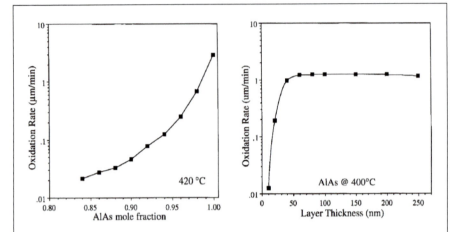

FIGURE 3.18 Left: Exponential dependence of the oxidation rate on Al molar fraction in the alloy. Right: Dependence of the oxidation rate on layer thickness. For $y > 60$ nm the plot is almost constant. (From Choquette, K.D. et al., *IEEE J. Sel. Top. Quant.*, 3, 916, 1997.)

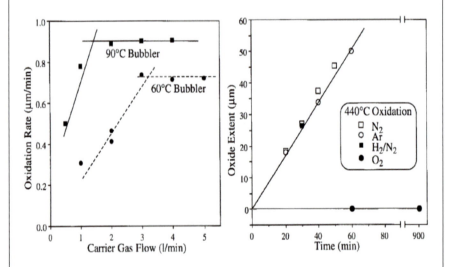

FIGURE 3.19 Impact of the carrier gas on oxidation. Left: the gas flow does not affect the process after a certain threshold, while a higher temperature ensures a larger oxidation rate. Right: the species of the carrier gas has a negligible effect apart from O_2, which completely suppresses the process. (From Choquette, K.D. et al., *IEEE J. Sel. Top. Quant.*, 3, 916, 1997.)

(Continued)

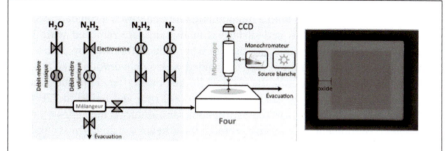

FIGURE 3.20 Left: scheme of oxidation machine. Right: microscope image showing the oxidation front advancing under a structure with low-Al content.

of expanding [54]. Oxidation must be complete as well, otherwise a considerable amount of strain accumulates at the interface between oxidized and unoxidized zones, causing severe delamination.

An oxidation system [44] consists of an oven in which samples are placed on a carbon holder and kept in vacuum conditions (Figure 3.20). The carbon holder supports temperature ramps as high as 40°C/min, allowing to reach the desired oxidation temperature (usually 390°C) in less than 1 hour. While the temperature is ramping, the atmosphere inside the chamber is kept constant at 500 mbar by balancing the injection of N_2 and N_2:H_2 dry gases and vacuum pumping. The humid gas is prepared by injecting in a controlled evaporator mixer at 95°C de-ionized water (liquid state) and N_2:H_2 gas. Once the desired oxidation conditions are met, the humid gas is injected in the chamber and the oxidation begins. The process can be monitored with an in-situ optical microscope, enabling the observation of the lateral oxidation front advancement (Figure 3.20).

The starting point of the AlGaAs-on-thick-AlOx fabrication is a MBE epitaxial wafer of the type (from bottom to top):

- GaAs wafer.
- 500 nm GaAs buffer.
- 90 nm transition AlGaAs layer with Al content linearly increasing from 5% to 98%.
- 1000 nm AlGaAs 98%.
- 90 nm transition AlGaAs layer with Al content linearly decreasing from 98% to 18%.
- from 200 to 400 nm AlGaAs 18%.
- PMMA resist layer to prevent oxidation over time.

Wafers grown in this way have a typical diameter of 5 cm. Starting from this, a sample of ~1 cm^2 is cleaved with a diamond tip and undergoes a standard cleaning procedure. Organic residues and surface contaminants are removed with 3 min ultrasound bath first in acetone and then in isopropanol, followed by 2 min in an O_2 plasma asher. The cleaning is then concluded with a dehydration bake at 120°C for 4 min for desorbing water on the surface. Immediately after, Ti-prime adhesion promoter is deposited on the surface, spun at 6000 rpm and baked at 120°C for 2 min in order to make the surface hydrophobic and improve the adhesion of the resist after development. In a former protocol the e-beam sensitive resist employed is negative-tone ma-N 2403 or 2401, that is a mixture of a polymeric bonding agent (novolac) and a photo-active compound, originally designed for UV photolithography, which nonetheless showed good performances in EBL too. Exposure of the resist to electrons makes polymer chains in the mixture cross-link with each other to form bigger chains, thus locally reducing the resist solubility in the developer. Few drops are collected with a pipette and deposited on the sample surface.

Spin coating at 3000 rpm leaves a 300 nm thick layer and a soft bake at 95°C for 1 min helps evaporating any residual trace of solvent in the resist layer. EBL is then performed, with a beam aperture of 10 μm, a good compromise between resolution, focusing ease, and patterning rate. The acceleration voltage is set at 20 kV, which again is a compromise between resolution and proximity effects. The typical dose suggested by the producing company for exposure is 120 μC/cm^2. This value can be slightly reduced depending on the geometry of the patterns to reduce proximity effects. After exposure the resist is developed in AZ 326 MIF, an alkali solution based on tetramethylammonium hydroxide (TMAH). The development is performed in multiple steps by dipping the sample in the developer for 10 s and then in deionized water for 90 s. Three or four steps are usually sufficient to obtain a clean result and resolve critical features.

Resist reflow is then performed by hot baking the sample at 150°C for 3 min. This step is crucial to reduce sidewall roughness. Structures are then defined with ICP etching, whose depth is controlled with an in-situ Fabry-Perot (FP) system based on a low-power CW laser centered at 633 nm.

After ICP, the residual resist is removed by leaving the sample in acetone at 40°C (near boiling point) for 20 min and then in isopropanol for 5 min. Oxidation during this time is negligible as the sample makes almost no contact with air. After the resist is removed, the sample is immediately put in the oxidation oven under vacuum ($\approx 10^{-2}$ mbar), and it is oxidized at 390°C for 30 min.

3.3.3 TEM CHARACTERIZATION

Samples fabricated with this protocol can be analyzed with a TEM, as in Ref. [53]. Prior to microscope observation of a thin cross sections, they are first encapsulated in platinum via evaporation and then cut with Focused Ion Beam (FIB). After such preparation, the specific sample described hereafter contains three nominally identical nanoantennas and is placed in a JEOL ARM 200F TEM. This microscope combines a cold field emission gun [55] and a CEOS hexapole spherical aberration corrector (CEOS GmbH) [56] to compensate for the spherical aberration of

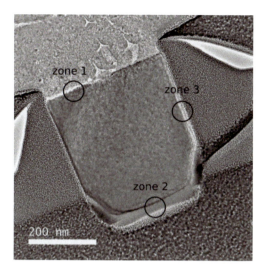

FIGURE 3.21 High-resolution TEM images of $Al_{0.98}Ga_{0.02}As$ nanoantennas. The light gray area is the AlO_x substrate, whose amorphous nature is confirmed. The nanoantenna shape slightly deviates from that of a cylinder and its final 100 nm resemble more a truncated cone.

the objective electron lens. Thanks to this TEM configuration a point resolution of 75 pm at 200 kV can be achieved [57]. High-resolution images of a single nanoantenna confirm its crystalline nature, as well as the amorphous nature of the AlOx layer, with the sporadic presence of polycrystalline islands due to the 0.02 Ga molar concentration in the $Al_{0.98}Ga_{0.02}As$ layer (Figure 3.21).

Moreover, a zoom in zones 2 and 3 highlights the presence of a thin (≈ 2 nm) AlOx passivation layer on the nanoantenna surface due to partial oxidation of the $Al_{0.18}Ga_{0.82}As$ layer. Conversely HR-TEM images also brought about the presence of an important source of nonideality [53]: despite the absence of any alteration of the crystalline matrix, the crystallographic directions of the three nanoantennas are neither parallel to each other nor to the GaAs substrate. The loss of this epitaxial relationship is due to the contraction of the $Al_{0.98}Ga_{0.02}As$ layer during wet oxidation (Figure 3.22).

More in detail, the tilt in the crystallographic directions of the nanoantennas was revealed with electron diffraction measurements, performed on the same machine with a double-tilt sample holder [57]. Using the GaAs substrate as a reference we determined the value of the tilt with respect to the [001] and [110] directions by comparing their electron diffraction maps in Fourier space. The measured values for these tilt angles are reported hereafter:

Zone	[001] tilt	[110] tilt
GaAs substrate	0.2°	0°
Nanopillar 1	5.7°	1.8°
Nanopillar 2	−3.1°	−0.9°
Nanopillar 3	0.7°	−1.3°

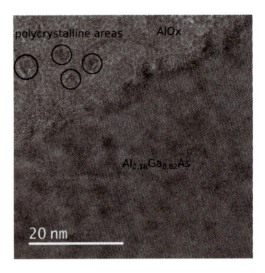

FIGURE 3.22 High-resolution zoom of zone 1.

The presence of these tilt angles has been proven to affect only slightly the nominal SHG response of these AlGaAs-on-AlO$_x$ monolithic pillars, highlighting their satisfactory use as an ideal test bed for nonlinear nanophotonics.

Despite the success of this fabrication procedure, ma-N e-beam resist limitations appear when attempting to fabricate more complex structures with gaps smaller than 50 nm. To demonstrate complex nanoscale structures, the use of a high-resolution e-beam resist called Hydrogen Silsesquioxane (HSQ) is more appropriate.

3.3.4 Ma-N vs HSQ in Nanofabrication

HSQ consists of HSiO clusters in a solvent solution (MIBK), and its exposure to the e-beam causes cross-linking of Si and O atoms in the clusters, eventually forming SiO$_x$, i.e., glass. Three different HSQ concentrations are commercially available: 6%, 2% and 1%. The advantage of this electron-resist is twofold: not only is its intrinsic resolution higher than that of ma-N series, but also its hardness, thus improving the ICP yield. An important challenge of HSQ implementation on III-V semiconductors regards the adhesion enhancement after development. Standard Ti-prime or HMDS treatment, used in standard ma-N based protocols, proved to be insufficient. Moreover, despite existing literature claiming improved adhesion of HSQ on III-V semiconductors with a cationic organic material (SurPass 3000) [58], we did not observe a reproducible enhancement. A solution to the adherence problem was found by depositing

3 nm of SiO_x with Plasma-Enhanced Chemical Vapor Deposition (PECVD) (see technical sheet: deposition). This PECVD step is performed right after the standard sample cleaning. Another possibility is to deposit a thin SiO_2 layer via Atomic Layer Deposition (ALD), which allows for greater precision than PECVD at the price of a longer processing time. Spin coating of 6% HSQ at 4000 rpm, followed by two hot plate bakings at 150°C and 200°C for 2 min each, allows obtaining a layer ~100 nm thick. As for e-beam lithography, the dose necessary to induce cross-linking in the resist molecular bonds at 20 kV is roughly one order of magnitude higher than the ma-N standard (~1800 $\mu C/c^2$), meaning that the average single-writing field patterning time increases form ~30 s to ~4 min. Development can achieved by dipping the sample in a KOH-based solution (AZ 400 K) for 1 min, plus rinsing in deionized water for 90 s. Alternatively, Tetramethilammonium Hydroxide (TMAH) protocols are also widely used. Unlike ma-N, HSQ development is single-step and no reflow is performed afterwards. Just before ICP, a CHF_3 plasma reactive ion etching can be performed to get rid of the SiO_2 buffer layer, which would greatly slow down the ICP step.

Particular attention must be paid to the process duration as HSQ etching rate is very similar to that of PECVD SiO_x. Thanks to this new protocol, structures impossible to fabricate with ma-N resist were achieved, such as pentamers with 30 nm gap (Figure 3.23).

The improved quality of dimers and pentamers demonstrates that HSQ is less sensible to secondary electron exposure, mitigating proximity effects with respect to ma-N. Another interesting comparison between the two resists can be made in considering nanoantennas surrounded by diffraction gratings. It is indeed possible to appreciate a net roughness decrease in the HSQ case, due to enhanced lateral resolution (Figure 3.24).

The lateral resolution improvement is even more evident in Atomic Force Microscope (AFM) characterization (Figure 3.25).

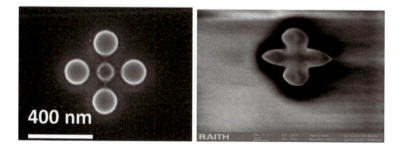

FIGURE 3.23 Left: top SEM image of a pentamer fabricated with HSQ lithographic mask. Right: attempt to fabricate the same structure using ma-N 2401.

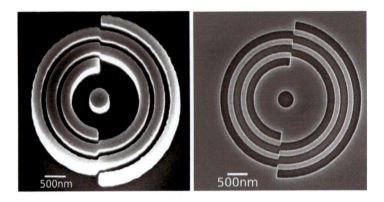

FIGURE 3.24 Left: tilted SEM image of a nanoantenna surrounded by a diffraction grating fabricated with ma-N 2401. Right: SEM image of the same structure with nominally identical parameters fabricated with HSQ (courtesy of Edmond Cambril). The latter presents both improved resolution and decreased lateral roughness.

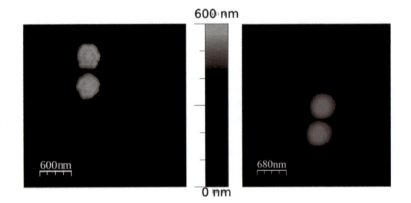

FIGURE 3.25 Left: AFM map of a dimer with 60 nm gap and a nominal radius of 225 nm fabricated with ma-N 2401. Right: dimer fabricated with HSQ technology. The nominal gap is the same (60 nm), while the nominal radius is slightly bigger (250 nm). The analysis is performed with WSxM software. (From Horcas, I., and Fernàndez, R., *Rev. Sci. Instrum.*, 78, 013705, 2007.)

3.3.5 Nonmonolithic Approaches

A possible strategy to achieve full control of harmonic radiation is to fabricate nanoantennas that can freely emit in both the forward and the backward directions. However, the monolithic semiconductor-on-insulator fabrication techniques do not always allow for this, as they require a nontransparent substrate for direct growth of semiconductors. As explained above, the growth on transparent substrates (e.g., glass) is avoided because it results in a high density of dislocations. Recent works have therefore been devoted to transfer semiconductor thin films onto a glass substrate, followed by the fabrication of micro- and nanostructures.

Formerly developed in the 1980s for solar cells [60], such process begins with the epitaxial growth of an active (Al)GaAs layer with low-Al content on top of a GaAs substrate, with a sacrificial AlAs layer sandwiched between them. The sample is then immersed in HF acid, which selectively attacks AlGaAs while leaving GaAs layers untouched, with an etching rate that sensibly grows with the Al content x in the AlGaAs alloy [61]. As the AlAs sacrificial layer is eaten by the acid, the active layer we are interested in detaches from the substrate and forms a thin film. Finally, the desired pattern is transferred on the active layer with standard lithographic processes (Figure 3.26). Such approach has been successfully applied to nanophotonics, with the demonstration of zero backscattering from GaAs nanoparticles [62].

The main drawback of this technique is that the interaction of HF with the AlAs sacrificial layer can leave some AlF_3 or $AlF(H_2O)$ impurities on the sample-film, which could be detrimental for some applications. To overcome this problem, the usage of lattice-matched sacrificial layers of materials such GaInP [63] or InAlP [64] has been shown to improve the quality of the fabricated films, with the surface mean square roughness decreasing from ~3 nm for AlAs, to less than 1 nm (Figure 3.27).

However, this method does not allow for the fabrication of high-resolution nano-structures with smooth surfaces and edges, which is crucial for the exploration of the bulk SHG from nanoantennas. Conversely, other AlGaAs nonmonolithic platforms were developed with the wafer-bonding technique. In this process two sufficiently flat and clean surfaces bond to each other with the help of an external applied pressure. In some cases this bonding is direct, not mediated by any surface activation. However, as AlGaAs does not stick well to glass, people have relied on the adhesive properties of Benzocyclobutene (BCB), a transparent polymer with a refractive index $n \approx 1.45$, which is compatible with optimal optical measurements.

A combination of BCB-mediated wafer-bonding and ELO-based peeling-off allowed to measure efficient SHG from AlGaAs nanoantennas embedded in a thin glass substrate. In Ref. [28] the nanoantenna pattern is first defined on an AlGaAs layer with EBL followed by vertical anisotropic etching. Afterwards, a Cl_2-based chemical treatment, combined by HF-etching of an AlAs sacrificial layer, makes the sample surface non-adhesive. The nanoantennas are then embedded in BCB, which mediates wafer-bonding to a glass substrate, and finally the sample is peeled-off the GaAs substrate as sketched in Figure 3.28.

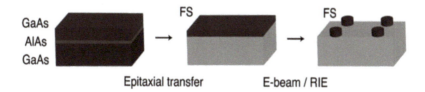

FIGURE 3.26 Fabrication of GaAs nanoparticles on fused silica (FS) via the epitaxial lift-off. (From Person, S. et al., *Nano Lett.*, 13, 1806, 2013.)

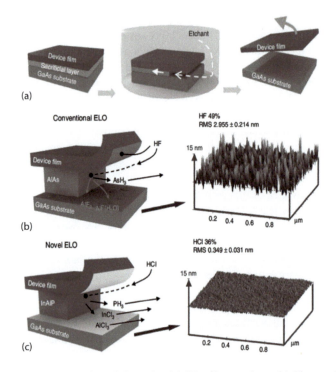

(a)

(b)

(c)

FIGURE 3.27 (a) Schematics of the epitaxial lift-off procedure. (b) The employment of AlAs as a sacrificial layer can lead to the formation of impurities, with a mean squared roughness of about 3 nm. (From Kongai, M. et al., *J. Cryst. Growth.*, 45, 277, 1978.) (c) Using sacrificial layers of different materials such as GaInP or InAlP causes a significant improvement, with a mean squared roughness less than one nanometer. (From Lee, K. et al., *J. Appl. Phys.*, 111, 033527, 2012; Cheng, C.W. et al., *Nat. Commun.*, 4, 1577, 2013.)

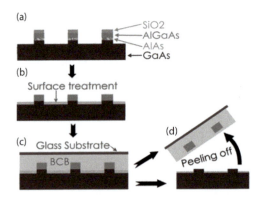

FIGURE 3.28 Fabrication procedure to obtain AlGaAs nanostructures embedded in glass. (a) The patterns are first defined on the AlGaAs layer. (b) While HF and Cl_2 chemical treatments are used to create a non-adhesive surface. (c) BCB is then deposited on the sample and is used as an intermediary for wafer bonding on a glass substrate. (d) Finally, the glass chip is peeled-off the GaAs substrate. (From Camacho-Morales, R. et al., *Nano Lett.*, 16, 7191, 2016.)

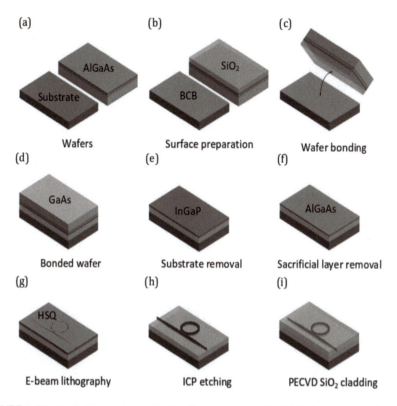

FIGURE 3.29 (a–i) Alternative wafer-bonding procedure, in which the wafer-bonding step is performed before the structure patterning. (From Pu, M. et al., *Optica* 3, 8, 2016.)

Alternatively, the inverse strategy can be adopted as well, that is to perform the wafer-bonding on a SiO_2 substrate first, and then pattern the AlGaAs layer at a later stage, as illustrated in Figure 3.29. This was done by M. Pu and coworkers, who have also employed a GaInP sacrificial layer rather than an AlAs one, with all the above-mentioned benefits [65]. This process enabled their demonstration of efficient frequency-comb generation in AlGaAs waveguide+microring systems.

3.4 METAL-DIELECTRIC AND METAL-SEMICONDUCTOR HYBRID NANOANTENNAS

As it was recalled above, metallic nanoparticles can exhibit high optical nonlinearity due to both strong local field enhancement and intrinsically high surface nonlinearity. However, since metallic structures absorb light, they tend to have a relatively low heat resistance to high power lasers and this could be a key limiting factor to nonlinear optics at the nanoscale. On the other hand, high-index dielectric nanoparticles can also provide significant focusing of light at the nanoscale, which is not limited to interfaces. Such nanoparticles offer unique opportunities for the study of nonlinear effects due to their very low losses as well as high light scattering efficiency. Importantly, dielectric nanostructures have multipolar characteristics of both electric and magnetic resonant modes that could aid the engineering of optical nonlinearity. The caveat of these benefits is a considerably lower field enhancement and the weaker intrinsic nonlinearity of the semiconductor materials used to construct these nanoparticles. Therefore, many believe that the best approach will incorporate the best of both worlds; i.e., hybrid metal-dielectric/semiconductor nanostructures.

For fabricating hybrid nanoantennas, unlike the processes described so far, multiple lithographies are necessary. In case of metallic structures placed above the dielectric/semiconductor ones, the workflow would generally consist of a first positive-tone lithography plus lift-off to define metallic nanostructures, followed by a negative-tone EBL and etching to reveal the dielectric/semiconductor nanostructures. Conversely, if both metallic and dielectric structures lie on the same substrate the processing order is inversed. To correctly align the mask of the second lithography to the first one, proper alignment marks are crucial. These consist of big arrows to easily find the to-be-patterned zone and crosses to finely align the EBL reference system to the sample (Figure 3.30).

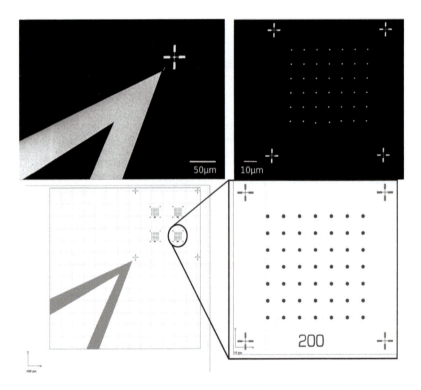

FIGURE 3.30 Alignment marks for hybrid dielectric-metal nanoantenna fabrication. A big arrow (left panels) is first used to find the structures. A set of four large crosses (bottom left panel) allows to perform a first coarse alignment. Smaller crosses (right panels) help to finely align each time the beam moves to pattern a new writing field. Top: SEM micrograph of fabricated Au alignment arrow and big crosses (left) and a nanoantenna array with small crosses (right). Bottom: structure schematics.

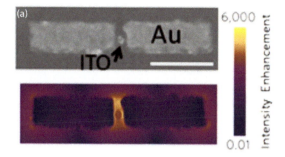

FIGURE 3.31 SEM image of a single nonlinear up-conversion system. The scale bar is 200 nm. (From Aouani, H. et al., *Nat. Nanotechnol.*, 9, 290–294, 2014.)

By exploiting this technique, a few years ago, a hybrid system was introduced, which can significantly increase the intensity of the nonlinear light. It was demonstrated that THG from an individual semiconductor indium tin oxide nanoparticle is significantly enhanced when coupled within a plasmonic gold dimer. The plasmonic dimer acts as a receiving optical antenna, confining the incident far-field radiation into a near field, localized at its gap; the ITO nanoparticle located at the plasmonic dimer gap acts as a localized nonlinear transmitter upconverting three incident photons. As can be seen in Figure 3.31, this hybrid nanodevice provides THG enhancements of up to 10^6 fold compared with an isolated ITO nanoparticle [19].

Alongside this, hybrid approaches have been applied as well to SOI and AlGaAs-on-AlOx platforms by integrating metal (Au) and semiconductor nanostructures on the same chip. These technological achievements allowed to demonstrate enhanced THG [66] and enhanced SHG [67], respectively (Figure 3.32). The role of the plasmonic nanoring in both works was to favor the excitation of an anapole mode at the fundamental frequency. While this mode is ideally radiation-less, it allows storing electromagnetic energy in the bulk of the dielectric nanoantenna, allowing for a considerable SHG/THG efficiency enhancement with respect to the bare pillar.

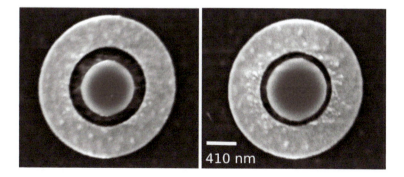

FIGURE 3.32 Fabricated AlGaAs-Au-on-AlOx nanostructures. Left: the gap between AlGaAs antenna and Au ring is 180 nm. Right: the gap is as small as 100 nm. The scalebar applies for both micrographs. (From Gili, V.F. et al., *Beilstein J. Nanotechnol.*, 9, 2306, 2018.)

LIST OF ACRONYMS

AFM	Atomic Force Microscope
Al	Aluminum
AlAs	Aluminum Arsenide
ALD	Atomic Layer Deposition
AlGaAs	Aluminum Gallium Arsenide
AlGaP	Aluminum Gallium Phosphide
AlOx	Aluminum oxide
Ar	Argon
As	Arsenic
Au	Gold
BCB	Benzocyclobutene
CHF$_3$	Fluoroform
CMOS	Complementary Metal-Oxide-Semiconductor
CW	Continuous-Wave
DFG	Difference-frequency generation
EBL	Electron-beam lithography
EELS	Electron Energy Loss Spectroscopy
ELO	Epitaxial Lift-Off
FP	Fabry-Perot
FIB	Focused ion beam
FWM	Four wave mixing
Ga	Gallium
GaAs	Gallium Arsenide
GaN	Gallium Nitride
GaP	Gallium Phosphide
Ge	Germanium
HF	Fluorhydric Acid
HR-TEM	High-Resolution Transmission Electron Microscope
HSQ	Hydrogen silsesquioxane
ICP	Inductively coupled
In	Indium
InAs	Indium Arsenide
InGaAs	Indium Gallium Arsenide
InP	Indium Phosphide
ITO	Indium tin oxide
LiNbO$_3$	Lithium Niobate
LNOI	LiNbO$_3$-on-insulator
MBE	Molecular beam epitaxy
MIBK	Metilisobutilchetone
MOCVD	Metal-organic chemical vapor deposition
NH$_3$	Ammonia
NIR	Near infrared
P	Phosphorus
PECVD	Plasma-enhanced chemical vapor deposition

PIC	Photonic integrated circuit
PMMA	Polymethyl methacrylate
PPLN	Periodically poled lithium niobate
Pt	Platinum
RF	Radio frequency
RHEED	Reflection high-energy electron diffraction
RIE	Reactive ion etching
SEM	Scanning electron microscope
SERS	Surface-enhanced Raman scattering
SFG	Sum-frequency generation
SHG	Second-harmonic generation
Si	Silicon
$SiCl_4$	Silicon Tetrachloride
SiH_4	Silane
SiOx	Silicon Oxide
SOI	Silicon-on-insulator
SOS	Silicon-on-sapphire
SPDC	Spontaneous parametric down-conversion
TEM	Transmission electron microscope
THG	Third-harmonic generation
Ti	Titanium
TMAH	Tetramethylammonium hydroxide
TPA	Two-photon absorption
ZnO	Zinc Oxide

REFERENCES

1. S. Wang, X. Ouyang, Z. Feng, Y. Cao, M. Gu, and X. Li, "Diffractive photonic applications mediated by laser reduced graphene oxides," *Optoelectronics and Advanced* 1, 170002 (2018).

2. P. Frigeri, L. Seravalli, G. Trevisi, and S. Franchi, "3.12: Molecular beam epitaxy: An overview," In P. Bhattacharya, R. Fornari, and H. Kamimura, *Comprehensive Semiconductor Science and Technology*, Vol. 3. Amsterdam, the Netherlands: Elsevier, pp. 480 (2011).

3. V. M. Donnelly, and A. Kornblit, "Plasma etching: Yesterday, today, and tomorrow," *Journal of Vacuum Science & Technology A: Vacuum, Surfaces, and Films* 31, 050825 (2013).

4. M. Munsch, N. S. Malik, E. Dupuy, A. Delga, J. Bleuse, J. M. Gérard, J. Claudon, N. Gregersen, and J. Mørk, "Dielectric GaAs antenna ensuring an efficient broadband coupling between an InAs quantum dot and a Gaussian optical beam," *Physical Review Letters* 111, 239902 (2013).

5. S. A. Maier, *Plasmonics: Fundamentals and Applications*. Boston, MA: Springer Science & Business Media (2007).

6. M. Rahmani, B. Luk'yanchuk, and M. Hong, "Fano resonance in novel plasmonic nanostructures," *Laser & Photonics Reviews* 7 (3), 329 (2012).

7. C. E. Talley, J. B. Jackson, C. Oubre, N. K. Grady, C. W. Hollars, S. M. Lane, T. R. Huser, P. Nordlander, and N. J. Halas, "Surface-enhanced Raman scattering from individual Au nanoparticles and nanoparticle dimer substrates," *Nano Letters* 5, 8 (2005).

8. M. Celebrano, X. Wu, M. Baselli, S. Großmann, P. Biagioni, A. Locatelli, C. De Angelis, et al. "Mode matching in multiresonant plasmonic nanoantennas for enhanced second harmonic generation," *Nature Nanotechnology* 10, 412 (2015).

9. T. Hanke, J. Cesar, V. Knittel, A. Trügler, U. Hohenester, A. Leitenstorfer, and R. Bratschitsch, "Tailoring spatiotemporal light confinement in single plasmonic nanoantennas," *Nano Letters* 12, 992 (2012).

10. F. Lu, W. Zhang, L. Huang, S. Liang, D. Mao, F. Gao, T. Mei, and J. Zhao, "Mode evolution and nanofocusing of grating-coupled surface plasmon polaritons on metallic tip," *Opto-Electronic Advances* 1, 6 (2018).

11. L. Chen, Y. Zhou, M. Wu, and M. Hong, "Remote-mode microsphere nano-imaging: New boundaries for optical microscopes," *Opto-Electronic Advance* 1, 170001 (2018).

12. P. Lalanne, S. Astilean, P. Chavel, E. Cambril, and H. Launois, "Blazed binary subwavelength gratings with efficiencies larger than those of conventional échelette gratings," *Optics Letters* 23, 14 (1998).

13. M. Khorasaninejad, W. T. Chen, R. C. Devlin, J. Oh, A. Y. Zhu, and F. Capasso, "Metalenses at visible wavelengths: Diffraction-limited focusing and subwavelength resolution imaging," *Science* 352, 6290 (2016).

14. A. Tittl, A. Leitis, M. Liu, F. Yesilkoy, D. Y. Choi, D. N. Neshev, Y. S. Kivshar, and H. Altug, "Imaging-based molecular barcoding with pixelated dielectric metasurfaces," *Science* 360, 1105 (2018).

15. A. I. Kuznetsov, A. E. Miroshnichenko, M. L. Brongersma, Y. S. Kivshar, and B. Luk'yanchuk, "Optically resonant dielectric nanostructures," *Science* 354, 6314 (2016).

16. X. Zhu, W. Yan, U. Levy, N. A. Mortensen, and A. Kristensen, "Resonant laser printing of structural colors on high-index dielectric metasurfaces," *Science Advances* 3, e1602487 (2017).

17. G. Grinblat, Y. Li, M. P. Nielsen, R. F. Oulton, and S. A. Maier, "Enhanced third harmonic generation in single germanium nanodisks excited at the anapole mode," *Nano Letters* 16, 4635 (2016).

18. H. Long, A. Chen, G. Yang, Y. Li, and P. Lu, "Third-order optical nonlinearities in anatase and rutile TiO_2 thin films," *Thin Solid Films* 517, 5601 (2009).

19. H. Aouani, M. Rahmani, M. Navarro-Cía, and S. A. Maier, "Third-harmonic-upconversion enhancement from a single semiconductor nanoparticle coupled to a plasmonic antenna," *Nature Nanotechnology* 9, 290 (2014).

20. H. Ye, Z. Su, F. Tang, Y. Bao, X. Lao, G. Chen, J. Wang, and S. Xu, "Probing defects in ZnO by persistent phosphorescence," *Opto-Electronic Advances* 1, 6 (2018).

21. R. W. Boyd, *Nonlinear Optics*. Burlington, VT: Academic Press, 3rd edition (2008).

22. B. Lu, H. Lan, and H. Liu, "Additive manufacturing frontier: 3D printing electronics," *Opto-Electronic Advances* 1, 1 (2018).

23. A. Nemati, Q. Wang, M. Hong, and J. Teng, "Tunable and reconfigurable metasurfaces and metadevices," *Opto-Electronic Advances* 1, 5 (2018).

24. L. Xu, M. Rahmani, K. Z. Kamali, A. Lamprianidis, L. Ghirardini, J. Sautter, R. Camacho-Morales, et al. "Boosting third-harmonic generation by a mirror-enhanced anapole resonator," *Light: Science and Applications* 7, 44 (2018).

25. M. R. Shcherbakov, P. P. Vabishchevich, A. S. Shorokhov, K. E. Chong, D. Y. Choi, I. Staude, A. E. Miroshnichenko, D. N. Neshev, A. A. Fedyanin, and Y. S. Kivshar "Ultrafast all-optical switching with magnetic resonances in nonlinear dielectric nanostructures," *Nano Letters* 15, 6985 (2015).

26. M. R. Shcherbakov, D. N. Neshev, B. Hopkins, A. S. Shorokhov, I. Staude, E. V. Melik-Gaykazyan, M. Decker, et al. "Enhanced third-harmonic generation in silicon nanoparticles driven by magnetic response," *Nano Letters* 14, 6488 (2014).

27. V. F. Gili, L. Carletti, A. Locatelli, D. Rocco, M. Finazzi, L. Ghirardini, I. Favero, C. Gomez, A. Lemâitre, M. Celebrano, C. De Angelis, and G. Leo, "Monolithic AlGaAs second-harmonic nanoantennas," *Optics Express* 24, 14 (2016).
28. R. Camacho-Morales, S. Kruk, M. Rahmani, L. Wang, L. Xu, D. Smirnova, A. Solntsev, et al. "Nonlinear generation of vector beams from AlGaAs nanoantennas," *Nano Letters* 16, 11, 7191 (2016).
29. S. Liu, M. B. Sinclair, S. Saravi, G. A. Keeler, Y. Yang, J. Reno, G. M. Peake, F. Setzpfandt, I. Staude, T. Pertsch, and I. Brener, "Resonantly enhanced second-harmonic generation using III-V semiconductor all-dielectric metasurfaces," *Nano Letters* 16, 9 (2016).
30. J. Cambiasso, G. Grinblat, Y. Li, A. Rakovich, E. Cortés, and S. A. Maier, "Bridging the gap between dielectric nanophotonics and the visible regime with effectively lossless gallium phosphide antennas," *Nano Letters* 17, 1219 (2017).
31. J. H. Epple, K. L. Chang, G. W. Pickrell, K. Y. Cheng, and K. C. Hsieh, "Thermal wet oxidation of GaP and $Al_{0.4}Ga_{0.6}P$," *Applied Physics Letters* 77, 1161 (2000).
32. J. Leuthold, C. Koos, and W. Freude, "Nonlinear silicon photonics," *Nature Photonics* 4, 535 (2010).
33. A. Boes, B. Corcoran, L. Chang, J. Bowers, and A. Mitchell, "Status and potential of lithium niobate on insulator (LNOI) for photonic integrated circuits," *Laser & Photonics Reviews* 12, 4 (2018).
34. C. Wang, Z. Li, M. H. Kim, X. Xiong, X. F. Ren, G. C. Guo, N. Yu, and M. Lončar, "Metasurface-assisted phase-matching-free second harmonic generation in lithium niobate waveguides," *Nature Communications* 8, 2098 (2017).
35. I. Shoji, T. Kondo, A. Kitamoto, M. Shirane, and R. Ito, "Absolute scale of second-order nonlinear-optical coefficients," *Journal of the Optical Society of America* 14, 2268 (1997).
36. D. E. Aspnes, S. M. Kelso, R. A. Logan and R. Bhat, "Optical properties of Al_xGa_{1-x} As," *Journal of Applied Physics* 60, 754 (1986).
37. K. Dolgaleva, W. C. Ng, L. Qian, and J. S. Aitchison, "Compact highly-nonlinear AlGaAs waveguides for efficient wavelength conversion," *Optics Express* 19, 13 (2011).
38. D. Duchesne, K. A. Rutkowska, M. Volatier, F. Légaré, S. Delprat, M. Chaker, D. Modotto, et al. "Second harmonic generation in AlGaAs photonic wires using low power continuous wave light," *Optics Express* 19, 13 (2011).
39. L. Carletti, D. Rocco, A. Locatelli, C. De Angelis, V. F. Gili, M. Ravaro, I. Favero et al. "Controlling second-harmonic generation at the nanoscale with monolithic AlGaAs-on-AlOx antennas," *Nanotechnology* 28, 114005 (2017).
40. J. M. Dallesasse, N. Holonyak, A. R. Sugg, T. A. Richard, and N. ElZein, "Hydrolyzation oxidation of $Al_xGa_{1-x}As/AlAs/GaAs$ quantum well heterostructures and superlattices," *Applied Physics Letters* 57, 2844 (1990).
41. K. L. Lear, K. D. Choquette, R. P. Schneider Jr., S. P. Kilcoyne, K. M. Geib, "Selectively oxidized vertical cavity surface emitting lasers with 50% power conversion efficiency," *Electronics Letters* 31, 3 (1995).
42. M. Savanier, C. Ozanam, L. Lanco, X. Lafosse, A. Andronico, I. Favero, S. Ducci, and G. Leo, "Near-infrared optical parametric oscillator in a III-V semiconductor waveguide," *Applied Physics Letters* 103, 261105 (2013).
43. F. Chouchane, G. Almuneau, N. Cherkashin, A. Arnoult, G. Lacoste, and C. Fontaine, "Local stress-induced effects on AlGaAs/AlOx oxidation front shape," *Applied Physics Letters* 105, 041909 (2014).
44. G. Leo, O. Stepanenko, and A. Lemaître, "Procédé de fabrication d'au moins une structure semi-conductrice présentant un empilement d'une ou de plusieurs couches d'Arséniure de Gallium-Aluminium," *French Patent* FR16 70043 (2016).

45. S. Liu, G. A. Keeler, J. L. Reno, M. B. Sinclair, and I. Brener, "III-V Semiconductor nanoresonators: A new strategy for passive, active, and nonlinear all-dielectric meta-materials," *Advanced Optical Materials* 4, 1457 (2016).

46. C. I. H. Ashby, J. P. Sullivan, K. D. Choquette, K. M. Geib, and H. Q. Hou, "Wet oxidation of AlGaAs: The role of hydrogen," *Journal of Applied Physics* 82, 3134 (1997).

47. C. I. H. Ashby, M. M. Bridges, A. A. Allerman, B. E. Hammons, and H. Q. Hou, "Origin of the time dependence of wet oxidation of AlGaAs," *Applied Physics Letters* 75, 73 (1999).

48. A. R. Sugg, N. Holonyak Jr., J. E. Baker, F. A. Kish, and J. M. Dallesasse, "Native oxide stabilization of AIAs-GaAs heterostructures," *Applied Physics Letters* 58, 1199 (1991).

49. S. K. Cheong, B. A. Bunker, T. Shibata, D. C. Hall, C. B. De Melo, Y. Luo, G. L. Snider, G. Kramer, and N. El-Zein, "Residual arsenic site in oxidized $Al_xGa_{1-x}As$ (x=0.96)," *Applied Physics Letters* 78, 2458 (2001).

50. C. I. H. Ashby, J. P. Sullivan, P. P. Newcomer, N. A. Missert, H. Q. Hou, B. E. Hammons, M. J. Hafich, and A. G. Baca, "Wet oxidation of $Al_xGa_{1-x}As$: Temporal evolution of composition and microstructure and the implications for metal-insulator-semiconductor applications," *Applied Physics Letters* 70, 2443 (1997).

51. B. E. Deal and A. S. Grove "General relationship for the thermal oxidation of silicon," *Journal of Applied Physics* 36, 3770 (1965).

52. K. D. Choquette, K. M. Geib, C. I. H. Ashby, R. D. Twesten, O. Blum, H. Q. Hou, D. M. Follstaedt, B. E. Hammons, D. Mathes, and R. Hull, "Advances in selective wet oxidation of AlGaAs alloys," *IEEE Journal of Selected Topics in Quantum Electronics* 3, 916 (1997).

53. V. F. Gili, L. Carletti, F. Chouchane, G. Wang, C. Ricolleau, D. Rocco, A. Lemaître, et al. "Role of the substrate in monolithic AlGaAs nonlinear nanoantennas," *Nanophoton* 7, 2 (2018).

54. O. Durand, F. Wyckzisk, J. Olivier, M. Magis, P. Galtier, A. De Rossi, M. Calligaro, et al. "Contraction of aluminum oxide thin layers in optical heterostructures," *Applied Physics Letters* 83, 2554 (2003).

55. Y. Kohno, E. Okunishi, T. Tomita, I. Ishikawa, T. Kaneyama, Y. Ohkura, Y. Kondo, and T. Isabell, "Long-range chemical orders in Au-Pd nanoparticles revealed by aberration-corrected electron microscopy," *Microscopy and Analysis Supplement* 24,S9 (2010).

56. M. Haider, S. Uhlemann, E. Schwan, H. Rose, B. Kabius and K. Urban, "Electron microscopy image enhanced," *Nature* 392, 768 (1998).

57. C. Ricolleau, J. Nelayah, T. Oikawa, Y. Kohno, N. Braidy, G. Wang, F. Hue, I. Florea, V. Pierron Bohnes, and D. Alloyeau, "Performances of an 80–200 kV microscope employing a cold-FEG and an aberration-corrected objective lens," *Journal of Electron Microscopy* 62, 283 (2012).

58. W. Erfurth, A. Thompson, and N. Ünal, "Electron dose reduction through improved adhesion by cationic organic material with HSQ resist on an InGaAs multilayer system on GaAs substrate," *Proceedings SPIE 8682, Advances in Resist Materials and Processing Technology* XXX, 8682, 1Z, March 29 (2013).

59. I. Horcas and R. Fernàndez, "WSXM: A software for scanning probe microscopy and a tool for nanotechnology," *Review of Scientific Instruments* 78, 013705 (2007).

60. M. Kongai, M. Sugimoto, and K. Takahashi, "High efficiency GaAs thin film solar cells by peeled film technology," *Journal of Crystal Growth* 45, 277 (1978).

61. X. S. Wu, L. A. Coldren, and J. L. Merz, "Selective etching characteristics of HF for $Al_xGa_{1-x}As$/GaAs," *Electronics Letters* 21, 13 (1985).

62. S. Person, M. Jain, Z. Lapin, J. J. Sáenz, G. Wicks, and L. Novotny, "Demonstration of zero optical backscattering from single nanoparticles," *Nano Letters* 13, 1806 (2013).

63. K. Lee, J. D. Zimmerman, X. Xiao, K. Sun, and S. R. Forrest, "Reuse of GaAs substrates for epitaxial lift-off by employing protection layers," *Journal of Applied Physics* 111, 033527 (2012).
64. C. W. Cheng, K. T. Shiu, N. Li, S. J. Han, L. Shi, and D. K. Sadana, "Epitaxial lift-off process for gallium arsenide substrate reuse and flexible electronics," *Nature Communications* 4, 1577 (2013).
65. M. Pu, L. Ottaviano, E. Semenova, and K. Yvind, "Efficient frequency comb generation in AlGaAs-on-insulator," *Optica* 3, 8 (2016).
66. T. Shibanuma, G. Grinblat, P. Albella, and S. A. Maier, "Efficient third harmonic generation from metal-dielectric hybrid nanoantennas," *Nano Letters* 17, 2647 (2017).
67. V. F. Gili, L. Ghirardini, D. Rocco, G. Marino, I. Favero, I. Roland, G. Pellegrini, et al. "Metal-dielectric hybrid nanoantennas for efficient frequency conversion at the anapole mode," *Beilstein Journal of Nanotechnology* 9, 2306 (2018).

4 Nonlinear Plasmonic Metasurfaces

Giovanni Sartorello, Alexey V. Krasavin,
Alexander E. Minovich, and Anatoly V. Zayats

CONTENTS

4.1 INTRODUCTION

Optical metamaterials have been developed to control light using assemblies of electromagnetically interacting nanostructures (sometimes called meta-atoms), often exploiting their resonant behavior. The dimensions, material, topology of meta-atoms, layout of their assemblies, as well as the nature and mechanism of their interaction with light vary widely, and so do the targeted spectral response and functionalities. Super- and hyper-lenses for high-resolution imaging [1,2], sensing and filtering [3], signal processing and nonlinearity enhancement [4–6] have been demonstrated, to name but a few. Recently, the metamaterial approach has been simplified to metasurfaces, two-dimensional subwavelength arrangements of meta-atoms which replace the necessity of complex three-dimensional nanostructuring. Metasurfaces can control the local phase of reflected and transmitted light and, therefore, emulate light reflection, transmission, and diffraction of a 3D medium with a 2D structure of, generally, subwavelength thickness. In order to achieve a strong effect of meta-atoms on the incident light, resonant optical modes need to be engineered. This can be achieved with a multipolar response provided by either high-index dielectrics or plasmonic materials [7–11]. In the former, the electromagnetic field at the resonances occupies the volume of nanostructures, whereas in the latter they are localized at the interface between structures and the surrounding dielectric medium. The latter plasmonic metasurfaces are the focus of this chapter.

Plasmonics studies the optical properties of free-carrier oscillations on metallic surfaces or nanostructures. The interaction of light and plasmonic structures gives rise to wavelength-dependent resonances, which strongly influence the behavior of the plasmonic systems, such as scattering, absorption as well as various nonlinear optical properties, including harmonic generation and Kerr-type nonlinearity. Free-carrier oscillations in confined geometries have characteristic frequencies determined by the specific system's physical characteristics (material, size, and shape), which can be exploited in plasmonic resonators to concentrate fields in local hotspots [12] in a wavelength- and geometry-dependent manner. Gold and silver nanoparticles, for example, have resonances in the visible, giving them a bright coloration. In addition to plasmonic resonances, other non-resonant geometric effects are possible in the light-nanoparticle interactions such as the "lightning rod" effect on sharp nanoscale tips and edges, a microscopic equivalent of the macroscopic tendency of metallic points to concentrate electric charge. The strong field enhancement that plasmonic systems exhibit makes them advantageous for nonlinear optics, as they can provide very high field magnitudes in very small volumes, allowing for strong nonlinear enhancement at the nanoscale.

Surface-enhanced Raman spectroscopy, which exploits field enhancement to increase sensitivity and allows for single-molecule detection, is one of the best known applications of plasmonics. The effect was originally observed on rough noble metal films [13], but has since been studied on nanostructured plasmonic systems such as nanorods [14] and nanotubes [15]. Nonlinear phenomena such as second-harmonic generation (SHG) can also be enhanced on roughened noble metals: enhanced SHG was first observed on rough silver in 1981, and simultaneously with SERS shortly after [16].

Enhanced harmonic generation has been observed in several designs, including nanocones, nanoscale apertures and dimers, nanocups, tapered waveguides, and many others [17,18]. Dimer nanoantennas, pairs of metallic cylinders or bars separated by a small gap, typically show strong field enhancement in the gap, which can be exploited for the study of second [19] and third [20,21] harmonic generation. Four-wave mixing (FWM) has also been reported on nanostructured surfaces [22].

In the context of nonlinear optics, plasmonic systems can exploit their intrinsic free-electron nonlinearities, but providing high field enhancement also important for enhancing the nonlinearity of other adjacent nonlinear materials [21]. This makes plasmonics suitable for all-optical, actively controlled signal processing systems. Such a system, based on waveguided surface plasmon polaritons (SPPs), was first proposed in 2004 [23] and SPP-based signal modulation and transmission techniques have since been brought to the femtosecond domain [24].

In this chapter, we give an overview of the principal applications of nonlinear plasmonic metasurfaces, including nonlinear generation, phase control, and all-optical modulation. We start in the next section by giving an overview of nonlinear behavior in optical metallic metasurfaces. In Section 4.3, we discuss the use of metasurfaces to study basic nonlinear optical phenomena, harmonics generation, and wave mixing. In Section 4.4, we discuss using nonlinear metasurfaces to modulate a light signal, and, in Section 4.5, we describe how to modulate nonlinear harmonic generation. Finally, we devote Section 4.6 to an overview of nonlinear holography.

4.2 NONLINEAR OPTICAL RESPONSE OF PLASMONIC METASURFACES

As there are several good reviews of nonlinearity of plasmonic systems and metasurfaces [10,11,17,25–27], we limit ourselves to recalling a few basic principles. In nonlinear systems, the n-order coherent nonlinear phenomena depend on the power n of the electric field, therefore the field enhancement effect found in plasmonic systems can greatly improve nonlinear generation. Thus, coherent phenomena such as harmonics generation and wave mixing are directly enhanced in plasmonic systems which support them, especially at the resonance. Kerr-type nonlinearities, which are caused by the light-induced change in permittivity, and allow for optical control of linear and nonlinear signals, are also enhanced. The more intense fields in plasmonic structures mean that a lower intensity of the control light is required to cause a nonlinear effect of a given magnitude. The field confinement can enhance effects due to the intrinsic nonlinearities of metals, but also enhance the nonlinear properties of adjacent materials.

The intrinsic nonlinear response of plasmonic systems consists of coherent and incoherent parts. The coherent part, responsible for harmonic generation, is related to an anharmonic response of the free carriers to the driving fields in conjunction with a symmetry breaking at the interface for SHG [28–30] and together with the ordinary bulk contribution in the case of THG. The incoherent part, producing Kerr nonlinearity, is related to the change in the optical properties due to the induced change of the energy distribution of the free carriers upon intraband and, if present, interband absorption, both happening at ultrafast timescales.

These general considerations apply to plasmonic metasurfaces, which can be used for the generation of coherent nonlinear effects as well as optical control of signals with schemes based on Kerr-type effects. Employing this mechanism, the linear optical response can be modulated, which can in turn be used to control the generated nonlinear effects, particularly harmonic generation. In addition, metasurfaces excel at phase control and come in a multitude of designs, which gives them great flexibility. Their subwavelength thickness and compactness mean that several of them can be used in parallel or cascaded in a small amount of space in an optical system, which could allow for integrated, modular, compact devices for light control and signal processing.

Plasmonic metasurfaces are excellent platforms for nonlinear generation. The metasurface itself, fabricated on a plain isotropic substrate, can be the source of all or most of the nonlinearity, or it might be used in conjunction with a designed nonlinear structure. For certain effects and designs, nonlinear generation happens within the subwavelength thickness of the metasurface, which is true, for example, for second-harmonic generation from metasurfaces fabricated on a homogeneous isotropic substrate. In other cases, the metasurface itself is not the only (or even the most important) source of nonlinearity, which may be by necessity (e.g., because metasurfaces are often fabricated on glass, which shows strong THG and FWM) or by design (e.g., because the metasurface is only one layer in a complex stack). Either way, the metasurface influences the generated nonlinear signals, which can be measured, for example, by comparison with a control sample or by performing polarization studies [31].

When dealing with nonlinear phenomena, symmetry considerations become paramount. The most frequently studied phenomena, those arising from second- and third-order nonlinearities, differ, as the former require noncentrosymmetry and the latter do not. Depending on the system symmetries, some or all of the elements of the third-rank tensor $\chi^{(2)}$ can be zero. In bulk media, these can be derived from the crystal group properties [32]. A metasurface, however, is a structured subwavelength material, and its effective $\chi^{(2)}$ reflects the symmetry of the component structures and their arrangement, not just the component materials.

The simplest way to achieve noncentrosymmetry is to break symmetry with a surface, which is why a second harmonic is obtained from thin films [30,33]. Suppose z is the direction perpendicular to the surface and there is isotropy at the surface xy. A general second-order susceptibility tensor $\chi_{ijk}^{(2)}$ has 27 elements in total, of which only 18 are independent because the element must not change if j and k are swapped. Adding isotropy in xy means that no element can have an odd number of indexes x or y. Moreover, the directions x and y are arbitrary and can be swapped. The remaining elements of ijk are $xxz = xzx = yyz = yzy$, $zxx = zyy$, and zzz. The effective $\chi^{(2)}$ of nonlinear systems at an interface is often written with subscripts "$\perp$" to denote the direction perpendicular to the surface (here, z) and "$\parallel$" for directions along it (here, x and y). This simplifies the list of elements of the tensor we just derived to $\chi_{\parallel\parallel\perp}^{(2)} = \chi_{\parallel\perp\parallel}^{(2)}$, $\chi_{\perp\parallel\parallel}^{(2)}$, and $\chi_{\perp\perp\perp}^{(2)}$. The major contribution to surface SHG comes from a depth of only a few atoms. Smaller contributions can also arise from bulk terms, due to field gradients, limited to the metal skin depth [25,34,35].

There are two different ways of engineering light interaction via the allowed tensor elements and achieving SHG on a metasurface. Under normally incident illumination and weak focusing, the metasurface must be noncentrosymmetric to generate SH. There must be lack of inversion symmetry in the metasurface general geometry, usually obtained by having individual noncentrosymmetric meta-atoms. However, even a metasurface comprising nothing but symmetric structures can generate significant SHG, either at higher diffraction orders (if the spacing is not subwavelength) [36], under oblique illumination [37], or by using strong focusing [19]. The latter two cases are in principle available for all metasurfaces. Conversely, SH tensor elements allowed by symmetry may be suppressed or enhanced for other reasons related to the meta-atom and arrangement geometries, specifically to how the local fields add up to yield the far-field SH [38–40].

Optimizing the geometry of plasmonic nanostructures for harmonic generation is an important part of the metasurface design. The most basic tool for predicting the nonlinear response of a structure is the Miller's rule which states that, under certain conditions, the quantity

$$\frac{\chi^{(2)}\left(\omega_1+\omega_2\right)}{\chi^{(1)}\left(\omega_1+\omega_2\right)\chi^{(1)}\left(\omega_1\right)\chi^{(1)}\left(\omega_2\right)} \tag{4.1}$$

is approximately constant in non-resonant materials. From this formula it follows that, for SHG

$$\chi^{(2)}\propto\left(\chi^{(1)}\left(\omega\right)\right)^2\chi^{(1)}\left(2\omega\right)$$

This method gives a good first approximation of the potential of the nanostructure to generate harmonics starting from the linear properties of materials, and the latter can be numerically simulated with relative ease. However, this rule is inaccurate for resonant structures, including metamaterials and metasurfaces [41]. This point is clearly illustrated in a work on SHG from rectangular C-shaped resonators whose shape is altered by changing the length and asymmetry ratio of the components (arm length divided by the sum of the length of all three segments) [42]. Experimental SH intensity is compared to the Miller's rule prediction, and the latter predicts peak generation at structure lengths close to the experiment, but not in the expected range of aspect ratios. A more accurate prediction can be made by integrating the product of nonlinear polarization (itself obtained from the linear polarization) and the SH mode. The overlap integral yields the effective nonlinear susceptibility of the system, providing an accurate estimate of the optimal dimensions for maximum SHG. Miller's rule, while a useful first-order approximation, fails to capture the necessary balance between large size (which provides more light absorption for nonlinear generation) and asymmetry (which provides the condition for nonvanishing $\chi^{(2)}$). This method can be applied to different geometries, though it is not as accurate with centrosymmetric particles [41].

4.3 HARMONIC GENERATION AND WAVE MIXING ON METASURFACES

4.3.1 SECOND-HARMONIC GENERATION

Resonances are essential to the linear and nonlinear capabilities of plasmonic nanostructures, including harmonics generation. A common approach is to excite a structure with fundamental wavelength at a strong absorption resonance, exploiting the strong field enhancement to increase nonlinear generation. Other resonances, if present, may then affect nonlinear emission to some degree. Plasmonic antennas resonant at both the fundamental and SH can enhance SHG by featuring field enhancement at both wavelengths, if there is a good spatial overlap of the modes [39,43,44]. However, pumping at a resonance is not a guarantee of optimal SH emission: as we pointed out, the way particle geometry influences the near field at different input polarizations also has an impact on the radiated SH intensity. Moreover, multiple resonances at both the fundamental and harmonic wavelengths do not always guarantee harmonics enhancement: on split-ring resonators, a resonance at the second harmonic was found to partially reabsorb the generated SH [45].

The strongest resonance on small, simple particles, such as bars and disks, is the electric dipolar one, due to simple back-and-forth free-carrier oscillations caused by the external fields. With larger size and more complex shapes, the contributions of higher-order electric multipoles and magnetic resonances become more important. The contributions of the various modes may be strongly influenced by roughness and fabrication defects. For example, multipolar effects were shown to significantly contribute to SHG from L-shaped particle arrays [46] but this contribution is mainly present due to fabrication imperfections. The values of the individual elements of the nonlinear response tensor for each multipolar contribution were obtained with polarization-sensitive techniques, and the dipolar response was found to dominate the theoretically allowed SH, with higher-order multipoles affecting tensor components which should be forbidden, but are present due to the surface defects [47]. It was later confirmed that the dipolar response dominates the SH emission when sample quality is high [48]. Unfortunately, as this example shows, defects are a common threat to the efficiency of a metasurface and the understanding of its behavior, while the high cost of state-of-the-art nanofabrication equipment puts a high bar to entry.

Magnetic dipole and electric multipole contributions become more significant in more complex structures. From the nanostructures designed to be noncentrosymmetric, the most widely used in nonlinear metasurfaces is the split-ring resonator (SRR). The split-ring resonator as a basic unit of the magnetic metamaterial (Figure 4.1a) appears in a 1999 paper by John Pendry and collaborators [49]. The authors proved that it is possible to build a material with a magnetic response at optical frequencies out of components made from nonmagnetic materials, and predicted the nonlinear enhancement properties of such an assembly. In conjunction with metallic wires, SRRs then became the basis of the first "left-handed" metamaterial, possessing a negative refractive index, first proposed [50] and demonstrated [51] in the microwave range. The SRRs used in the experiment were square (a design later implemented at shorter wavelengths) and the unit cell contained two nested SRRs, whereas optical metasurfaces tend to use only one. In general, the shape and design of the individual ring affect the position of the resonances [52], and spectral features which appear in that design may not appear in another [53]. Moreover, the distribution of edges and corners affects the location of intensity hot spots, and the shape influences the interaction between elements of the metasurface. Whatever the specifics of the design, SHG on SRRs is allowed by noncentrosymmetry and enhanced by the magnetic resonance. Square SRRs were used as the fundamental unit of the first NIR and mid-IR metasurface [54], which was followed up with a scaled-down design for the visible and NIR [55]. This work resulted in the first visible/NIR nonlinear plasmonic metamaterial (Figure 4.1b), an array of 220 nm square SRRs with a magnetic resonance at the wavelength of 1.5 μm, capable of efficient SHG [56] and THG [57]. The contribution of the Lorentz force to SH enhancement is important, but a rigorous analysis of the nonlinear response based on the hydrodynamic model later showed this contribution to be negligible relative to the purely electric one given by oscillating currents caused by the charge displacement induced by the electric field [58].

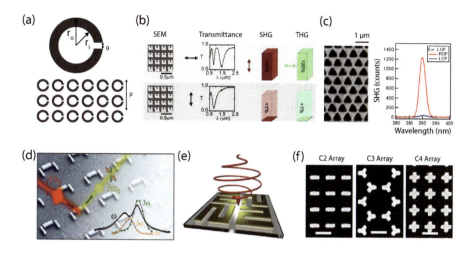

FIGURE 4.1 Nonlinearity of plasmonic metasurfaces. (a) Schematic of a split-ring reso-nator of inner radius r_i, outer radius r_0, and gap size g, by itself (top) and in an array of period p (bottom). (b) SHG and THG from a magnetic metamaterial. Arrows indicate the orientation of the electric field. SH is emitted with polarization orthogonal to the funda-mental, and TH with the same polarization as the fundamental. Emission is only efficient under illumination with polarization across the gap. (Reproduced from Klein M. et al., *Optics Express*, 15, 5238, 2007. With permission.) (c) SHG with circular polarization gen-erated by a nanohole array. (Reproduced from Konishi, K. et al., *Phys. Rev. Lett.*, 112, 5, 135502, 2014. With permission.) (d) THG from U-shaped structures. (Reproduced from Metzger, B. et al., *ACS Photonics*, 1, 471–476, 2014. With permission.) (e) SHG metasur-face for nonlinear chiral optical studies. (Reproduced from Valev, V.K. et al., *Adv. Mater.*, 26, 4074–4081, 2014. With permission.) (f) Arrays of meta-atoms with two-, three-, and fourfold rotational symmetry. (Reproduced from Chen, S.M. et al., *Phys. Rev. Lett.*, 113, 5, 033901, 2014. With permission.)

Another avenue for obtaining multipole-dependent SH emission is to build mul-tilayered structures. Both magnetic and electric quadrupolar resonances were shown to coexist and contribute to SHG in 140 nm, three-layer metal-dielectric-metal disks [59,60]. The use of both electric and magnetic modes is also typical for dielec-tric metasurfaces.

Although most metasurfaces we discuss are composed of isolated nanostructures on a dielectric substrate, the inverse design, with voids patterned in a metallic film (hole arrays), is also suitable for metamaterials, according to Babinet's principle relating antennas and their complementary screens [61,62]. Shortly after SRRs were first used for SHG, the complementary SRR (CSRR) geometry was also tested, and a patterned film of SRR-shaped voids was shown to generate SH almost as effi-ciently as its more conventional counterpart [63]. SHG was more recently studied on triangular hole arrays (Figure 4.1c), showing that a circularly polarized fundamen-tal light incident on a metasurface with a threefold rotational symmetry generates SH with the opposite handedness, as predicted by the symmetry-derived selection rules [64]. The same selection rules forbid SHG for other even- and odd-fold rotational symmetries, a good example of the fact that breaking of centrosymmetry is a necessary, but not sufficient condition for SHG.

4.3.2 PHASE CONTROL, THIRD-HARMONIC GENERATION, AND FOUR-WAVE MIXING

The most straightforward way of creating a metasurface is to design meta-atoms and, arrange them in arrays with regular spacing and orientation. However, the arrangement is another degree of freedom which can be exploited to tailor the linear and nonlinear properties of the system. For example, although THG can be obtained from isotropic materials, its properties can be controlled by the interaction of nearby nanostructures. Meta-atoms formed by nanobars arranged in a U-shape (with the bars not in contact with each other) emit resonance-enhanced TH largely polarized parallel to the polarization of the fundamental light (Figure 4.1d). By moving the bottom bar, TH polarized orthogonal to the polarization of the fundamental light is enhanced, as the destructive interference that prevented it in the symmetric case is reduced [21]. Chiral metasurfaces commonly make use of particular meta-atom arrangements. Nonlinear chiral metasurfaces can be built by taking noncentrosymmetric building blocks and arranging them rotated at right angles (Figure 4.1e) [65], or displaced from the simple square grid [66]. Illuminated with circularly polarized light, they can generate SH light with a strong chiral dependence, such as SH emission has the complementary contrast patterns when the input handedness is swapped, that presents a more pronounced behavior than in the linear regime. Arranging SRRs on a square grid rotated by 180 degrees in contiguous regions allows precise control of the nonlinear diffraction, which is used in planar nonlinear lenses [67], and nonlinear beam shaping [68]. By individually rotating neighboring meta-atoms, the local geometric (Pancharatnam-Berry) phase imparted on the generated SH can be controlled, which in turn allows full control of the nonlinear wavefront [69].

Under linearly polarized illumination, THG is allowed from meta-atoms of any shape, and dependent on the effective size of the particle as seen by the incident electric field. When using circularly polarized light, however, selection rules dependent on the metasurface's symmetries must be obeyed. Under normal-incidence illumination, nanostructures with n-fold rotational symmetries (Figure 4.1f) allow harmonics of order $m = nl \pm 1$, where l is a positive integer and $+ (-)$ means that the emitted harmonic is co-polarized (cross-polarized) with the fundamental light. For example, C2 metasurfaces can generate TH with the same and opposite handedness as the fundamental, C3 structures can only generate SH with the opposite handedness, and C4 structures can only generate TH, also with the opposite handedness. In an experimental test of the selection rules, under illumination with right-handed circular polarization (RCP), C2-symmetric nanobars generate TH with both RCP and LCP, albeit weaker, C3 trefoils generate no TH, and C4 nanocrosses generate only LCP TH [70]. Such structures can be used to build nonlinear metasurfaces that exploit the geometric phase to locally control the phase of the generated harmonic, which in turn allows the control of the harmonic's diffraction [71]. Structures with rotational symmetries can be designed in chiral variants, removing the mirror symmetry by adding additional features. Triskelions and gammadions are the C3 and C4 examples. They exhibit large circular dichroism between their response to RCP and LCP in their SH and TH spectra, respectively [72].

Most nonlinear metasurfaces are fabricated on a plain substrate (such as fused silica, other glasses, sapphire, etc.) often with the aid of a thin adhesion layer (such as indium tin oxide or tantalum pentoxide). More elaborate designs are possible by fabricating the metasurface on top of a multilayered structure. Arrays of gold nanodisks on an ITO/gold double layer were shown to be good THG generators by careful tuning of coupled magnetic and SPP resonances [73], and a silicon heterostructure was shown to enhance nonlinear generation by orders of magnitude by coupling plasmonic modes in L-shaped nanostructures with intersubband transitions in a quantum-well layer [74].

The metasurface has remained the most common plasmonic metamaterial because at the nanoscale it is far easier to fabricate two-dimensional structures and because subwavelength thickness is an attractive property for compact systems. However, by their nature, 2D metasurfaces cannot achieve properties such as well-defined bulk negative refractive index [75]. This is the domain of more complex 3D optical metamaterials [76–78].

In addition to harmonics generation, wave mixing processes are also commonly studied in nonlinear metasurfaces. Suitably designed $Au/SiO_2/Au$ nanodisks have been shown to be excellent FWM generators: they show greatly enhanced nonlinear generation induced by plasmonic field confinement, by about 10 times when compared to a gold film, while also exhibiting, like the gold film, negative refraction of the FWM as derived by phase matching conditions [79]. Similarly to harmonics, resonant enhancement is possible for FWM, which was shown in nanodisk clusters [80].

4.4 MODULATION OF LIGHT WITH METASURFACES

4.4.1 OPTOMECHANICAL MODULATION

Metasurfaces have shown an incredible ability to shape the wavefront of light in an engineered manner, independently in both amplitude and in phase. They have also been used, in numerous approaches for the active control of light. The optical response of metasurfaces can be highly dependent on the coupling between the meta-atoms, which was exploited by fabricating the metasurface on an elastic (PDMS) substrate and stretching it. It was experimentally shown that 6% stretching is enough for two-fold modulation of the transmittance of a nanohole array in a gold film [81], while a complete switch of a holographic image was demonstrated with 24% stretching of a nanorod-based metasurface [82]. Similar approaches based on changing the geometrical configuration of the metasurface by a voltage-induced electrostatic force [83,84], Lorentz force, or Joule-heating-based thermal expansion of a current-supporting structure [85] have been also implemented. In the latter approach, the metasurface was made up of horizontal metallic wires with a certain degree of movement in the vertical direction, and mechanical resonances excited by the control electrical signal were additionally utilized. In the static regime, applying a variable voltage to different wires, it was theoretically shown that it is possible to steer or focus the incident light signal upon reflection [86,87]. A thermally induced transmission change in the tens of percent was obtained, exploiting a thermo-optical effect, in a hybrid semiconductor/metallic metasurface comprising silicon and gold nanoantennas [88], while introducing ferromagnetic materials, e.g., cobalt, allows the control of the polarization state of the reflected light by the application of a magnetic field [89].

4.4.2 ELECTRO-OPTICAL MODULATION

Although the above approaches provide efficient signal control, they are relatively slow, operating either in a static regime or at modulation speeds in the Hz to kHz range of the mechanical or thermal processes they exploit. Higher modulation speeds are achievable utilizing electro-optic (EO) effects, related to molecular or electron dynamics. At mid-IR frequencies, this can be done through the giant quantum-confined Stark effect in a quantum-well structure placed in-between a metasurface produced by an array of cross-shaped hole nanoresonators in a metallic film, which acts as an electrode, and the other metal electrode below (Figure 4.2a) [90].

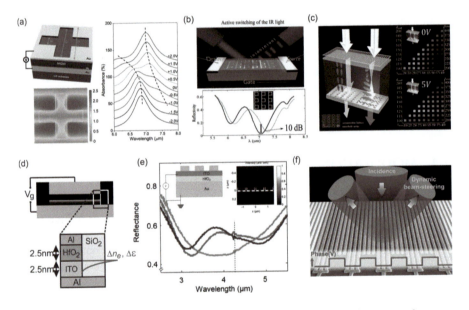

FIGURE 4.2 Electrically controlled metasurfaces. (a) A schematic of a metasurface produced by cross-opening resonators in a metallic film interfacing a quantum well structure. The plot shows the metasurface absorption change under the application of a control voltage. (Reproduced from Lee, J. et al., *Adv. Opt. Mater.*, 2, 1057–1063, 2014. With permission.) (b) Operating principle and modulation performance of a graphene-based electrically driven plasmonic metasurface. (Reproduced from Dabidian, N. et al., *ACS Photonics*, 2, 216–227, 2015. With permission.) (c) Changing the color of the transmitted light by an electrically-controlled metasurface operating in conjunction with a liquid crystal cell. (Reproduced from Lee, Y. et al., *ACS Photonics*, 4, 1954–1966, 2017. With permission.) (d, e) Modulation of the absorbance and (f) light steering by a plasmonic metasurface upon electrically controlled charge accumulation in a TCO layer. (Reproduced from [d] Krasavin, A.V. and Zayats, A.V., *Phys. Rev. Lett.*, 109, 053901, 2012; [e] Park, J. et al., *Sci. Rep.*, 5, 9, 15754, 2015; [f] Huang, Y.W. et al., *Nano Lett.*, 16, 5319–5325, 2016. With permission.)

The applied electric field modifies the electronic quantum states in the quantum wells, which leads to a shift in the intersubband transition between them and consequently to the change in the material optical properties. It was experimentally shown that sweeping the voltage from −1 V to 1 V leads to a shift in the corresponding optical transition wavelength from 6.5 to 7.5 μm. As the strongest changes in the optical parameters of the quantum-well system happen near this resonant wavelength, the plasmonic resonance of the cross-shaped nanoresonators was adjusted to the intersubband transition frequency at a zero bias, which consequently leads to Rabi oscillations between the material and plasmonic excitations. Overall, the voltage sweep from 0 to 5 V causes ~30% metasurface absorbance change at the wavelength of 7.5 μm, while the measured modulation speed approaches the MHz range, limited only by the RC time constant of the implemented circuit.

Graphene is a very promising electro-optic material whose optical properties can be naturally and strongly modified through the change of its Fermi level by the application of a control voltage (Figure 4.2b) [91,92]. At the same time, due to an atomically thin thickness of the graphene layer, its overall nonlinear response is rather weak and elaborate engineering the metasurface is needed to achieve high sensitivity to the nonlinear change and obtain a prominent modulation contrast. An example of this is a metasurface built out of electrically connected coupled nanorod/C-shaped antenna structures placed on the top of a graphene layer and supporting two narrow high-Q Fano resonances [91]. It shows a tenfold modulation of the metasurface reflectivity at a wavelength of 7 μm, though with a substantial applied voltage of −390 V. Graphene-based electro-optical structures were also experimentally implemented for electric control of a phase of the reflected wave [93], while the possibility of dynamic focusing was theoretically examined [94,95].

The near-IR and visible frequency ranges the traditional approach for electro-optical modulation is to use liquid crystals [96,97] and chromophore-based materials, frequently regarded as electro-optical polymers because the chromophore molecules are placed in a polymer matrix [98]. As an example of the former, a metasurface based on a rectangular array of nanoholes in a metallic film was used to control color states in conjunction with a liquid crystal (Figure 4.2c) [96]. The different periods in the two principal axes of the array correspond to two different resonant wavelengths of the extraordinary transmission for the light polarizations along these directions. A liquid crystal cell adjacent to the metasurface was used for an electrically driven continuous sweep of the polarization of the incident white light and therefore the continuous change of the transmission peak at one resonant transmission wavelength to that at another. Since the resonance peaks are quite broad, this leads to the change of the color of the transmitted light over the visible spectral range from 500 to 650 nm as the control voltage sweeps from 1.6 to 2.5 V. It has also been proposed to fill the gaps between the metallic components of an MIM-based metasurface grating with an EO chromophore [98]. Utilizing the interplay between two resonant MIM grating modes, a 15-dB extinction ratio was obtained for realistic EO parameters of the chromophore under 10 V electrical modulation. As the nonlinear response time of the polymer can originate from the electronic polarization taking place at a picosecond timescale, the metasurface modulation speed is defined by its RC time constant and theoretically can reach the GHz range. The possibility of an electrical control of the color of the reflected light along with its polarization state and the angle of reflection using a chromophore-based EO material has been also studied [99].

Although from the physical point of view electrically induced changes in the material bulk refractive indices are very high, reaching the values of a few percent, when implemented in nanoscale structures and meta-atoms, they are rather low in practice, as the mode does not have the distance to accumulate the nonlinear change that the propagating wave has. This is especially true for plasmonic nanocomponents in the visible spectral range, where extreme mode localization is combined with rather high losses. At the opposite end, graphene can offer a larger nonlinear response, but over just an atomically thin material layer. There is however an alternative in the form of studying electro-optic effects in transparent conductive oxides (TCOs), which are highly doped, or even degenerate semiconductors, e.g., indium tin oxide (ITO) or aluminum zinc oxide (AZO). These materials combine drastic unity-scale refractive index changes happening exactly at the nanoscale dimensions of highly localized optical modes [100,101]. In this approach, the metallic nanostructures, in our case electrically connected meta-atoms, forming one electrode, are placed next to the second, gating electrode, with the gap between them filled with a sandwich of dielectric and TCO layers (Figure 4.2d). When applying a control voltage, electrons supplied by the adjacent metallic electrode stop at the TCO/dielectric boundary and produce an accumulation layer, screening the applied potential. The electron density in the layer rises orders of magnitude, drastically changing the local plasma frequency, and through it, the complex local dielectric function, by tens or even hundreds percent [100,101]. If the opposite bias is applied, the device can similarly work in a depletion regime. The thickness of the layer is defined by the Thomas-Fermi screening length and is of the order of 1 nm [101], which can exactly fit the nanoscale dimensions of the engineered plasmonic modes on the metasurface. Thus, this method offers a drastic electro-optic and/or electro-absorption effect, orders of magnitude larger than in bulk counterparts and particularly suitable for applications at the nanoscale.

Absorption change in ITO was used to demonstrate electrically-driven light modulation in the near-IR region. The MIM sandwich described above was realised with the metasurface made of a metallic line grating, with ITO itself working as a control electrode (Figure 4.2e) [102]. The optical response of the metasurface is generally defined by the resonant excitation of Fabry-Perot MIM modes localized in the ITO/dielectric gap, with the resonant wavelength defined by the size of the cavity. As it was stressed above, in this approach the field of the mode is highly localized in the gap, which makes its excitation very sensitive to the optical properties of ITO, which in turn can be drastically changed by application of the voltage. To enhance the electro-optic effect, the width of the strip and the initial electron concentration in ITO were chosen such that the resonant wavelength of the Fabry-Perot mode matched the wavelength at which ITO has an epsilon-near-zero (ENZ) behaviour, where the real part of the Drude-model permittivity crosses zero to become metallic at longer wavelengths. Such wavelength match makes the plasmonic mode and the material resonance in ITO hybridized, producing two absorption peaks, one red- and another blue-shifted with respect to the initial resonant wavelength of 4 µm. The operational wavelength of the metasurface modulator is matched to be in-between the absorption peaks, so its reflectance is high, around 90%. (The nature of the ITO material resonance at the ENZ point was not specified in that work, but we assume that it is due to the continuity of the normal component of the electric displacement in the gap. At the ENZ wavelength, the electric field becomes very large in the ITO layer, which is an indication of the corresponding high density of the electromagnetic states, which effectively creates nonlinear response.) When a +5V voltage is applied, the depletion region is produced, removing the ENZ operation regime, so the initial Fabry-Perot resonance is restored, resulting now in a transmission dip at the operational wavelength (Figure 4.2e). Thus, the reflectance is reduced, and the optical signal is modulated. The 3-dB modulation cut-off frequency was measured to be 125 kHz, which is in a general agreement with an estimation based on the ITO resistivity and the capacitance of the device. The same approach was used to modulate not only the reflectivity of the metasurface, but also the phase of the reflected wave [103,104]. The important difference implemented in Ref. [103] was that the voltage was applied not to the ITO layer continuously extending over the metasurface, but to the metallic strips themselves. This gave the opportunity to apply different voltages to a periodic subset of the strips, thus producing an amplitude/ phase-change diffraction grating, which led to the experimental realization of electrically driven beam steering (Figure 4.2f) [103]. Alternative methods for the realization of electro-optic metasurfaces utilize the refractive index change upon carrier injection in Si [105,106] or in InSb [107].

4.4.3 ALL-OPTICAL MODULATION

In all-optical nonlinear metasurfaces the intensity and/or polarization state of the transmitted and/or reflected light are modulated by a control light signal. All-optical interactions are usually extremely fast, offering a vital technology for future ultrafast all-optical telecommunications and computing systems. Usually they utilize Kerr-type effects, in which the transmission or reflection properties of the signal light are controlled by the changes in the complex refractive index (or complex effective refractive index) of the medium. These can be changes in the real part of the refractive index, usually leading to the shifts of the optical resonances of the meta-atoms, or changes in the imaginary part, leading to the modulation of the absorption. Here, however, a lot depends on the nature of the nonlinear effect involved. There is usually a trade-off between the magnitude of the effect and its speed. For example, nonlinearities produced by thermo-optical effects are large but slow, while the nonlinearities produced by electronic polarization are fast, but weak [32]. At the same time, with the development of nanotechnology, new types of structures, which can drastically enhance nonlinearities, become available, while with the development of laser technologies, particularly ultrafast ones, ever increasing speeds of nonlinear effects become accessible.

We start our discussion of all-optical effects looking at optomechanical metasurfaces. Upon interaction of the meta-atoms with the control light, usually in a resonant regime, the charge and current distributions interact with the local electromagnetic field, and the meta-atoms experience torque. If the meta-atoms are not fixed, e.g., if they lie on an elastic support, this leads to their displacement and the reconfiguration of the metasurface structure. The subsequent change in the metasurface properties causes a change in the reflection and transmission of the probe light. Control of the polarization state of light can be obtained through optically-induced chirality in meta-molecules produced by two meta-atoms supporting dipolar optical resonances and symmetrically positioned at an angle to each other [108] (Figure 4.3a).

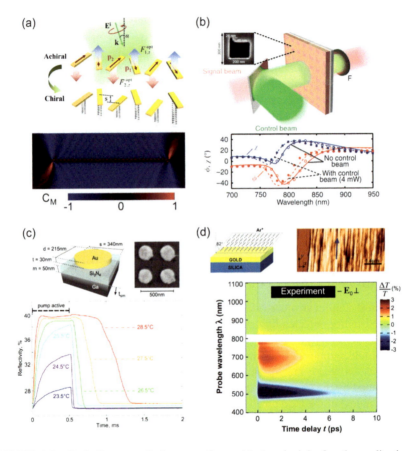

FIGURE 4.3 Optically controlled metasurfaces. (a) A principle for the realization of optically induced stereoscopic chirality and the resulting LCP/RCP asymmetry in scattering of linearly polarized probe light, defined by the polarization of the pump light. (Reproduced from Liu, M.K. et al., *Adv. Opt. Mater.*, 5, 8, 1600760, 2017. With permission.) (b) Optically-induced change of the polarization state of light (polarization azimuth rotation angle ϕ and ellipticity angle χ) transmitted though a nonlinear plasmonic metasurface functionalized with an EO polymer. (Reproduced from Ren, M.X. et al., *Light Sci. Appl.*, 6, 5, e16254, 2017. With permission.) (c) All-optical modulation of the reflectivity of Ga-based nonlinear metasurfaces. (Reproduced from Waters, R.F. et al., *Appl. Phys. Lett.*, 107, 4, 081102, 2015. With permission.) (d) Optically induced modulation of transmission through a metasurfaces produced by an ensemble of flat-lying metallic nanorods, driven through heating of the free electron gas by a femtosecond light pulse. (Reproduced from Della Valle, G. et al., *Phys. Rev. B*, 91, 235440, 2015. With permission.)

Upon simultaneous excitation of the symmetric and asymmetric modes in this coupled system, possible because of broad resonances, with pump polarization breaking the metamolecule symmetry, the two meta-atoms experience an out-of-plane force in the opposite directions, which results in the opposite displacement and an overall stereoscopic chiral structure of the metamolecule. The probe light experiences the induced 3D chirality and becomes predominantly right or left-hand polarized (Figure 4.3a). The handedness of the polarization depends on the induced chirality sign, which is defined by the polarization of the pump beam. Furthermore, it was found that the coupling effects in the meta-atom chain or in the metasurface array can enhance the effect. As it was discussed above, although the changes are quite dramatic, the modulation speeds here are not as high. For example, it was estimated that the maximum bandwidth of a generic optomechanical system operating in the frequently occurring hysteresis (bistability) regime is a few MHz [109]. The reason for this is a drastic decrease in the transformation speed as the system approach the point of switching between the two states, defined by the gathering nonlinear dynamics. Also, sometimes, for the optomechanical modulation to be efficient, the resonant excitation of the meta-atom mechanical modes is needed, thus the pump light rather than having a desired modulation profile has a harmonic modulation with a frequency of the utilized mechanical resonance [110].

A more conventional approach for all-optical modulation is based on pump-induced changes in the molecular configuration of chromophores, which are frequently dispersed in a polymer matrix to obtain nonlinear polymers. An optically controlled modulation of the polarization state of the transmitted light was demonstrated with a metasurface made of an array of L-shaped slits in a metallic film covered with a layer of ethyl red doped PMMA [111] (Figure 4.3b). A 532 nm control laser beam having a power of just 4 mW focused to a 9 μm spot resulted in a change of the refractive index of the nonlinear polymer, providing a ~10 nm shift of the metasurface plasmonic resonance, which led to changes in the polarization azimuthal rotation and ellipticity angles of the transmitted probe light of tens of degrees. The switching speed of the device was measured to be of the order or tens of Hz, which might be improved by the choice of the nonlinear material or through the implementation of additional optical effects, enhancing the device performance. Another example of pump-probe modulation approach is hybridization of plasmonic resonances in a metamolecule made from two triangular antennas, leading to metamaterial-induced transparency (MIT), which results in narrow transmission resonances in an ITO/graphene/Au-metasurface/ZnO-nanoparticles/multilayer-graphene nonlinear sandwich [112]. A resonance shift of about 20 nm leads to a change of about 15% in the probe light transmission with pump intensities of about 1.5 kW/cm², three times smaller than in the previous work, and with remarkable 1.96 ps (fast component) and 42.3 ps ("slow" component) switching times. The fast component was attributed to the third-order nonlinearity of graphene, while the "slow" component was attributed to electron gas heating in gold (see below the detailed discussion of this effect) with the subsequent charge injection into ITO. Also, in the MIT window a substantial slow-light effect was observed, although in our view this did not have a direct impact on the modulation, as it was observed not at the MIT wavelength, but at the wavelengths of the steepest transmission slopes. Another approach for realization of efficient all-optical meta-surface switches based on nonlinear polymers or semiconductors utilizes optical bistability [113,114].

A remarkable all-optical switching mechanism is based on light-induced nanoscale structural phase transitions. Initially proposed for modulation of propagating SPP signals [23,115] using phase transitions in gallium, this approach was extended to nonlinear all-optical metasurfaces [116]. Gallium possesses two structural phases with drastically different electrical and optical properties, a low-electron-density α-phase and a metallic liquid phase. The transformation between the phases can be induced by an optical pulse. Moreover, due to a 19 nm skin depth of the metallic phase, only a nanometer-scale phase change is needed to completely change the optical characteristics perceived by incident light or a meta-atom mode. Such an effect was used in an optically controlled nanodisc-based metasurface interfacing gallium through a 50 nm Si_3N_4 spacer [116] (Figure 4.3c). A relative change of about 50% in the meta-surface transmission was observed upon illumination with 500 μs pump pulses at a wavelength of 1310 nm with an intensity of just a few kW/cm². The required pump intensity and the relaxation time to the initial α-phase (which defines the modulation speed, measured to be in the range of kHz range) were found to be dependent on how close the metasurface temperature is to the melting point of gallium (29.8°C), as expected. Optically induced control of metasurface transmission was also realized in GST-based plasmonic metasurfaces, but in a non-volatile fashion, making such a metasurface reconfigurable, rather than dynamic [117]. Vanadium dioxide can be another material offering dynamic optical switching between two phases with essentially different optical properties, but the realization of the effect with this material in a metasurface geometry has not yet been shown. Temperature-induced phase control is one more avenue for the modulation of an optical signal and its polarization state in particular [118].

So far in the discussion of all-optical modulation of light by metasurfaces upon a light-induced change in the refractive index of the constituting material, dielectrics or semiconductors have been considered, while the role of the plasmonic meta-atoms was to (a) provide an optical effect (usually resonant) whose magnitude can be easily modulated by the above change, and (b) to enhance the local optical fields in order to increase the nonlinear response. Metals, however, themselves have extremely large and ultrafast third-order nonlinearity related to optically induced change in the energy distribution of electrons in the conduction band. Upon illumination with a femtosecond control light pulse conduction band electrons are excited to much higher energy levels, to become what is known as "hot electrons" [119]. In addition, if the photon energy of the pump light is higher than the energy of the intraband transitions, the electrons from the valence d-band are excited into the conduction band, affecting the energy distribution as well as the overall number of free electrons. Thus, a non-Fermi distribution of highly non-thermalized electrons is created in the conduction band. Then, at the characteristic timescale of ~100 fs, upon electron-electron scattering, hot electrons transfer their energy to the rest of the electron pool and Fermi distribution is established again, but with a greatly elevated electron temperature which can reach thousands of degrees. Such drastic changes in the electron distribution manifest themselves with optically induced nonlinear changes of the metal permittivity [6,120]. This approach has led to active polarization control [121] as well as the realization of on-demand nonlinear metamaterials with engineered spectral [40,122] and temporal characteristics [123]. Furthermore, such Kerr-type nonlinearity is ultrafast, existing for timescales of a few picoseconds, during which the electrons thermalize to the room temperature, transferring their energy to phonons through electron-phonon scattering. As the heat capacity of the phonon pool is much higher than the heat capacity of the electron gas, the final rise of the lattice temperature is of a order of a few degrees centigrade only.

A nonlinear plasmonic metasurface utilizing electronic nonlinearity in gold for all-optical switching of the transmitted optical signal was fabricated using Ar-beam sputtering of a gold film at a grazing angle. As a result, a dense assembly of parallel, flat-lying nanowires with quite a broad distribution of widths and lengths was obtained, with lengths in a 10–100 nm range and widths of 5–15 nm [120]. (Figure 4.3d). As it follows from symmetry considerations, the metasurface possesses an anisotropic optical response: for the incident polarization along the nanowire direction the wavelength dependence of the transmission is quite flat and can be approximated by that of a thin gold film, while for the orthogonal polarization across the nanowires there is a transmission dip produced by the transverse plasmonic resonances of nanowires with various widths present in the ensemble. This was found to be effectively approximated by the response of an ensemble of nanowires with fixed length of 110 nm and width of 8.5 nm, but with increased losses. Such approximations are required for the further estimation of the nonlinear response. A femtosecond pump-probe technique was employed to study the optically induced change in the metasurface transmission. An ultrafast modulation of the probe transmission of the order of 1% was observed, with expected relaxation time of a few picoseconds (Figure 4.3d). To understand the experimental results an extended two-temperature model (with one temperature describing the temperature of the electron gas and the other temperature describing the temperature of the phonons), taking into account all the above-described energy exchange processes, was implemented. The temperature dynamics of the electron gas defined the optical properties of the metal, which in turn determined the change in transmission of the approximated structure for the two probe polarizations. It was found that the transmission modulation in the region of 450–550 nm, an increase at shorter wavelengths and a decrease at longer wavelengths, observed for both polarizations, is related to the change in interband absorption caused by the smoother Fermi distribution at the high temperature of the free-electron gas. A similar change in the transmission behavior present only for the probe polarization perpendicular to the nanowires is attributed to the shift of the transverse plasmonic resonance of the nanowires due to the change of the optical response of the electron gas at high temperatures and consequently the change in the optical properties of the metal. At the increased pump fluence of 9 mJ/cm^2, a 15% optically induced relative change in the metasurface transmission was demonstrated. The Kerr-type electronic nonlinearity was also demonstrated in conductive oxides, which are considered an important alternative material for linear, and now nonlinear, plasmonics [124].

In a recent example of optical switching in reflection, it was proposed an approach to build a nonlinear metasurface, based on gold/silica multilayer elements (essentially quantum wells) coupled to an underlying gold film underneath a dielectric spacer. Such a material is predicted to behave as a mirror at low incident powers, and as a grating at high incident powers [125].

4.5   ACTIVE CONTROL OF NONLINEAR GENERATION AND WAVE MIXING

The ultrafast control of optical signals is a building block for more complex signal processing systems. Plasmonic systems are suitable for the fast control of signals, both electro-optically on a sub-ns scale [126] and all-optically on a sub-ps scale [24,127]. The concepts of nonlinear harmonic generation and light-induced ultrafast control of optical properties can be combined to demonstrate all-optical control of nonlinear harmonic generation on metasurfaces. These metasurfaces generate optical harmonics (or wave mixing effects) with certain efficiencies, which depend on the optical properties of the system and on the permittivity of the constituting materials. As was discussed above, in plasmonic systems the presence of resonances is paramount, while their shape and spectral position can be affected by a change in permittivity. This change in permittivity can be induced all-optically by pump light, via the Kerr effect. Pump-probe setups can be employed for all-optical control, switching and modulation in plasmonic metamaterials and metasurfaces, from large, controllable and ultrafast (sub-picosecond) changes in transmittance [6,128–131]. to nonlinear harmonic generation, and particularly SHG, THG, and FWM.

This can be demonstrated on a system consisting of arrays of cut disks (Figure 4.4a) [53]. Symmetry considerations for these structures are largely the same as for more conventional square SRRs. They belong to the C_{1v} crystal group (they are symmetric only by reflection about the plane that bisects them). A 200 fs probe pulse with a central wavelength of 1310 nm, matching the strongest resonance, generates second and third harmonics whose magnitude can be controlled all-optically by using a 200 fs pump pulse with a central wavelength of 1028 nm to modify the metal permittivity. Modulation amplitudes of 20% of the pump SH, 10% of the probe SH, and 15% for the probe TH were obtained. These changes are associated with the change in magnitude, width, and position of the resonances under illumination, with both beams contributing to affect each others nonlinear signals. The timescales of the nonlinear harmonics modulation are those characteristic of pump-probe technique and the underlying Kerr nonlinearity mechanism: the rise or fall at a zero pulse delay is sharp, reaching maximum modulation within one pulse width (200 fs), while the relaxation back to the ground state lasts a period in the order of the characteristic carrier cooling time, which is usually several picoseconds in plasmonic systems of this kind. Therefore, it is more advantageous for a system built around this phenomenon to exploit the leading control pulse edge for the modulation.

The alternative to all-optical control is active electrical control. This has been shown in the steady-state regime with THG from U-shaped three-bar assemblies similar to structures used for THG before, built on an Au-doped $SrTiO_3$ layer on silicon and positioned between two electrodes (Figure 4.4b) [132]. THG conversion efficiency is very high, and introducing a bias voltage changes the permittivity of the metasurface, which results in a change in the optical properties. As a consequence, TH emission intensity decreases away from 0 V bias, with a maximum 12% modulation.

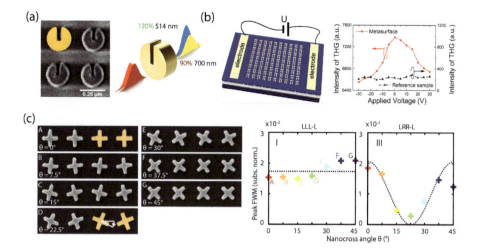

FIGURE 4.4 Active control of nonlinear generation. (a) Cut disks (left) generate SH for both input beams. If the beams overlap (right) the SH from both of them is modulated. (Reproduced from Sartorello, G. et al., *ACS Photonics*, 3, 1517–1522, 2016. With permission.) (b) A metasurface with an applied voltage shows a marked dependence of THG on the bias. (Reproduced from Gong, Z. et al., *Opt. Mater.*, 60, 552–558, 2016. With permission.) (c) Nanocross metasurfaces with varying meta-atom orientation angle (SEM images) generate the same FWM in co-polarized cases (left graph) but have a dependence on the meta-atom angle induced by the geometric phase in cross-polarized cases (right graph). (Reproduced from Li, G. et al., *Laser Photon. Rev.*, 12, 1800034, 2018. With permission.)

FWM is another interesting phenomenon, useful in the context of optical parametric amplification in laser systems and multiplexing in optical signal transmission. As was shown previously, C4 nanocrosses offer a good platform for the study of THG and its relationship to the polarization and geometric phase. These results were extended to FWM, another third-order phenomenon, by looking at its selection rules and the effect of geometric phase on its intensity under circular polarization (Figure 4.4c) [31]. The selection rules can be summed up by saying that, if FWM is at the frequency $2\omega_1 - \omega_2$, then the handedness of FWM is the same as that of the ω_2 beam. The cross-polarized cases are not suppressed for the array, unlike for the substrate, as can be derived by a symmetry analysis of the metasurface. To clarify the effect of the geometric phase, arrays were fabricated with nanocrosses rotated toward each other, in pairs, by an angle θ in 7.5° increments. As the phase factor acquired by a nanocross is $e^{i4\theta}$, it can be derived that the FWM intensity is modulated by $\cos^2(4\theta)$.

These methods for light control are interesting on their own, but their true potential might emerge if they are integrated together. Assemblies of ultrathin metasurfaces could exploit different nonlinear effects, modulated all-optically by the Kerr effects and controlled by the use of the local phase profiles in order to implement various optical processing functions.

4.6   NONLINEAR METASURFACE HOLOGRAPHY

Phase and amplitude modulation by metasurface elements can be used for optical holograms [10,133]. The subwavelength size of a metasurface unit cell leads to the suppression of diffraction orders, thus the term "zero-order hologram" is an appropriate description for the operation principle of metasurface holograms. The small unit cell size also results into large viewing angles for the metasurface-based holograms. Moreover, geometric features of the metasurface elements can introduce additional degrees of freedom to the holographic effect such as polarization, wavelength, spin angular momentum, and wavevector (angle of incidence). They give the possibility to encode multiple holographic images (multiplexing) that can be selectively reconstructed by certain illumination conditions.

Nonlinear processes can also be implemented in the design of the metasurface holograms. SHG is the most widely utilized effect in nonlinear metasurface holography. It is mainly used for encoding multiple images in both fundamental and SH frequencies. When the geometric phase method is implemented for the modulation, the phase values for the cross-polarized diffracted fundamental beam are $2\sigma\varphi(\mathbf{r})$, where σ is the spin (helicity) sign of the fundamental wave, and $\varphi(\mathbf{r})$ is the angle of rotation of the metasurface element at point $\mathbf{r}$. The SHG beam has phase modulations $\sigma\varphi(\mathbf{r})$ for co-polarized and $3\sigma\varphi(\mathbf{r})$ for cross-polarized components. Following this principle, gold split-ring resonators were used to encode three different images of letters "X," "R," and "L" (Figure 4.5a), which were reconstructed simultaneously in the in cross-polarized signal at the fundamental frequency and in the generated SH in both σ and $-\sigma$ helicity states [134]. Here, SHG occurs due to the metal nonlinearity of the split-rings elements. Simultaneous encoding of multiple images becomes possible because of the different multiplication factors (1, 2, and 3) between the phase value and the meta-atom orientation in the output frequency/ polarization channels. Thus, two elements whose rotation angles differ by π impose the same phase modulation on the fundamental wave (cross-polarization) while the phase in the generated harmonic beam will be different. For cross-polarized SHG, two meta-atoms rotated by $2/3\pi$ give the same phase, but for the co-polarized SHG channel, the phase values for the same elements are different. This provides enough degrees of freedom to encode three different images in one phase hologram.

The encoding of image information using an SHG metasurface hologram has also been demonstrated [135]. The hidden data can be only observed in the SH frequency output, when the structure is illuminated by the fundamental wave. The illumination by light of SH frequency does not reveal the encoded image, nor does any incoherent and/or unpolarized illumination, thus providing a useful method for security holography. Metasurface elements with a three-fold symmetry were used in the design of the hologram. Selection rules prohibit radiation of co-polarized SHG, thus, only the cross-polarized components remain in the output. The metasurface elements are utilized in pairs, where the structures are rotated relative to each other. When the rotation angle reaches $\pi/3$, the phase difference between SHG signals becomes π and, as a result of destructive interference, the area looks dark at the SH frequency. When the rotation angle is $0°$, the SHG signal reaches maximum. The metasurface elements form pixels and the bright pixels form an image that becomes visible in the generated SH light (Figure 4.5b).

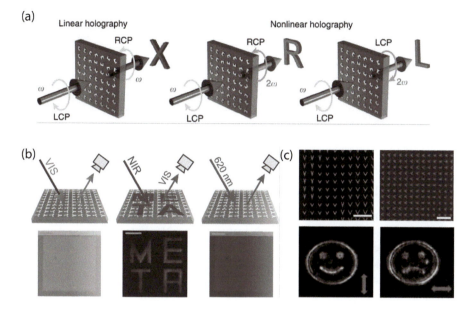

FIGURE 4.5 Metasurface holography. (a) Metasurface hologram encoding multiple images. Left: linear mode. Centre: SHG image observed in cross-polarized light. Right: same, but in co-polarized light. (Reproduced from Ye, W.M. et al., *Nat. Commun.*, 7, 7, 11930, 2016. With permission.) (b) Nonlinear metasurface under different illumination conditions. Left: illumination by visible, incoherent, unpolarized light does not reveal the hidden image. Center: illumination by 1240 nm circularly polarized fundamental light shows the hidden text in the SHG signal. Right: illumination by 620 nm circularly polarized SH does not detect the hidden image. Scale bars: 20 μm. (Reproduced from Walter, F. et al., *Nano Lett.*, 17, 3171–3175, 2017. With permission.) (c) Two-layer THG hologram. Top left: a metasurface layer for vertical polarization. Top right: a metasurface layer for horizontal polarization. Bottom left: the THG image displayed under vertical fundamental polarization. Bottom right: same, but with horizontal polarization. Scale bars: 1 μm. (Reproduced from Almeida, E. et al., *Nat. Commun.*, 7, 7, 12533, 2016. With permission.)

A nonlinear metasurface that operates in TH frequencies has also been demonstrated [133]. V-shaped antennas were used as metasurface elements (Figure 4.5c). The metasurface operates for linearly polarized light with polarization oriented along the axis of meta-atom symmetry. The phase shift acquired by the transmitted wave depends on the length of the antenna arms and the angle between them. The V-element parameters where chosen in such a way that the holographic signal is minimized for the polarization perpendicular to the angle bisector. The phase shift imposed by the elements in THG is three times larger than that of the fundamental wave. Utilizing a set of metasurface elements with an equal THG amplitude and discretized phase steps, a phase hologram was encoded and printed in the first metasurface layer. In the second layer, a hologram with a different image was imprinted with V-antennas turned by 90°. By changing the polarization direction, it is possible to achieve the activation of different layers of the hologram and therefore switching between the encoded images.

Numerous metasurface holographic designs have been demonstrated to date, but there are remaining challenges on the way of implementation of the nanophotonic structures for holography. First of all there is a nanofabrication issue: electron beam lithography and focused ion beam techniques are expensive and slow, thus only the improvement of nanoprinting methods could provide methods suitable for mass production. Second, the principles of dynamic holography, which can provide such applications as holographic displays for TV and smartphones, still remain under-developed. However, nonlinear metasurfaces utilizing Kerr effect can be one more route to achieve all-optical dynamic control.

4.7   CONCLUSIONS AND OUTLOOK

Research now in progress will determine if practical, fast, and power-efficient signal processing with light is possible with metasurfaces. Nonlinear effects are widely exploited in these systems, as they offer a broad selection of methods to manipulate light. Plasmonic and hybrid plasmonic-dielectric systems provide a flexible platform for designing metasurfaces well suited to nonlinear enhancement, so it is likely they will play an important role in our possible "more photonic" future. There are, as always, numerous challenges to overcome. Traditional plasmonic systems are greatly affected by losses, and the search for novel materials to use in nanostructures and metamaterials has been a topic of intensive research, with new nonlinear materials proposed, including those exploiting epsilon-near-zero properties [136,137] as well as plasmonic properties of highly-doped semiconductors [138].

While signal transmission over long distances is already dominated by fiber optics and optical interconnects are replacing electronic hardware in data centres, a fully photonic system able to process digitally encoded data would require the development of true optical logic gates. At the same time, this is very hard to do in prac-tice: logic gates must operate in sequence without reflections back to the previous elements and without distorting the signal excessively, they must not be affected by losses and must restore signal quality to prevent it from degrading into noise [139].

With plasmonics, metasurfaces and nonlinear materials have found new uses for light, and researchers have gained extensive experience in studying and designing opti-cal properties by combining many different materials. If the challenges are overcome, this knowledge can be applied to the design of optical devices with unprecedented capa-bilities, to open further opportunities for light in the sciences and in daily life.

REFERENCES

1. J. B. Pendry, "Negative refraction makes a perfect lens," *Phys. Rev. Lett.* **85**, 3966–3969 (2000).
2. Z. W. Liu, H. Lee, Y. Xiong, C. Sun, and X. Zhang, "Far-field optical hyperlens magni-fying sub-diffraction-limited objects," *Science* **315**, 1686–1686 (2007).
3. A. V. Kabashin, P. Evans, S. Pastkovsky, W. Hendren, G. A. Wurtz, R. Atkinson, R. Pollard, V. A. Podolskiy, and A. V. Zayats, "Plasmonic nanorod metamaterials for biosensing," *Nat. Mater.* **8**, 867–871 (2009).

4. A. D. Neira, G. A. Wurtz, and A. V. Zayats, "All-optical switching in silicon photonic waveguides with an epsilon-near-zero resonant cavity invited," *Photonics Res.* **6**, B1–B5 (2018).

5. A. D. Neira, G. A. Wurtz, P. Ginzburg, and A. V. Zayats, "Ultrafast all-optical modulation with hyperbolic metamaterial integrated in Si photonic circuitry," *Opt. Express* **22**, 10987–10994 (2014).

6. G. A. Wurtz, R. Pollard, W. Hendren, G. P. Wiederrecht, D. J. Gosztola, V. A. Podolskiy, and A. V. Zayats, "Designed ultrafast optical nonlinearity in a plasmonic nanorod metamaterial enhanced by nonlocality," *Nat. Nanotechnol.* **6**, 107–111 (2011).

7. D. Smirnova, and Y. S. Kivshar, "Multipolar nonlinear nanophotonics," *Optica* **3**, 1241–1255 (2016).

8. N. Meinzer, W. L. Barnes, and I. R. Hooper, "Plasmonic meta-atoms and metasurfaces," *Nat. Photonics* **8**, 889–898 (2014).

9. B. Sain, C. Mieier, and T. Zentgraf, "Nonlinear optics in all-dielectric nanoantennas and metasurfaces: A review," *Adv. Photon.* **1**, 024002 (2019).

10. G. Li, S. Zhang, and T. Zentgraf, "Nonlinear photonic metasurfaces," *Nat. Rev. Mater.* **2**, 14, 17010 (2017).

11. A. E. Minovich, A. E. Miroshnichenko, A. Y. Bykov, T. V. Murzina, D. N. Neshev, and Y. S. Kivshar, "Functional and nonlinear optical metasurfaces," *Laser Photon. Rev.* **9**, 195–213 (2015).

12. M. I. Stockman, L. N. Pandey, and T. F. George, "Inhomogeneous localization of polar eigenmodes in fractals," *Phys. Rev. B* **53**, 2183–2186 (1996).

13. A. Campion, and P. Kambhampati, "Surface-enhanced Raman scattering," *Chem. Soc. Rev.* **27**, 241–250 (1998).

14. K. Fukami, M. L. Chourou, R. Miyagawa, A. M. Noval, T. Sakka, M. Manso-Silvan, R. J. Martin-Palma, and Y. H. Ogata, "Gold nanostructures for surface-enhanced Raman spectroscopy, prepared by electrodeposition in porous silicon," *Materials* **4**, 791–800 (2011).

15. X. B. Xu, H. F. Li, D. Hasan, R. S. Ruoff, A. X. Wang, and D. L. Fan, "Near-field enhanced plasmonic-magnetic bifunctional nanotubes for single cell bioanalysis," *Adv. Funct. Mater.* **23**, 4332–4338 (2013).

16. C. K. Chen, A. R. B. Decastro, and Y. R. Shen, "Surface-enhanced 2nd-harmonic generation," *Phys. Rev. Lett.* **46**, 145–148 (1981).

17. M. Kauranen, and A. V. Zayats, "Nonlinear plasmonics," *Nat. Photonics* **6**, 737–748 (2012).

18. S. Takahashi, and A. V. Zayats, "Near-field second-harmonic generation at a metal tip apex," *Appl. Phys. Lett.* **80**, 3479–3481 (2002).

19. G. F. Walsh, and L. Dal Negro, "Enhanced second harmonic generation by photonic-plasmonic fano-type coupling in nanoplasmonic arrays," *Nano Lett.* **13**, 3111–3117 (2013).

20. T. Hanke, J. Cesar, V. Knittel, A. Trugler, U. Hohenester, A. Leitenstorfer, and R. Bratschitsch, "Tailoring spatiotemporal light confinement in single plasmonic nanoantennas," *Nano Lett.* **12**, 992–996 (2012).

21. B. Metzger, T. Schumacher, M. Hentschel, M. Lippitz, and H. Giessen, "Third harmonic mechanism in complex plasmonic fano structures," *ACS Photonics* **1**, 471–476 (2014).

22. J. Renger, R. Quidant, N. van Hulst, and L. Novotny, "Surface-enhanced nonlinear four-wave mixing," *Phys. Rev. Lett.* **104**, 4, 046803 (2010).

23. A. V. Krasavin, and N. I. Zheludev, "Active plasmonics: Controlling signals in Au/Ga waveguide using nanoscale structural transformations," *Appl. Phys. Lett.* **84**, 1416–1418 (2004).

24. K. F. MacDonald, Z. L. Samson, M. I. Stockman, and N. I. Zheludev, "Ultrafast active plasmonics," *Nat. Photonics* **3**, 55–58 (2009).

25. J. Butet, P. F. Brevet, and O. J. F. Martin, "Optical second harmonic generation in plasmonic nanostructures: From fundamental principles to advanced applications," *ACS Nano* **9**, 10545–10562 (2015).

26. S. Keren-Zur, L. Michaeli, H. Suchowski, and T. Ellenbogen, "Shaping light with nonlinear metasurfaces," *Adv. Opt. Photonics* **10**, 309–353 (2018).

27. A. Krasnok, M. Tymchenko, and A. Alu, "Nonlinear metasurfaces: A paradigm shift in nonlinear optics," *Mater. Today* **21**, 8–21 (2018).

28. A. V. Krasavin, P. Ginzburg, and A. V. Zayats, "Free-electron optical nonlinearities in plasmonic nanostructures: A review of the hydrodynamic description," *Laser Photon. Rev.* **12**, 24, 1700082 (2018).

29. N. Bloembergen, R. K. Chang, S. S. Jha, and C. H. Lee, "Optical second-harmonic generation in reflection from media with inversion symmetry," *Phys. Rev.* **174**, 813–822 (1968).

30. J. E. Sipe, V. C. Y. So, M. Fukui, and G. I. Stegeman, "Analysis of second-harmonic generation at metal surfaces," *Phys. Rev. B* **21**, 4389–4402 (1980).

31. G. Li, G. Sartorello, S. Chen, L. H. Nicholls, K. F. Li, T. Zentgraf, S. Zhang, and A. V. Zayats, "Spin and geometric phase control four-wave mixing from metasurfaces," *Laser Photon. Rev.* **12**, 1800034 (2018).

32. R. W. Boyd, *Nonlinear Optics* (Academic Press, 2008).

33. J. Rudnick, and E. A. Stern, "Second-harmonic radiation from metal surfaces," *Phys. Rev. B* **4**, 4274–4290 (1971).

34. D. Krause, C. W. Teplin, and C. T. Rogers, "Optical surface second harmonic measurements of isotropic thin-film metals: Gold, silver, copper, aluminum, and tantalum," *J. Appl. Phys.* **96**, 3626–3634 (2004).

35. F. X. Wang, F. J. Rodriguez, W. M. Albers, R. Ahorinta, J. E. Sipe, and M. Kauranen, "Surface and bulk contributions to the second-order nonlinear optical response of a gold film," *Phys. Rev. B* **80**, 4, 233402 (2009).

36. N. I. Zheludev, and V. I. Emel'yanov, "Phase matched second harmonic generation from nanostructured metallic surfaces," *J. Opt. A-Pure Appl. Opt.* **6**, 26–28 (2004).

37. M. D. McMahon, R. Lopez, R. F. Haglund, E. A. Ray, and P. H. Bunton, "Second-harmonic generation from arrays of symmetric gold nanoparticles," *Phys. Rev. B* **73**, 4, 041401 (2006).

38. R. Czaplicki, J. Makitalo, R. Siikanen, H. Husu, J. Lehtolahti, M. Kuittinen, and M. Kauranen, "Second-harmonic generation from metal nanoparticles: Resonance enhancement versus particle geometry," *Nano Lett.* **15**, 530–534 (2015).

39. G. Marino, P. Segovia, A. V. Krasavin, P. Ginzburg, N. Olivier, G. A. Wurtz, and A. V. Zayats, "Second-harmonic generation from hyperbolic plasmonic nanorod metamaterial slab," *Laser Photon. Rev.* **12**, 1700189 (2018).

40. B. Wells, A. Y. Bykov, G. Marino, M. E. Nasir, A. V. Zayats, and V. A. Podolskiy, "Structural second-order nonlinearity in plasmonic metamaterials," *Optica* **5**, 1502–1507 (2018).

41. J. Butet, and O. J. F. Martin, "Evaluation of the nonlinear response of plasmonic metasurfaces: Miller's rule, nonlinear effective susceptibility method, and full-wave computation," *J. Opt. Soc. Am. B* **33**, A8-A15 (2016).

42. K. O'Brien, H. Suchowski, J. Rho, A. Salandrino, B. Kante, X. Yin, and X. Zhang, "Predicting nonlinear properties of metamaterials from the linear response," *Nat. Mater.* **14**, 379–383 (2015).

43. K. Thyagarajan, S. Rivier, A. Lovera, and O. J. F. Martin, "Enhanced second-harmonic generation from double resonant plasmonic antennae," *Opt. Express* **20**, 12860–12865 (2012).

44. M. Celebrano, X. Wu, M. Baselli, S. Grossmann, P. Biagioni, A. Locatelli, C. De Angelis, et al., "Mode matching in multiresonant plasmonic nanoantennas for enhanced second harmonic generation," *Nat. Nanotechnol.* **10**, 412–417 (2015).

45. F. B. P. Niesler, N. Feth, S. Linden, and M. Wegener, "Second-harmonic optical spectroscopy on split-ring-resonator arrays," *Opt. Lett.* **36**, 1533–1535 (2011).

46. S. Kujala, B. K. Canfield, M. Kauranen, Y. Svirko, and J. Turunen, "Multipole interference in the second-harmonic optical radiation from gold nanoparticles," *Phys. Rev. Lett.* **98**, 4, 167403 (2007).

47. M. Zdanowicz, S. Kujala, H. Husu, and M. Kauranen, "Effective medium multipolar tensor analysis of second-harmonic generation from metal nanoparticles," *New J. Phys.* **13**, 12, 023025 (2011).

48. R. Czaplicki, M. Zdanowicz, K. Koskinen, J. Laukkanen, M. Kuittinen, and M. Kauranen, "Dipole limit in second-harmonic generation from arrays of gold nanoparticles," *Opt. Express* **19**, 26866–26871 (2011).

49. J. B. Pendry, A. J. Holden, D. J. Robbins, and W. J. Stewart, "Magnetism from conductors and enhanced nonlinear phenomena," *IEEE Trans. Microw. Theory Tech.* **47**, 2075–2084 (1999).

50. D. R. Smith, W. J. Padilla, D. C. Vier, S. C. Nemat-Nasser, and S. Schultz, "Composite medium with simultaneously negative permeability and permittivity," *Phys. Rev. Lett.* **84**, 4184–4187 (2000).

51. R. A. Shelby, D. R. Smith, and S. Schultz, "Experimental verification of a negative index of refraction," *Science* **292**, 77–79 (2001).

52. K. Aydin, I. Bulu, K. Guven, M. Kafesaki, C. M. Soukoulis, and E. Ozbay, "Investigation of magnetic resonances for different split-ring resonator parameters and designs," *New J. Phys.* **7**, 15, 168 (2005).

53. G. Sartorello, N. Olivier, J. J. Zhang, W. S. Yue, D. J. Gosztola, G. P. Wiederrecht, G. Wurtz, and A. V. Zayats, "Ultrafast optical modulation of second- and third-harmonic generation from cut-disk-based metasurfaces," *ACS Photonics* **3**, 1517–1522 (2016).

54. S. Linden, C. Enkrich, M. Wegener, J. F. Zhou, T. Koschny, and C. M. Soukoulis, "Magnetic response of metamaterials at 100 terahertz," *Science* **306**, 1351–1353 (2004).

55. C. Enkrich, M. Wegener, S. Linden, S. Burger, L. Zschiedrich, F. Schmidt, J. F. Zhou, T. Koschny, and C. M. Soukoulis, "Magnetic metamaterials at telecommunication and visible frequencies," *Phys. Rev. Lett.* **95**, 4, 203901 (2005).

56. M. W. Klein, C. Enkrich, M. Wegener, and S. Linden, "Second-harmonic generation from magnetic metamaterials," *Science* **313**, 502–504 (2006).

57. M. W. Klein, M. Wegener, N. Feth, and S. Linden, "Experiments on second- and third-harmonic generation from magnetic metamaterials," *Opt. Express* **15**, 5238–5247 (2007).

58. C. Ciracì, E. Poutrina, M. Scalora, and D. R. Smith, "Origin of second-harmonic generation enhancement in optical split-ring resonators," *Phys. Rev. B* **85**, 201403(R) (2012).

59. S. Kruk, M. Weismann, A. Y. Bykov, E. A. Mamonov, I. A. Kolmychek, T. Murzina, N. C. Panoiu, D. N. Neshev, and Y. S. Kivshar, "Enhanced magnetic second-harmonic generation from resonant metasurfaces," *ACS Photonics* **2**, 1007–1012 (2015).

60. I. A. Kolmychek, A. Y. Bykov, E. A. Mamonov, and T. V. Murzina, "Second-harmonic generation interferometry in magnetic-dipole nanostructures," *Opt. Lett.* **40**, 3758–3761 (2015).

61. F. Falcone, T. Lopetegi, M. A. G. Laso, J. D. Baena, J. Bonache, M. Beruete, R. Marques, F. Martin, and M. Sorolla, "Babinet principle applied to the design of metasurfaces and metamaterials," *Phys. Rev. Lett.* **93**, 4, 197401 (2004).

62. T. Zentgraf, T. P. Meyrath, A. Seidel, S. Kaiser, H. Giessen, C. Rockstuhl, and F. Lederer, "Babinet's principle for optical frequency metamaterials and nanoantennas," *Phys. Rev. B* **76**, 4, 033407 (2007).

63. N. Feth, S. Linden, M. W. Klein, M. Decker, F. B. P. Niesler, Y. Zeng, W. Hoyer, et al., "Second-harmonic generation from complementary split-ring resonators," *Opt. Lett.* **33**, 1975–1977 (2008).

64. K. Konishi, T. Higuchi, J. Li, J. Larsson, S. Ishii, and M. Kuwata-Gonokami, "Polarization-controlled circular second-harmonic generation from metal hole arrays with threefold rotational symmetry," *Phys. Rev. Lett.* **112**, 5, 135502 (2014).

65. V. K. Valev, J. J. Baumberg, B. De Clercq, N. Braz, X. Zheng, E. J. Osley, S. Vandendriessche, et al., "Nonlinear superchiral meta-surfaces: Tuning chirality and disentangling non-reciprocity at the nanoscale," *Adv. Mater.* **26**, 4074–4081 (2014).

66. R. Kolkowski, L. Petti, M. Rippa, C. Lafargue, and J. Zyss, "Octupolar plasmonic meta-molecules for nonlinear chiral watermarking at subwavelength scale," *ACS Photonics* **2**, 899–906 (2015).

67. N. Segal, S. Keren-Zur, N. Hendler, and T. Ellenbogen, "Controlling light with meta-material-based nonlinear photonic crystals," *Nat. Photonics* **9**, 180–184 (2015).

68. S. Keren-Zur, O. Avayu, L. Michaeli, and T. Ellenbogen, "Nonlinear beam shaping with plasmonic metasurfaces," *ACS Photonics* **3**, 117–123 (2016).

69. M. Tymchenko, J. S. Gomez-Diaz, J. Lee, N. Nookala, M. A. Belkin, and A. Alu, "Gradient nonlinear pancharatnam-berry metasurfaces," *Phys. Rev. Lett.* **115**, 5, 207403 (2015).

70. S. M. Chen, G. X. Li, F. Zeuner, W. H. Wong, E. Y. B. Pun, T. Zentgraf, K. W. Cheah, and S. Zhang, "Symmetry-selective third-harmonic generation from plasmonic metac-rystals," *Phys. Rev. Lett.* **113**, 5, 033901 (2014).

71. G. X. Li, S. M. Chen, N. Pholchai, B. Reineke, P. W. H. Wong, E. Y. B. Pun, K. W. Cheah, T. Zentgraf, and S. Zhang, "Continuous control of the nonlinearity phase for harmonic generations," *Nat. Mater.* **14**, 607–612 (2015).

72. S. M. Chen, F. Zeuner, M. Weismann, B. Reineke, G. X. Li, V. K. Valev, K. W. Cheah, N. C. Panoiu, T. Zentgraf, and S. Zhang, "Giant nonlinear optical activity of achiral origin in planar metasurfaces with quadratic and cubic nonlinearities," *Adv. Mater.* **28**, 2992–2999 (2016).

73. H. Liu, G. X. Li, K. F. Li, S. M. Chen, S. N. Zhu, C. T. Chan, and K. W. Cheah, "Linear and nonlinear fano resonance on two-dimensional magnetic metamaterials," *Phys. Rev. B* **84**, 6, 235437 (2011).

74. J. Lee, M. Tymchenko, C. Argyropoulos, P. Y. Chen, F. Lu, F. Demmerle, G. Boehm, M. C. Amann, A. Alu, and M. A. Belkin, "Giant nonlinear response from plasmonic metasurfaces coupled to intersubband transitions," *Nature* **511**, 65-U389 (2014).

75. C. M. Soukoulis, and M. Wegener, "Past achievements and future challenges in the development of three-dimensional photonic metamaterials," *Nat. Photonics* **5**, 523–530 (2011).

76. N. Liu, H. C. Guo, L. W. Fu, S. Kaiser, H. Schweizer, and H. Giessen, "Three-dimensional photonic metamaterials at optical frequencies," *Nat. Mater.* **7**, 31–37 (2008).

77. G. Dolling, M. Wegener, and S. Linden, "Realization of a three-functional-layer nega-tive-index photonic metamaterial," *Opt. Lett.* **32**, 551–553 (2007).

78. M. B. Ross, M. G. Blaber, and G. C. Schatz, "Using nanoscale and mesoscale anisot-ropy to engineer the optical response of three-dimensional plasmonic metamaterials," *Nat. Commun.* **5**, 11, 4090 (2014).

79. S. Palomba, S. Zhang, Y. Park, G. Bartal, X. B. Yin, and X. Zhang, "Optical negative refraction by four-wave mixing in thin metallic nanostructures," *Nat. Mater.* **11**, 34–38 (2012).

80. Y. Zhang, F. Wen, Y. R. Zhen, P. Nordlander, and N. J. Halas, "Coherent fano res-onances in a plasmonic nanocluster enhance optical four-wave mixing," *Proc. Natl. Acad. Sci. U.S.A.* **110**, 9215–9219 (2013).

81. D. Yoo, T. W. Johnson, S. Cherukulappurath, D. J. Norris, and S. H. Oh, "Template-stripped tunable plasmonic devices on stretchable and rollable substrates," *ACS Nano* **9**, 10647–10654 (2015).

82. S. C. Malek, H. S. Ee, and R. Agarwal, "Strain multiplexed metasurface holograms on a stretchable substrate," *Nano Lett.* **17**, 3641–3645 (2017).
83. J. Y. Ou, E. Plum, J. F. Zhang, and N. I. Zheludev, "An electromechanically reconfigurable plasmonic metamaterial operating in the near-infrared," *Nat. Nanotechnol.* **8**, 252–255 (2013).
84. P. Cencillo-Abad, J. Y. Ou, E. Plum, and N. I. Zheludev, "Electro-mechanical light modulator based on controlling the interaction of light with a metasurface," *Sci. Rep.* **7**, 7, 5405 (2017).
85. J. Valente, J. Y. Ou, E. Plum, I. J. Youngs, and N. I. Zheludev, "A magneto-electro-optical effect in a plasmonic nanowire material," *Nat. Commun.* **6**, 6, 7021 (2015).
86. P. Cencillo-Abad, E. Plum, E. T. F. Rogers, and N. I. Zheludev, "Spatial optical phase-modulating metadevice with subwavelength pixelation," *Opt. Express* **24**, 18790–18798 (2016).
87. P. Cencillo-Abad, N. I. Zheludev, and E. Plum, "Metadevice for intensity modulation with sub-wavelength spatial resolution," *Sci. Rep.* **6**, 7, 37109 (2016).
88. J. Ward, K. Z. Kamali, L. Xu, G. Q. Zhang, A. E. Miroshnichenko, and M. Rahmani, "High-contrast and reversible scattering switching via hybrid metal-dielectric metasurfaces," *Beilstein J. Nanotechnol.* **9**, 460–467 (2018).
89. G. Armelles, A. Cebollada, F. Garcia, A. Garcia-Martin, and N. de Sousa, "Far- and near-field broad-band magneto-optical functionalities using magnetoplasmonic nanorods," *ACS Photonics* **3**, 2427–2433 (2016).
90. J. Lee, S. Jung, P. Y. Chen, F. Lu, F. Demmerle, G. Boehm, M. C. Amann, A. Alu, and M. A. Belkin, "Ultrafast electrically tunable polaritonic metasurfaces," *Adv. Opt. Mater.* **2**, 1057–1063 (2014).
91. N. Dabidian, I. Kholmanov, A. B. Khanikaev, K. Tatar, S. Trendafilov, S. H. Mousavi, C. Magnuson, R. S. Ruoff, and G. Shvets, "Electrical switching of infrared light using graphene integration with plasmonic fano resonant metasurfaces," *ACS Photonics* **2**, 216–227 (2015).
92. S. Dutta-Gupta, N. Dabidian, I. Kholmanov, M. A. Belkin, and G. Shvets, "Electrical tuning of the polarization state of light using graphene-integrated anisotropic metasurfaces," *Philos. Trans. R. Soc. A-Math. Phys. Eng. Sci.* **375**, 11, 20160061 (2017).
93. M. C. Sherrott, P. W. C. Hon, K. T. Fountaine, J. C. Garcia, S. M. Ponti, V. W. Brar, L. A. Sweatlock, and H. A. Atwater, "Experimental demonstration of > 230 degrees phase modulation in gate-tunable graphene-gold reconfigurable mid-infrared metasurfaces," *Nano Lett.* **17**, 3027–3034 (2017).
94. Z. B. Li, K. Yao, F. N. Xia, S. Shen, J. G. Tian, and Y. M. Liu, "Graphene plasmonic metasurfaces to steer infrared light," *Sci. Rep.* **5**, 9, 12423 (2015).
95. S. R. Biswas, C. E. Gutierrez, A. Nemilentsau, I. H. Lee, S. H. Oh, P. Avouris, and T. Low, "Tunable graphene metasurface reflectarray for cloaking, illusion, and focusing," *Phys. Rev. Appl.* **9**, 12, 034021 (2018).
96. Y. Lee, M. K. Park, S. Kim, J. H. Shin, C. Moon, J. Y. Hwang, J. C. Choi, H. Park, H. R. Kim, and J. E. Jang, "Electrical broad tuning of plasmonic color filter employing an asymmetric-lattice nanohole array of metasurface controlled by polarization rotator," *ACS Photonics* **4**, 1954–1966 (2017).
97. O. Buchnev, N. Podoliak, M. Kaczmarek, N. I. Zheludev, and V. A. Fedotov, "Electrically controlled nanostructured metasurface loaded with liquid crystal: Toward multifunctional photonic switch," *Adv. Opt. Mater.* **3**, 674–679 (2015).
98. J. Q. Zhang, Y. Kosugi, A. Otomo, Y. Nakano, and T. Tanemura, "Active metasurface modulator with electro-optic polymer using bimodal plasmonic resonance," *Opt. Express* **25**, 30304–30311 (2017).

99. J. J. Guo, Y. Tu, L. L. Yang, R. W. Zhang, L. L. Wang, and B. P. Wang, "Electrically tunable gap surface plasmon-based metasurface for visible light," *Sci. Rep.* **7**, 7, 14078 (2017).

100. J. A. Dionne, K. Diest, L. A. Sweatlock, and H. A. Atwater, "Plasmostor: A metal-oxide-Si field effect plasmonic modulator," *Nano Lett.* **9**, 897–902 (2009).

101. A. V. Krasavin, and A. V. Zayats, "Photonic signal processing on electronic scales: Electro-optical field-effect nanoplasmonic modulator," *Phys. Rev. Lett.* **109**, 053901 (2012).

102. J. Park, J. H. Kang, X. G. Liu, and M. L. Brongersma, "Electrically tunable epsilon-near-zero (enz) metafilm absorbers," *Sci. Rep.* **5**, 9, 15754 (2015).

103. Y. W. Huang, H. W. H. Lee, R. Sokhoyan, R. A. Pala, K. Thyagarajan, S. Han, D. P. Tsai, and H. A. Atwater, "Gate-tunable conducting oxide metasurfaces," *Nano Lett.* **16**, 5319–5325 (2016).

104. G. K. Shirmanesh, R. Sokhoyan, R. A. Pala, and H. A. Atwater, "Dual-gated active metasurface at 1550 nm with wide (> 300 degrees) phase tunability," *Nano Lett.* **18**, 2957–2963 (2018).

105. Y. L. Sun, Y. H. Ling, T. J. Liu, and L. R. Huang, "Electro-optical switch based on continuous metasurface embedded in Si substrate," *AIP Adv.* **5**, 8, 117221 (2015).

106. K. Y. Lee, J. W. Yoon, S. H. Song, and R. Magnusson, "Multiple p-n junction subwavelength gratings for transmission-mode electro-optic modulators," *Sci. Rep.* **7**, 8, 46508 (2017).

107. P. P. Iyer, M. Pendharkar, and J. A. Schuller, "Electrically reconfigurable metasurfaces using heterojunction resonators," *Adv. Opt. Mater.* **4**, 1582–1588 (2016).

108. M. K. Liu, D. A. Powell, R. Guo, I. V. Shadrivov, and Y. S. Kivshar, "Polarization-induced chirality in metamaterials via optomechanical interaction," *Adv. Opt. Mater.* **5**, 8, 1600760 (2017).

109. S. Viaene, V. Ginis, J. Danckaert, and P. Tassin, "Do optomechanical metasurfaces run out of time?" *Phys. Rev. Lett.* **120**, 6, 197402 (2018).

110. J. Y. Ou, E. Plum, J. Zhang, and N. I. Zheludev, "Giant nonlinearity of an optically reconfigurable plasmonic metamaterial," *Adv. Mater.* **28**, 729–733 (2016).

111. M. X. Ren, W. Wu, W. Cai, B. Pi, X. Z. Zhang, and J. J. Xu, "Reconfigurable metasurfaces that enable light polarization control by light," *Light Sci. Appl.* **6**, 5, e16254 (2017).

112. C. C. Lu, X. Y. Hu, K. B. Shi, Q. Hu, R. Zhu, H. Yang, and Q. H. Gong, "An actively ultrafast tunable giant slow-light effect in ultrathin nonlinear metasurfaces," *Light Sci. Appl.* **4**, 9, e302 (2015).

113. Z. Q. Huang, A. Baron, S. Larouche, C. Argyropoulos, and D. R. Smith, "Optical bistability with film-coupled metasurfaces," *Opt. Lett.* **40**, 5638–5641 (2015).

114. B. Memarzadeh, and H. Mosallaei, "Engineering optical bistability in a multimaterial loop metasurface," *J. Opt. Soc. Am. B* **31**, 1539–1543 (2014).

115. A. V. Krasavin, K. F. MacDonald, N. I. Zheludev, and A. V. Zayats, "High-contrast modulation of light with light by control of surface plasmon polariton wave coupling," *Appl. Phys. Lett.* **85**, 3369–3371 (2004).

116. R. F. Waters, P. A. Hobson, K. F. MacDonald, and N. I. Zheludev, "Optically switchable photonic metasurfaces," *Appl. Phys. Lett.* **107**, 4, 081102 (2015).

117. A. K. U. Michel, P. Zalden, D. N. Chigrin, M. Wuttig, A. M. Lindenberg, and T. Taubner, "Reversible optical switching of infrared antenna resonances with ultrathin phase-change layers using femtosecond laser pulses," *ACS Photonics* **1**, 833–839 (2014).

118. Y. Zhu, X. Y. Hu, H. Yang, and Q. H. Gong, "Switchable cross-polarization conversion in ultrathin metasurfaces," *J. Opt.* **17**, 8, 105101 (2015).

119. J. B. Khurgin, "How to deal with the loss in plasmonics and metamaterials," *Nat. Nanotechnol.* **10**, 2–6 (2015).

120. G. Della Valle, D. Polli, P. Biagioni, C. Martella, M. C. Giordano, M. Finazzi, S. Longhi, L. Duò, G. Cerullo, and F. Buatier de Mongeot, "Self-organized plasmonic metasurfaces for all-optical modulation," *Phys. Rev. B* **91**, 235440 (2015).

121. L. H. Nicholls, F. J. Rodriguez-Fortuno, M. E. Nasir, R. M. Cordova-Castro, N. Olivier, G. A. Wurtz, and A. V. Zayats, "Ultrafast synthesis and switching of light polarization in nonlinear anisotropic metamaterials," *Nat. Photonics* **11**, 628–633 (2017).

122. A. D. Neira, N. Olivier, M. E. Nasir, W. Dickson, G. A. Wurtz, and A. V. Zayats, "Eliminating material constraints for nonlinearity with plasmonic metamaterials," *Nat. Commun.* **6**, 7757 (2015).

123. L. H. Nicholls, T. Stefaniuk, M. E. Nasir, F. J. Rodriguez-Fortuno, G. A. Wurtz, and A. V. Zayats, "Designer photonic dynamics by using non-uniform electron temperature distribution for on-demand all-optical switching times," *Nat. Commun.* **10**, 8, 2967 (2019).

124. J. Kim, E. G. Camemolla, C. DeVaul, A. M. Shaltout, D. Faccio, V. M. Shalaev, A. V. Kildishev, M. Ferrera, and A. Boltasseva, "Dynamic control of nanocavities with tunable metal oxides," *Nano Lett.* **18**, 740–746 (2018).

125. Y. Xiao, H. Qian, and Z. Liu, "Nonlinear metasurface based on giant optical kerr response of gold quantum wells," *ACS Photonics* **5**, 1654–1659 (2018).

126. A. Melikyan, L. Alloatti, A. Muslija, D. Hillerkuss, P. C. Schindler, J. Li, R. Palmer, et al., "High-speed plasmonic phase modulators," *Nat. Photonics* **8**, 229–233 (2014).

127. R. B. Davidson, A. Yanchenko, J. I. Ziegler, S. M. Avanesyan, B. J. Lawrie, and R. F. Haglund, "Ultrafast plasmonic control of second harmonic generation," *ACS Photonics* **3**, 1477–1481 (2016).

128. K. M. Dani, Z. Y. Ku, P. C. Upadhya, R. P. Prasankumar, S. R. J. Brueck, and A. J. Taylor, "Subpicosecond optical switching with a negative index metamaterial," *Nano Lett.* **9**, 3565–3569 (2009).

129. M. Ren, B. Jia, J. Y. Ou, E. Plum, J. Zhang, K. F. MacDonald, A. E. Nikolaenko, J. Xu, M. Gu, and N. I. Zheludev, "Nanostructured plasmonic medium for terahertz bandwidth all-optical switching," *Adv. Mater.* **23**, 5540–5544 (2011).

130. M. Abb, Y. D. Wang, C. H. de Groot, and O. L. Muskens, "Hotspot-mediated ultrafast nonlinear control of multifrequency plasmonic nanoantennas," *Nat. Commun.* **5**, 8, 4869 (2014).

131. P. J. Guo, R. D. Schaller, J. B. Ketterson, and R. P. H. Chang, "Ultrafast switching of tunable infrared plasmons in indium tin oxide nanorod arrays with large absolute amplitude," *Nat. Photonics* **10**, 267 (2016).

132. Z. Gong, C. Li, X. Hu, H. Yang, and Q. Gong, "Active control of highly efficient thirdharmonic generation in ultrathin nonlinear metasurfaces," *Opt. Mater.* **60**, 552–558 (2016).

133. E. Almeida, O. Bitton, and Y. Prior, "Nonlinear metamaterials for holography," *Nat. Commun.* **7**, 7, 12533 (2016).

134. W. M. Ye, F. Zeuner, X. Li, B. Reineke, S. He, C. W. Qiu, J. Liu, Y. T. Wang, S. Zhang, and T. Zentgraf, "Spin and wavelength multiplexed nonlinear metasurface holography," *Nat. Commun.* **7**, 7, 11930 (2016).

135. F. Walter, G. X. Li, C. Meier, S. Zhang, and T. Zentgraf, "Ultrathin nonlinear metasurface for optical image encoding," *Nano Lett.* **17**, 3171–3175 (2017).

136. M. Z. Alam, I. De Leon, and R. W. Boyd, "Large optical nonlinearity of indium tin oxide in its epsilon-near-zero region," *Science* **352**, 795–797 (2016).

137. N. Kinsey, C. DeVault, J. Kim, M. Ferrera, V. M. Shalaev, and A. Boltasseva, "Epsilon-near-zero Al-doped ZnO for ultrafast switching at telecom wavelengths," *Optica* **2**, 616–622 (2015).

138. R. M. Cordova-Castro, M. Casavola, M. van Schilfgaarde, A. V. Krasavin, M. A. Green, D. Richards, and A. V. Zayats, "Anisotropic plasmonic CuS nanocrystals as a natural electronic material with hyperbolic optical dispersion," *ACS Nano* **13**, 6550–6560 (2019).

139. D. A. B. Miller, "Are optical transistors the logical next step?" *Nat. Photonics* **4**, 3–5 (2010).

5 Quantum Nonlinear Metasurfaces

Alexander N. Poddubny, Dragomir N. Neshev, and Andrey A. Sukhorukov

CONTENTS

5.1 INTRODUCTION

The recent advent of optical metasurfaces, ultra-thin structures composed of nanoresonators facilitating strong light concentration [1–3], opens the path toward efficient frequency conversion [4–7] in a material thousand times thinner than a human hair [8]. There is a remarkable capacity to individually shape each nanoresonator of the metasurface to spatially and spectrally control the conversion process with unprecedented nanoscale resolution, facilitating an ultimate flexibility to selectively convert, focus, and image different colors with a single metasurface. The strong enhancement of the nonlinear processes in dielectric nanoresonators is largely due to the absence of material absorption and the excitation of Mie-type bulk resonances [2]. The highest conversion efficiency to date has been achieved employing III–V semiconductor nanostructures, such as AlGaAs which is a non-centrosymmetric material with high quadratic nonlinear susceptibility. In particular, second-harmonic generation efficiencies up to 10^{-4} have been recently demonstrated [8–12], six orders of magnitude higher than in plasmonics. Such capabilities can make an immense fundamental and practical impact, including the quantum state generation in nonlinear metasurfaces as we discuss in this chapter.

The quantum state of correlated photon pairs is the essential building block for photon entanglement [13,14], which underpins many quantum applications, including secure networks, enhanced measurement and lithography, and quantum information processing [15]. One of the most versatile techniques for the generation of correlated photons is the process of spontaneous parametric down-conversion (SPDC) [16,17], see Figure 5.1. It allows for an arbitrary choice of energy and momentum correlations between the generated photons, robust operation at room temperature, as well as for spatial and temporal coherence between simultaneously pumped multiple SPDC sources. Alternative approaches based on atom-like single-photon emitters, such as solid-state fluorescent atomic defects [18], quantum dots [19,20], and 2D host materials [21,22], have reached a high degree of frequency indistinguishability, purity and brightness [19,20]. However, this comes with the expense of operation at cryogenic temperatures and lack of spatial coherence between multiple quantum emitters. These features might limit possible applications and reduce the potential for device scalability. Furthermore, the small size of the atomic sources often requires complex schemes aimed at coupling to optical nanoantennas and improving the photon extraction efficiency [21].

The miniaturization of SPDC quantum light sources to micro and nanoscale dimensions is a continuing quest, as it enables denser integration of functional quantum devices. Traditionally, bulky cm-sized crystals were utilized for SPDC, entailing the difficulty of aligning multiple optical elements after the SPDC crystal, while offering relatively low photon-pair rates [17]. As a first step of miniaturization, SPDC was realized in low-index-contrast waveguides, which allowed confining light down to several square micrometers transversely to the propagation direction, significantly enhancing the conversion efficiency [23]. However, this approach still

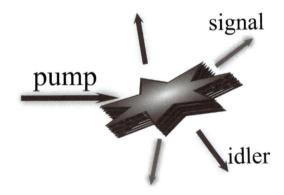

FIGURE 5.1 General concept of biphoton generation in an SPDC process. (After Poddubny, A.N. et al., *Phys. Rev. Lett.*, 117, 123901, 2016.)

requires centimeters of propagation length, which makes the on-chip integration with other elements challenging [24]. The introduction of high-index contrast wave-guides and ring resonators allowed for shrinking the sizes necessary for SPDC to millimeters [25], and to tens of micrometers [26]. However, further miniaturization down to the metasurfaces composed of nanoresonators requires conceptually different approaches.

The generation of quantum light with nonlinear nanoresonators, acting both as sources of quantum states and nanoantennas shaping the emitted photons, has only been reported last year [27]. Such nanoscale multi-photon quantum sources offer an unexplored avenue for applications of highly indistinguishable and spatially reconfigurable quantum states, through the spatial multiplexing of coupled nanoantennas on metasurfaces.

The chapter is organized as follows. First, in Section 5.2 we outline a general quantum theory of spontaneous photon-pair generation in arbitrary nonlinear photonic structures, including nanoresonators and metasurfaces, which provides an explicit analytical solution for the photon state expressed through the classical Green function. In the following Section 5.3 we formulate the correspondence between the quantum photon-pair generation and classical sum-frequency process in nonlinear media, and discuss its application in various contexts, including waveguide circuits and nanostructures. Then, in Section 5.4 we present the first experimental results demonstrating photon-pair generation in a single nonlinear nanoantenna. Finally, in Section 5.5 we present conclusions and outlook toward the generation of quantum entangled images with nonlinear metasurfaces and emerging opportunities for applications.

5.2 GREEN FUNCTION THEORY

In this section we outline the general theoretical approach to for two-photon generation via spontaneous parametric down-conversion (SPDC) and spontaneous four-wave mixing (SFWM). Section 5.2.1 presents the derivation of the two-photon wavefunction. In Section 5.2.2 we discuss the photon heralding efficiency. In Section 5.2.3 we show, how our results can be reduced to those known in literature for the specific case of coupled dispersionless waveguides. Finally, in Section 5.2.4 we consider the generation of entangled plasmon-photon pairs.

5.2.1 TWO-PHOTON WAVEFUNCTION

The photon-pair generation is described by the Hamiltonian [29]

$$H_{NL} = \frac{1}{2} \int \frac{d\omega_1 d\omega_2}{(2\pi)^2} \int d^3r E_\alpha^\dagger(\omega_1, r) E_\beta^\dagger(\omega_2, r) \Gamma_{\alpha\beta}(r) + \text{H.c.} \qquad (5.1)$$

where E is the electric field operator, $\alpha, \beta = x, y, z$, and $\Gamma_{\alpha\beta}$ is the generation matrix. We consider two possibilites [30], spontaneous parametric down-conversion (SPDC) due to $\chi^{(2)}$ nonlinear susceptibility and spontaneous four-wave mixing (SFWM) governed by $\chi^{(3)}$ nonlinearity, when

$$\Gamma_{\alpha\beta}(\mathbf{r}) = \begin{cases} \chi^{(2)}_{\alpha\beta\gamma}(\mathbf{r};\omega_1,\omega_2;\omega_p)\varepsilon_{p,\gamma}(\mathbf{r})e^{-i\omega_p t}, \\\\ \chi^{(3)}_{\alpha\beta\gamma\delta}(\mathbf{r};\omega_1,\omega_2;\omega_p,\omega_p)\varepsilon_{p,\gamma}(\mathbf{r})\varepsilon_{p,\delta}(\mathbf{r})e^{-2i\omega_p t}, \end{cases} \tag{5.2}$$

Here, ε_p is the classical pump at frequency ω_p, and $\gamma, \delta = x, y, z$ are the Cartesian indices.

We explicitly introduce the *sensors that detect the quantum electromagnetic field* [31] to find the experimentally measurable quantities. The sensors are modeled as signal (s) and idler (i) two-level systems with the Hamiltonians

$$H_{i,s} \equiv H_{i,s}^{(0)} + V_{i,s} = \hbar\omega_{i,s}a_{i,s}^{\dagger}a_{i,s} - \hat{\mathbf{d}}_{i,s}\cdot\mathbf{E}(\mathbf{r}_{i,s}), \tag{5.3}$$

with the resonant energies $\hbar\omega_s$ and $\hbar\omega_i$, respectively. Here, $a_{s,i}^{\dagger}$ are the corresponding exciton creation and $\hat{\mathbf{d}}_{i,s} = a\mathbf{d}_{i,s}^* + a^{\dagger}\mathbf{d}_{i,s}^*$ are the dipole momentum operators. The detected two-quantum state is $|\Psi\rangle = a_i^{\dagger}a_s^{\dagger}|0\rangle$ with both detectors excited by the photon-pair.

The direct approach to describe the two-photon generation would be to expand the electric field operator over the set of eigenmodes of the linear problem, see the left part Figure 5.2. However, this technique is impractical for nanostructured meta-materials. The eigenmodes are poorly defined due to the presence of ohmic and radiative losses as well as the dispersion. Instead, we describe the linear electromagnetic problem by the classical Green tensor,

$$\left[\text{rot rot} - \left(\frac{\omega}{c}\right)^2 \varepsilon(\omega,\mathbf{r}) \right] G(\mathbf{r},\mathbf{r}',\omega) = 4\pi \left(\frac{\omega}{c}\right)^2 \hat{1}\delta(\mathbf{r}-\mathbf{r}'), \tag{5.4}$$

that explicitly accounts for arbitrary strong ohmic losses and mode dispersion. The *classical* Green function allows to circumvent the calculation of eigenmodes in macroscopic *quantum* electrodynamics. The Green function method was previously

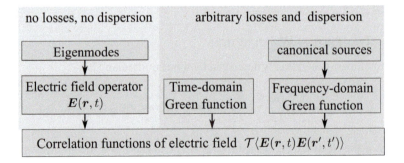

FIGURE 5.2   Scheme of different approaches to calculate the two-photon wavefunction.

applied [32] to describe the spontaneous two-photon emission (STPE) [33] from a single atom. However, the current problem is quite distinct from STPE, because nonlinear spontaneous wave-mixing acts as a coherent spatially extended source.

In the Green function method the two-photon wavefunction $|\Psi\rangle$ can be equivalently obtained using the time-dependent perturbation theory or the time-independent one, see middle and right parts in Figure 5.2, respectively. The first approach follows the methodology of the quantum field and condensed matter theories and allows for a more compact derivation provided that the Feynman diagram technique [34] is used. The second one requires only the knowledge of standard quantum mechanics but has to be supplemented with the local source quantization scheme for the electric field [35]. Below we will outline the derivation using both methods and demonstrate their equivalence.

5.2.1.1 Time-Dependent Perturbation Technique

Formally, the process of photon-pair generation, propagation, and detection can be described by the scattering matrix element $S_{is} = \langle \Psi | U | 0 \rangle$, where U is the evolution operator [36],

$$U = T e^{-i \int_{-\infty}^{\infty} W(\tau) d\tau}. \tag{5.5}$$

Here, T denotes the time-ordered product and

$$W(\tau) = e^{iH^{(0)}\tau/\hbar}(V_i + V_s + H_{\mathrm{NL}})e^{-iH^{(0)}\tau/\hbar}, \quad H^{(0)} = H_i^{(0)} + H_s^{(0)} + H_{\mathrm{lin}} \tag{5.6}$$

Is the operator describing the generation and detection in the interaction representation and $H^{(0)}$ is the sum of Hamiltonian of noninteracting photons $H_i^{(0)}$ and the detector Hamiltonians. Our goal is to calculate the scattering matrix element in the lowest nonvanishing order of the time-dependent perturbation theory. Since two photons have to be generated and each of them has to be absorbed, we need to consider the third order process in the operator in Eq. (5.6), given by the expansion

$$S_{is} = \left(\frac{-i}{\hbar}\right)^3 \frac{1}{3!} \int_{-\infty}^{\infty}\int_{-\infty}^{\infty}\int_{-\infty}^{\infty} d\tau_1 d\tau_2 d\tau_3 \, T\langle \Psi | W(\tau_1)W(\tau_2)W(\tau_3)|0\rangle. \tag{5.7}$$

The diagrammatic representation of Eq. (5.7) is shown in Figure 5.3a. Explicitly, the matrix element reads

$$T\langle \Psi | W(\tau_1)W(\tau_2)W(\tau_3)|0\rangle = \frac{1}{2}d_i^* d_s^* e^{i\tau_1\omega_i + i\tau_2\omega_s - Ni\omega_p\tau_3}$$

$$\times \sum_{\alpha\beta}\Gamma_{\alpha\beta}(r_0)\int d^3r_0 \, T\langle E_{\sigma_i}(r_i,\tau_1)E_{\sigma_s}(r_s,\tau_s)E_\alpha(r_0,\tau_3)E_\beta(r_0,\tau_3)\rangle \tag{5.8}$$

$$+ \text{ permutations of } \tau_1,\tau_2,\tau_3,$$

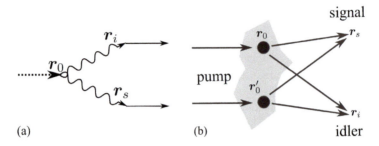

FIGURE 5.3 (a) Feynman diagram for Eq. (5.7), describing generation of two photons in the point r_0 and their detection in the points r_i and r_f. Dotted line corresponds to the classical pump, curved lines describe photon Green functions and solid line are the detectors. (b) Graphic representation of the same process and Eq. (5.12) for the two-photon wave function. (After Poddubny, A.N. et al., *Phys. Rev. Lett.*, 117, 123901, 2016.)

where $N = 1(2)$ for SPDC (SWFM). As usual, the factor $1/3!$ in Eq. (5.7) cancels out with the combinatorial factor arising from six different permutations of the times τ_1, τ_2, τ_3 [34]. Hence, the scattering matrix element is determined only by the correlation function in the second line of Eq. (5.8).

The correlation functions in Eq. (5.8) are calculated by using the equivalent alternative definition of the Green function from Eq. (5.4) in the time domain [34]

$$G_{\alpha\beta}(\boldsymbol{r},\boldsymbol{r}',t-t') = \frac{\mathrm{i}}{\hbar} T\langle E_\alpha(t,\boldsymbol{r})E_\beta(t',\boldsymbol{r}')\rangle. \tag{5.9}$$

The function in Eq. (5.9) is related to the frequency-dependent function in Eq. (5.4) by the Fourier transform $G(t) = \int d\omega e^{-i\omega t} G(\omega)/2\pi$. Once the Green function is introduced, the correlation function in Eq. (5.8) can be calculated by means of the Wick's theorem,

$$T\langle E_{\sigma_i}(\boldsymbol{r}_i,\tau_1)E_{\sigma_f}(\boldsymbol{r}_s,\tau_2)E_\alpha(\boldsymbol{r}_0,\tau_3)E_\beta(\boldsymbol{r}_0,\tau_3)\rangle$$

$$= T\langle E_{\sigma_i}(\boldsymbol{r}_i,\tau_1)E_\alpha(\boldsymbol{r}_0,\tau_3)\rangle\langle E_{\sigma_s}(\boldsymbol{r}_s,\tau_2)E_\beta(\boldsymbol{r}_0,\tau_3)\rangle + \tag{5.10}$$

$$T\langle E_{\sigma_i}(\boldsymbol{r}_i,\tau_1)E_{\sigma_s}(\boldsymbol{r}_s,\tau_2)\rangle\langle E_\alpha(\boldsymbol{r}_0,\tau_3)E_\beta(\boldsymbol{r}_0,\tau_3)\rangle.$$

Using the Green function definition in Eq. (5.9) we find the scattering matrix element in the form

$$S_{is} = -2\pi i\delta(\hbar\omega_i + \hbar\omega_s - N\hbar\omega_{\text{pump}})T_{is} \tag{5.11}$$

and obtain

$$T_{is}(\mathbf{r}_i,\omega_i,\mathbf{d}_i;\mathbf{r}_s,\omega_s,\mathbf{d}_s) = \sum_{\alpha\beta,\sigma_i,\sigma_s} d^*_{i,\sigma_i} d^*_{s,\sigma_s}$$

$$\times \int d^3r_0 G_{\sigma_i\alpha}(\mathbf{r}_i,\mathbf{r}_0,\omega_i) G_{\sigma_s\beta}(\mathbf{r}_s,\mathbf{r}_0,\omega_s)\Gamma_{\alpha\beta}(\mathbf{r}_0). \tag{5.12}$$

By construction the two-photon transition amplitude T_{is} has the meaning of the complex wave function fully defining the pure two-photon state.

Equation (5.12) is the central result of our study. The form of Eq. (5.12) clearly represents the interference between the spatially entangled photons generated in the different points of space $\mathbf{r}_0$ [37], as schematically illustrated in Figure 5.3b. The coincidence rate, which defines simultaneous detection of two photons at different positions in space, is found as:

$$W_{is} = (2\pi/\hbar)\delta(\hbar\omega_i + \hbar\omega_s - N\hbar\omega_p)|T_{is}|^2. \tag{5.13}$$

5.2.1.2 Time-Independent Local Source Quantization Scheme

In this approach instead of Eq. (5.7) in the time domain we write [36] for the two-photon wavefunction

$$T_{is} = \lim_{\varepsilon\to 0}\langle\Psi|V\frac{1}{N\hbar\omega_p - H_0 + i\varepsilon}V\frac{1}{N\hbar\omega_p - H_0 + i\varepsilon}V|0\rangle, \tag{5.14}$$

In the frequency domain, where $N = 1,2$ for SPDC (SFWM), respectively. The Hamiltonian H_{lin}, describing the *linear propagation of the generated photons*, is written in the *local source quantization scheme* [35] as

$$H_{\text{lin}} = \int d^3r \int_0^\infty \sum_{\alpha=x,y,z} d\omega\hbar\omega f^\dagger_\alpha(\mathbf{r},\omega)f_\alpha(\mathbf{r},\omega), \tag{5.15}$$

where $f_\alpha(\mathbf{r},\omega)$ are the canonical bosonic source operators for the quantum electric field [35]:

$$\mathbf{E}(\mathbf{r}) = \int_0^\infty \frac{d\omega}{2\pi}\mathbf{E}(\mathbf{r},\omega) + \text{H.c.}, \tag{5.16}$$

$$\widehat{\mathbf{E}}(\omega) = i\sqrt{\hbar}\int d^3r' G_{\alpha\beta}(\mathbf{r},\mathbf{r}',\omega)\sqrt{\text{Im}\varepsilon(\omega,\mathbf{r}')}f_\beta(\mathbf{r}',\omega).$$

Now we directly substitute Eq. (5.16) into Eq. (5.14). The averaging yields the two products of the Green functions, that are evaluated using the identity [35]

$$\text{Im}G_{\beta\beta'}(\mathbf{r},\mathbf{r}'') = \frac{1}{4\pi}\int d^3r' \text{Im}[\varepsilon(\mathbf{r}')]G_{\beta\alpha}(\mathbf{r},\mathbf{r}')G^*_{\beta'\alpha'}(\mathbf{r}'',\mathbf{r}'), \tag{5.17}$$

And the integration over frequency ω' is performed using the Kramers-Kronig relations,

$$\lim_{\varepsilon \to 0} \int_0^\infty \frac{d\omega'}{\pi} \mathrm{Im} G_{\alpha\beta}(\boldsymbol{r},\boldsymbol{r}') \left(\frac{1}{\omega' - \omega_s - \mathrm{i}\varepsilon} + \frac{1}{\omega_s + \omega' - \mathrm{i}\varepsilon} \right) = G_{\alpha\beta}(\boldsymbol{r},\boldsymbol{r}'). \qquad (5.18)$$

The final result yields again Eq. (5.12) in full agreement with the time-dependent technique. The time-dependent approach relies directly on the Green function from Eq. (5.9) while the time-independent one is built on the electric field expansion in Eq. (5.16). Their equivalence can be verified by substituting Eq. (5.16) into Eq. (5.9).

5.2.2 HERALDING EFFICIENCY

The role of ohmic and radiative losses becomes especially important for the two-photon generation process. After an entangled photon pair is generated, a single absorption event is sufficient to fully destroy it, and leave only one photon in a mixed state. Hence, the problem of detection of individual photons in a lossy system where entangled photon pairs are generated, provides an interesting question. Given that a signal photon is detected, did it correspond to the photon-pair where the idler photon has been emitted or to the pair, where it has been absorbed? The relative magnitude of the first term provides the heralding efficiency, characterizing the robustness of the two-photon generation setup to the ohmic losses.

In order to evaluate the heralding efficiency we start with the calculation of the total single-photon count rate. Similarly to the two-photon wavefunction, the rate can be rigorously obtained using the time-dependent perturbation theory in the framework of the Keldysh diagram technique or using time-independent approach with the canonical local sources [28]. Another equivalent and very compact derivation, given below, is based on the semiclassical expression for the signal polarization combined with the fluctuation-dissipation theorem. We use the standard result of the quantum photodetector theory for the signal photon count rate [38,39].

$$W_s(\boldsymbol{r}_s,\sigma_s) = \frac{1}{\hbar} \sum_{\sigma_s \sigma_{s'}} d_{s\sigma_s} d_{s',\sigma_{s'}}^* \int_{-\infty}^\infty \langle E_{\sigma_s}(\boldsymbol{r}_s,\tau) E_{\sigma_{s'}}(\boldsymbol{r}_s,0) \rangle e^{\mathrm{i}\omega_s \tau} d\tau, \qquad (5.19)$$

where the averaging is performed over the states of the electromagnetic field E. In order to calculate the field of signal photons, acting upon the detector, we introduce the dielectric polarization as a variation of the density of the nonlinear Hamiltonian in Eq. (5.1) over the electric field

$$P_\alpha(\boldsymbol{r}_0,t) = -\frac{\delta H_{\mathrm{NL}}}{\delta E_s} = -\Gamma_{\alpha\beta}(\boldsymbol{r}_0) E_{i,\beta}(\boldsymbol{r}_0,t). \qquad (5.20)$$

Next, we determine the electric field induced by this polarization

$$E_{\sigma_s}(\mathbf{r}_s,t) = \int_{-\infty}^{t} dt' G_{\sigma_s\alpha}(\mathbf{r}_s,\mathbf{r}_0,t-t')P_\alpha(\mathbf{r}_0,t'),$$ (5.21)

where G is the time-dependent retarded Green function. Using Eqs. (5.20), (5.21) and the fluctuation-dissipation theorem [40] for the electric field components in the form [34]

$$\langle E_\beta(\mathbf{r},t)E_{\beta'}(\mathbf{r}',t')\rangle = \int_{-\infty}^{\infty} \frac{d\omega}{2\pi} \hbar \mathrm{Im} G^R_{\beta\beta'}(\omega,\mathbf{r}_1,\mathbf{r}_2) \mathrm{sign}\omega e^{-i\omega(t-t')}$$ (5.22)

we obtain the signal photon count rate

$$W_s(\mathbf{r}_s) = \frac{2}{\hbar} \iint d^3 r_0' d^3 r_0'' \sum_{\sigma_s \sigma_{s'}} d_{s,\sigma_s} d^*_{s,\sigma_{s'}} \mathrm{Im} G_{\beta\beta'}(\mathbf{r}_0',\mathbf{r}_0'',\omega_p - \omega_s)$$

(5.23)

$$\times \Gamma_{\alpha\beta}(\mathbf{r}_0')\Gamma^*_{\alpha'\beta'}(\mathbf{r}_0'')G_{\sigma_s,\alpha}(\mathbf{r}_s,\mathbf{r}_0',\omega_s)G^*_{\sigma_{s'},\alpha'}(\mathbf{r}_s,\mathbf{r}_0'',\omega_s).$$

It is instructive to separate the rate of single-photon counts in Eq. (5.23) into two contributions,

$$W_s(\mathbf{r}_s) = W_s^{(\mathrm{rad})}(\mathbf{r}_s) + W_s^{(\mathrm{Ohmic})}(\mathbf{r}_s).$$ (5.24)

The first term corresponds to the photon pairs with the idler photon emitted into the far-field, and the second term describes the pairs with the idler photon absorbed in the medium, see Figure 5.4a and b, respectively. The ratio of the first term to the total single-photon count rate will provide the far-field heralding efficiency. Explicitly, the radiative term reads

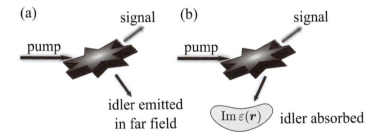

FIGURE 5.4 Illustration of the photon pairs with idler photon (a) emitted and (b) absorbed. The relative value of the pairs (a), corresponding to the first term in Eq. (5.24), provides the heralding efficiency.

$$W_s^{(\mathrm{rad})}(\mathbf{r}_s, \sigma_s) = \frac{c}{2\pi\hbar\omega_i} \sum_{\sigma_s, \sigma_{s'}, \beta} d_{s,\sigma_s} d_{s,\sigma_{s'}}^* \oint dS_{i,\beta} e_{\beta\sigma_i\sigma_{i'}}$$

$$\mathrm{Re}\tilde{T}_{is}(\mathbf{r}_s, \mathbf{r}_i, \sigma_s, \sigma_i)\bar{T}_{is}^*(\mathbf{r}_s, \mathbf{r}_i, \sigma_{s'}, \sigma_{i'}),$$

$$\tilde{T}_{is}(\mathbf{r}_s, \mathbf{r}_i, \sigma_s, \sigma_i) = \int d^3 r_0 G_{\sigma_i\alpha}(\mathbf{r}_i, \mathbf{r}_0, \omega_i) G_{\sigma_s\beta}(\mathbf{r}_s, \mathbf{r}_0, \omega_s) \Gamma_{\alpha\beta}(\mathbf{r}_0), \tag{5.25}$$

$$\bar{T}_{is}(\mathbf{r}_s, \mathbf{r}_i, \sigma_s, \sigma_i) = \frac{c}{i\omega_i} \sum_{\beta\gamma} e_{\sigma_i\beta\gamma} \frac{\partial \tilde{T}(\mathbf{r}_s, \mathbf{r}_i, \sigma_s, \gamma)}{\partial x_{i,\gamma}}, \omega_i = N\omega_p - \omega_s,$$

where $e_{\beta\sigma_i\sigma_{i'}}$ is the Levi-Civita tensor. The second term, corresponding to the pairs where the idler photon has been absorbed, has the form

$$W_s^{(\mathrm{Ohmic})}(\mathbf{r}_s) = \frac{\hbar}{4\pi^2 |d_i|^2} \sum_{\sigma_i} \int d\omega_i \int d\mathbf{r}_i W_{is}(\mathbf{r}_i, \omega_i, d_i; \mathbf{r}_s, \omega_s, \sigma_s) \mathrm{Im}\,\varepsilon(\mathbf{r}_i, \omega_i), \tag{5.26}$$

and can be expressed via the joint two-photon count rate

$$W_{is}(\mathbf{r}_i, \omega_i, d_i; \mathbf{r}_s, \omega_s, \sigma_s) = \frac{2\pi}{\hbar} \delta(\hbar\omega_i + \hbar\omega_s - N\hbar\omega_p) |T_{is}(\mathbf{r}_i, \omega_i, d_i; \mathbf{r}_s, \omega_s, e_{\sigma_s} d_s)|^2. \tag{5.27}$$

The volume integration in Eq. (5.26) is performed only over the lossy region where $\mathrm{Im}\,\varepsilon \neq 0$. If the integral is extended over the whole volume and is regularized at $r \to \infty$ by adding infinitesimal losses to the permittivity as

$$\varepsilon(\mathbf{r}_i) \to \varepsilon(\mathbf{r}_i) + i\Delta\varepsilon, \tag{5.28}$$

the result reduces to the total rate $W_s(\mathbf{r}_s)$ in Eq. (5.23), rather than to $W_s^{(\mathrm{Ohmic})}(\mathbf{r}_s)$:

$$W_s(\mathbf{r}_s) = \frac{\hbar}{4\pi^2 |d_i|^2} \sum_{\sigma_i} \int d\omega_i \lim_{\Delta\varepsilon \to 0} \int_{\mathrm{tot}} d\mathbf{r}_i W_{is}(\mathbf{r}_i, \omega_i, d_i; \mathbf{r}_s, \omega_s, \sigma_s) \mathrm{Im}\,\varepsilon(\mathbf{r}_i, \omega_i). \tag{5.29}$$

By definition, the heralding efficiency is the ratio of the heralded counts, corresponding to the pairs where the idler photon has been emitted into the far-field to the total single-photon rate. As such, this quantity should characterize the generated photon pairs but not the specific detection setup. However, the single-photon count rates in Eqs. (5.23), (5.25), (5.26) and the two-photon count rate in Eq. (5.27) by construction depend on the efficiencies of the detectors, determined by the dipole momentum matrix elements $d_{i,s}$. In order to determine the heralding efficiency, it is necessary to explicitly calibrate the detectors to the flux of the photons. The procedure is outlined in [28] and the calibrated heralding efficiency reads

$$QE = \frac{\tilde{W}_s^{(rad)}}{\tilde{W}_s},$$

(5.30)

$$\tilde{W}_s^{(rad)} = \frac{c}{4\pi\omega_i}\oint dS_{i,\alpha} e_{\alpha\sigma_i\sigma_{i'}} \mathrm{Re}\tilde{T}_{is}(\mathbf{r}_s,\mathbf{r}_i,\sigma_s,\sigma_i)\tilde{T}_{is}^*(\mathbf{r}_s,\mathbf{r}_i,\sigma_{s'},\sigma_{i'}),$$

$$\tilde{W}_s = \iint d^3r_0' d^3r_0'' \mathrm{Im}G_{\beta\beta'}(\mathbf{r}_0',\mathbf{r}_0'',\omega_p-\omega_s)\Gamma_{\alpha\beta}(\mathbf{r}_0')\Gamma_{\alpha'\beta'}^*(\mathbf{r}_0'') \times$$

$$G_{\sigma_s,\alpha}(\mathbf{r}_s,\mathbf{r}_0',\omega_s)G_{\sigma_s,\alpha'}^*(\mathbf{r}_s,\mathbf{r}_0'',\omega_s).$$

In the case of planar geometry, the Green functions in Eq. (5.30) are to be replaced by their Fourier components along x,y and the unit area interval by the unit wave vector interval, $1/dS \rightarrow dk_x dk_y /(2\pi)^2$.

5.2.3   CORRESPONDENCE TO THE RESULTS FOR WAVEGUIDES

Now we consider an example of one-dimensional dispersionless waveguides with weak losses, see Figure 5.5. We will show how our general Green function theory allows one to recover the results known in literature [41,42].

The Green function for the one-dimensional scalar problem reads

$$G(x,x') = \frac{i(\omega/c)^2}{2q}e^{iq|x-x'|}$$

(5.31)

and satisfies the inhomogeneous Helmholtz equation

$$\left(\frac{d^2}{dx^2}+q^2\right)G(x,x') = -4\pi\left(\frac{\omega}{c}\right)^2\delta(x-x').$$

(5.32)

Here, q is the wave propagation constant (photon wave vector), that can be complex for lossy medium. The results in [41] implied (i) weak losses and (ii) one-way propagation. Using these additional assumptions for the Green function in Eq. (5.31) we obtain

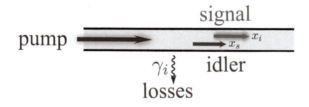

FIGURE 5.5   Illustration of the two-photon generation in a nonlinear lossy waveguide.

$$G(x, x') = ige^{i\beta(x-x')-\gamma(x-x')}\theta(x - x'), \quad g = \frac{(\omega/c)^2}{2\beta}. \tag{5.33}$$

Here, we explicitly introduced the real and imaginary parts of the propagation constant $q = \beta + i\gamma$, neglected the imaginary part of q in the factor g and took into account only the term in the factor $e^{iq|x-x'|} = \theta(x - x')e^{iq(x-x')} + \theta(x' - x)e^{iq(x'-x)}$, corresponding to the waves propagating along the positive direction of the real axis. We present the generation matrix for the SPDC process as

$$\Gamma(x_0) = \frac{\chi}{2}\varepsilon_p e^{-\gamma_p x_0}\theta(x - x_0), \tag{5.34}$$

where without the loss of generality we assumed that the real part of the propagation constant for the pump wave is set to zero. The two-photon wave function determined by Eq. (5.12) then reads

$$T_{is}(x_i, \omega_i; x_s, \omega_s) = -d_i^* d_s^* g_i^* g_s^* \chi \varepsilon_p \times \tag{5.35}$$

$$\int_0^{\min(x_i, x_s)} dx_0 e^{(i\beta_s - \gamma_s)(x_s - x_0)} e^{(i\beta_i - \gamma_i)(x_i - x_0)} e^{-\gamma_p x_0}.$$

For $x_i = x_s$ this answer for $T(x_s, x_s)$ exactly matches the solution in Eq. (5.9) of the differential Eq. (5) in [41]. In order to determine the single-photon count rate we use Eq. (5.29) and take into account that for small losses $\text{Im}\,\varepsilon(\omega_i)$ is proportional to γ_i. Next, we present the integral in Eq. (5.29) as

$$\gamma_i \int_0^\infty d\omega_i \int_0^\infty dx_i W_{is}(x_i, \omega_i; x_s, \omega_s) = \gamma_i \int_0^{x_s} dx_i \,|T_{is}(x_i, x_s)|^2 + \gamma_i \int_{x_s}^\infty dx_i \,|T_{is}(x_i, x_s)|^2, \tag{5.36}$$

where the two-photon wave function in Eq. (5.35) has the form

$$T_{is}(x_i, x_s) = \begin{cases} e^{(i\beta_s - \gamma_s)(x_s - x_i)} T_{is}(x_i, x_i), & (x_i < x_s), \\ e^{(i\beta_i - \gamma_i)(x_i - x_s)} T_{is}(x_s, x_s), & (x_i > x_s). \end{cases} \tag{5.37}$$

Substituting Eq. (5.37) into Eq. (5.36) we find

$$W_s(x_s) \propto 2\gamma_i \int_0^{x_s} dx_i \,|T_{is}(x_i, x_s)|^2 + |T_{is}(x_s, x_s)|^2. \tag{5.38}$$

The first term in Eq. (5.38) exactly corresponds to the term $\tilde{I}_s(z)$ in Eq. (10) of [41], i.e. the contribution from the states, where the idler photon has been absorbed.

The second term in Eq. (5.38) matches the term $I_s^{(0)}$ in Eq. (10) of [41]. It is the contribution of the states where the idler photon has still not been absorbed up to the point x_s.

5.2.4 APPLICATION FOR SURFACE PLASMONS

We now apply our general theory to layered metal-dielectric plasmonic structures. First, we analyze the degenerate spontaneous four-wave mixing for the metallic layer of the thickness $d_{silver} = 20$ nm on top of the nonlinear dielectric, see Figure 5.6a–c. Due to the translational symmetry, the total in-plane momentum **k** of the photons and plasmons is conserved, i.e. $k_{i,\alpha} + k_{s,\alpha} = 2k_{p,\alpha}$ for $\alpha = x, y$. The most interesting situation is realized for oblique pump incidence, giving rise to four different regimes when (a) both signal and idler, (b) only idler, (c) neither signal nor idler, and (d) only signal in-plane wave vectors lie outside the corresponding light cone

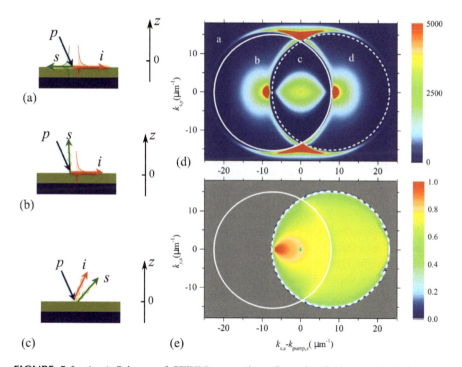

FIGURE 5.6 (a–c) Scheme of SFWM generation of a pair of (a) entangled plasmons, (b) photon entangled with plasmon, and (c) entangled photons in the gold/nonlinear dielectric structure. (d) Color map of the two-photon detection probability $|T(\mathbf{k}_i, \mathbf{k}_s)|^2$ in the reciprocal space vs. the in-plane wave vector components (arb.un.) in TM polarization ($\mathbf{d}_{i,s} \propto \mathbf{k} \times \mathbf{k} \times \hat{\mathbf{z}}$) at $z_i = z_s = 100$ nm. The signal (solid) and idler (dashed) light lines are plotted in white color. The letters a–d mark the near- and far-field signal and idler generation regimes. (e) Efficiency of signal heralding by far-field idler photons, Eq. (5.39). For all plots $\hbar\omega_i \approx \hbar\omega_s \approx \hbar\omega_p \approx 3$ eV, $\varepsilon_{diel} = 2$, $d_{silver} = 20$ nm, pump is TM polarized, $k_{p,x} = 0.5\omega_p/c$. (After Poddubny, A.N. et al., *Phys. Rev. Lett.*, 117, 123901, 2016.)

boundaries $\omega_{i,s}/c$. The first three situations are schematically shown in Figure 5.6a–c. Two-photon generation occurs in case (c), while (b) and (d) correspond to plasmon generation heralded by the far-field photon.

We perform numerical simulations considering isotropic dielectric with electronic $\chi^{(3)}$ nonlinearity tensor as [30]: $\chi_{\alpha\beta\gamma\delta} = \chi_0(\delta_{\alpha\beta}\delta_{\gamma\delta} + \delta_{\alpha\delta}\delta_{\beta\delta} + \delta_{\alpha\gamma}\delta_{\beta\delta})$. We plot the Fourier transform of the two-photon detection amplitude $|T_{is}(\boldsymbol{k}_s, z_i, z_s)|^2$ for $z_i = z_s = 100$ nm above the structure, defined as $T(\boldsymbol{k}_i) = \int dx dy \exp(-ik_x x - ik_y y)T(x, y)$, which characterizes the signal-idler generation efficiency in all different regimes. The relevant Fourier transforms of the Green functions were evaluated analytically following [43]. Silver permittivity has been taken from [44] and includes the losses and dispersion. The overall map of the correlations resembles that for the generation of the polarization-entangled photons from a bulk nonlinear uniaxial crystal [17]: it shows strong maxima at the intersections of the signal and idler light cone boundaries. However, contrary to the bulk, the calculated map reflects the two-quantum correlations of both photons and plasmons. In the region (c) the shown signal can be directly measured from the far-field photon-photon correlations. For the chosen 30° pump incidence angle the bright spot in the region (b) of Figure 5.6d corresponds to the signal photons emitted in the normal direction. The near-field signal in the regions (a), (b), (d) can be recovered by using the grating to outcouple the plasmons to the far-field [45] or with the near-field scanning optical microscopy setup [46].

The bright spot in the map Figure 5.6d for $k_{s,x} - k_{p,x} \approx 10\,\mu\text{m}^{-1}$ reveals the resonantly enhanced plasmonic emission heralded by the normally propagating idler photons. The heralding efficiency in Eq. (5.16), adopted for the planar geometry, reads

$$QE = \sum_{z_i=-L,L d_i=\hat{x},\hat{y},\hat{z}} \sum \frac{c\cos\theta_i}{2\pi\hbar\omega_i} \frac{|T_{is}(\boldsymbol{k}_s, z_s, z_i, \boldsymbol{d}_i)|^2}{W_s(\boldsymbol{k}_s)}, \qquad (5.39)$$

where $\cos\theta_i = \sqrt{1-(ck_i/\omega_i)^2}$. The summation over z_i in Eq. (5.39) accounts for the total idler photon flux through the surfaces $z_i = \pm L$ above and below the nonlinear structure. The calculated values of the signal heralding, shown in Figure 5.6e, are remarkably high. They reach almost 100% in the case when both signal and idler photon are in the far-field, see the bright spot at $k_{s,x} - k_{p,x} \approx -5\,\mu\text{m}^{-1}$. In the case of signal plasmons the heralding efficiency is uniform and about 70%. We note that the results in Figure 5.6e correspond to the internal heralding efficiency, calculated for the plane pump wave. The external quantum heralding efficiency has to account also for the plasmonic losses due to the propagation from the pump spot to the near-field detector, which can be optimized in the actual experimental setup.

5.3 QUANTUM-CLASSICAL CORRESPONDENCE

The characterization of the two-photon state generated by a structured nonlinear system is a hard experimental task [47–50]. It requires a number of measurements and resources that increase quadratically with system size, making the characterization of states generated by large waveguide circuits impractical or very time consuming. Additionally, integrated circuits are typically patterned on wafers and produced in large numbers, and efficient techniques for fast quality checking of device performance are urgently needed. Here we present an efficient method for the characterization of two-photon states generated from arbitrary optical structures with quadratic nonlinearity that has both fundamental and practical importance for the development of future integrated quantum photonics technologies.

A practical approach for predicting the biphoton state produced from a nonlinear device using only classical detectors and laser sources was proposed in [51] based on the concept of stimulated emission tomography (SET). This technique takes advantage of the analogy between a spontaneous nonlinear process and its classical stimulated counterpart, i.e. difference-frequency generation or stimulated four-wave mixing. It was applied experimentally for spectral characterization of two-photon states with an accuracy unobtainable with single-photon detection methods [52–55], and fast reconstruction of the density matrix of entangled photon sources [56,57].

However, for large optical networks SET becomes a challenging task as discussed in [58], which prevented its experimental realizations for multi-mode path entangled states. The reason is that one would need to precisely inject the seed beam in the individual supermodes supported by the structure. A possible workaround is to inject the seed beam in each single channel individually and to perform a transform through supermode decomposition to obtain quantum predictions. In either case, one requires complete knowledge of the linear light dynamics inside the whole structure, making SET a multi-step procedure prone to errors and not applicable to "black-box" circuits. Additionally, the analogy between a spontaneous nonlinear process and its stimulated counterpart is strictly valid only in the limit of zero propagation losses [59]. This assumption poses a fundamental limit for the characterization of integrated waveguide circuits. Indeed, the effect of scattering losses, due to impurities or surface and side-wall roughness, becomes prominent with increasing miniaturization of photonic devices. Sum-frequency generation (SFG), the reverse process of SPDC, was identified in [59] as the ideal approach for characterizing second-order nonlinear circuits in presence of losses. Nevertheless, the method was formulated only for a single and homogeneous waveguide, posing a stringent restriction for the characterization of more complex devices.

Here, we overcome the limitations of the previous proposals by establishing a rigorous equivalence between the biphoton wavefunction in the undepleted pump regime and the sum-frequency field generated by classical wave-mixing in the reverse direction of SPDC. Our theory analysis is generalized by the use of the Green function method [28,60], and holds for arbitrarily complex second-order nonlinear circuits and in presence of propagation losses. More importantly, the SFG-SPDC analogy can be expressed in any measurement basis, providing a simple experimental tool for the characterization of any "black-box" $\chi^{(2)}$-nonlinear process, see Figure 5.7.

5.3.1 Quantum SPDC-SFG Reciprocity Relationship

We start with proving the general identity for the correspondence between the SPDC process and the sum-frequency generation (SFG) processes in the reversed geometry. This identity will generalize the Lorentz reciprocity theorem in linear electromagnetism [61]

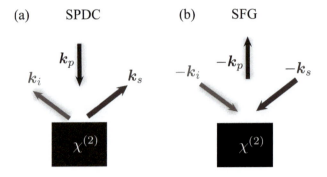

FIGURE 5.7 Schematic illustration of the reciprocal (a) spontaneous parametric down-conversion and (b) sum-frequency generation processes.

$$\int d^3 r\, \boldsymbol{P}_1(\boldsymbol{r}) \cdot \boldsymbol{E}_2(\boldsymbol{r}) = \int d^3 r\, \boldsymbol{P}_2(\boldsymbol{r}) \cdot \boldsymbol{E}_1(\boldsymbol{r}) \tag{5.40}$$

That links the electric field distribution $\boldsymbol{E}_1(\boldsymbol{r})$ and $\boldsymbol{E}_2(\boldsymbol{r})$, induced by the polarization distributions $\boldsymbol{P}_1$ and $\boldsymbol{P}_2$, respectively. The biphoton wavefunction in the SPDC regime, following from Eq. (5.12), reads

$$\Psi(\boldsymbol{r}_s,\boldsymbol{r}_i,\sigma_s,\sigma_i,\omega_s,\omega_i) = \int d^3 r_0 G_{\sigma_s \alpha}(\boldsymbol{r}_s,\boldsymbol{r}_0;\omega_s) G_{\sigma_i \beta}(\boldsymbol{r}_i,\boldsymbol{r}_0;\omega_i) \chi^{(2)}_{\alpha\beta\gamma} E_{p,\gamma}(\boldsymbol{r}_0). \tag{5.41}$$

On the other hand, the nonlinear wave at the sum-frequency $\omega_p = \omega_i + \omega_s$ generated from the waves $\boldsymbol{E}_s(\boldsymbol{r}_i)$, $\boldsymbol{E}_i(\boldsymbol{r}_i)$ in the nonlinear structure can be presented as

$$E_{SFG,\sigma_p}(\boldsymbol{r}_p) = \int d^3 r_0 G_{\sigma_p \gamma}(\boldsymbol{r}_p,\boldsymbol{r}_0) \chi^{(2)}_{\alpha\beta\gamma}(\boldsymbol{r}_0) E_{s,\alpha}(\boldsymbol{r}_0) E_{i,\beta}(\boldsymbol{r}_0). \tag{5.42}$$

Inspired by linear reciprocity relationship Eq. (5.40) we introduce the polarizations $P_{i,s,p}(\boldsymbol{r})$ inducing the correspondent waves $E_{s,i,p}(\boldsymbol{r})$,

$$E_\nu(\boldsymbol{r}) = \int d^3 r_0 \hat{G}(\boldsymbol{r},\boldsymbol{r}_0;\omega_p) P_\nu(\boldsymbol{r}_0), \quad \nu = i,s,p. \tag{5.43}$$

This allows us to rewrite Eq. (5.41) and Eq. (5.42) as

$$\Psi(\boldsymbol{r}_s,\boldsymbol{r}_i,\sigma_s,\sigma_i,\omega_s,\omega_i) = \int d^3 r_0 \int d^3 r_p G_{\sigma_s \alpha}(\boldsymbol{r}_s,\boldsymbol{r}_0) G_{\sigma_i \beta}(\boldsymbol{r}_i,\boldsymbol{r}_0) \chi^{(2)}_{\alpha\beta\gamma}(\boldsymbol{r}_0) \times$$
$$G_{\gamma\sigma_p}(\boldsymbol{r}_0,\boldsymbol{r}_p) P_{p,\sigma_p}(\boldsymbol{r}_p), \tag{5.44}$$

and

$$E_{SFG,\sigma_p}(\mathbf{r}_p;\omega_i + \omega_s) = \int d^3 r_0 \int d^3 r_i \int d^3 r_s G_{\sigma_p\gamma}(\mathbf{r}_p,\mathbf{r}_0)\chi_{\alpha\beta\gamma}^{(2)}(\mathbf{r}_0)\times$$
$$G_{\alpha\sigma_s}(\mathbf{r}_0,\mathbf{r}_s)G_{\alpha\sigma_i}(\mathbf{r}_0,\mathbf{r}_i)E_{s,\sigma_s}(\mathbf{r}_s)E_{i,\sigma_i}(\mathbf{r}_i). \tag{5.45}$$

We have omitted the frequency arguments in the Green functions for the sake of brevity. In the reciprocal structure the Green functions satisfy the reciprocity property

$$G_{\alpha\beta}(\mathbf{r}_1,\mathbf{r}_2) = G_{\beta\alpha}(\mathbf{r}_2,\mathbf{r}_1), \tag{5.46}$$

that is equivalent to Eq. (5.40). Comparing Eqs. (5.44) and (5.45) with the help of Eq. (5.46), we establish the general reciprocity relationship between SPDC and SFG processes in the form

$$\iint d^3 r_i d^3 r_s \Psi(\mathbf{r}_s,\mathbf{r}_i,\sigma_s,\sigma_i,\omega_s,\omega_i)P_{i,\sigma_i}(\mathbf{r}_i)P_{s,\sigma_s}(\mathbf{r}_i)$$
$$= \int d^3 r_p E_{SFG,\gamma}(\mathbf{r}_p;\omega_i + \omega_s)P_{p,\gamma}(\mathbf{r}_p). \tag{5.47}$$

5.3.2 SPDC-SFG Correspondence for a Localized Nonlinear Source

Now we apply the general SPDC-SFG reciprocity relationship Eq. (5.47) to the case of photon-pair generation from a localized nonlinear source. We start with calculating electric field of the structure illuminated by two plane waves, the "idler" one with the wave vector $-\mathbf{k}_i$ and the "signal" one with the wave vector $-\mathbf{k}_s$, see Figure 5.7b. To this end we assume that both waves are generated by point dipoles located in the far-field zone in the points $\mathbf{r}_{i,s} \parallel \mathbf{k}_{i,s}$, and having the unit amplitudes $\mathbf{d}_i^* \perp \mathbf{k}_i$ and $\mathbf{d}_s^* \perp \mathbf{k}_s$, respectively. In this case the linear fields at idler and signal frequencies can be written as

$$\mathbf{E}_{i,s}(\mathbf{r}) = G(\mathbf{r},\mathbf{r}_{i,s};\omega_{i,s})\mathbf{d}_{i,s}^* \tag{5.48}$$

where G is the electromagnetic Green function tensor satisfying the equation

$$\text{rot rot}\, G(\mathbf{r},\mathbf{r}';\omega) = \left(\frac{\omega}{c}\right)^2 \varepsilon(\mathbf{r})G(\mathbf{r},\mathbf{r}';\omega) + 4\pi\left(\frac{\omega}{c}\right)^2 \delta(\mathbf{r}-\mathbf{r}'). \tag{5.49}$$

Far away from the nonlinear structure, where $\varepsilon = 1$, the Green function reduces to the free Green function

$$G_{\alpha\beta} = \left[\left(\frac{\omega}{c}\right)^2 + \frac{\partial^2}{\partial x_\alpha \partial x_\beta}\right]\frac{e^{i\omega|r-r'|/c}}{|\mathbf{r}-\mathbf{r}'|}. \tag{5.50}$$

The locally-plane idler and signal waves, incident upon the structure, are found by substituting Eq. (5.50) into Eq. (5.48) and assuming $r_{i,s} \gg r$. Hence, we find

$$E_{i,s}^{(0)} = \frac{e^{i\omega_{i,s}r_{i,s}/c}}{r_{i,s}} q_{i,s}^2 d_{i,s}^*,$$

(5.51)

where $q_{i,s} = \omega_{i,s}/c$. The corresponding time-averaged fluxes are found as

$$\Phi_{i,s} = \frac{c}{2\pi} |E_{i,s}^{(0)}|^2 = \frac{c}{2\pi} \frac{q_{i,s}^4}{r_{i,s}^2} |d_{i,s}|^2.$$

(5.52)

The nonlinear SFG field is found as a convolution of the $\chi^{(2)}$ susceptibility, the incident fields, and the Green function at the sum-frequency:

$$E_\alpha^{(SFG)}(\mathbf{r}_p \leftarrow \mathbf{r}_i, \mathbf{d}_i^*; \mathbf{r}_s, \mathbf{d}_s^*)$$

(5.53)

$$= \int d^3r' G_{\alpha\beta}(\mathbf{r}_p, \mathbf{r}') \chi_{\gamma\delta,\beta}^{(2)} G_{\gamma\nu}(\mathbf{r}', \mathbf{r}_i) G_{\delta\mu}(\mathbf{r}', \mathbf{r}_s) d_{i,\nu}^* d_{s,\mu}^*.$$

Here, $\alpha, \beta, \gamma, \delta, \mu, \nu$ are the Cartesian indices and the frequency arguments of the Green functions are omitted for the sake of brevity. It is instructive to present the Green functions in the following way

$$G_{\alpha\beta}(\mathbf{r}, \mathbf{r}') = q^2 \frac{e^{iqr}}{r} g_{\alpha\beta}\left(\frac{\mathbf{r}}{r}, \mathbf{r}'\right) \text{for } r \gg c/\omega, r \gg r',$$

(5.54)

where the dimensionless scattering amplitudes $g_{\alpha\beta}\left(\frac{\mathbf{r}}{r}, \mathbf{r}'\right) \equiv g_{\alpha\beta}(\mathbf{k}, \mathbf{r}')$ describe the conversion between the near-field at the point $\mathbf{r}'$ and the plane wave propagating in the direction $\mathbf{r}/r \parallel \mathbf{k}$. We also use the Lorentz reciprocity property

$$G_{\alpha\beta}(\mathbf{r}, \mathbf{r}') = G_{\beta\alpha}(\mathbf{r}', \mathbf{r})$$

(5.55)

Applying Eqs. (5.54) and (5.55) to Eq. (5.53) we rewrite the sum-frequency wave as

$$E_\alpha^{(SFG)}(\mathbf{r}_p \leftarrow \mathbf{r}_i, \mathbf{d}_i^*; \mathbf{r}_s, \mathbf{d}_s^*) =$$

(5.56)

$$\frac{q_i^2 q_s^2 q_p^2}{r_i r_s r_p} e^{i(q_i r_i + q_s r_s + q_p r_p)} \int d^3r' g_{\alpha\beta}(\mathbf{k}_p, \mathbf{r}') \chi_{\gamma\delta,\beta}^{(2)} g_{\nu\gamma}(\mathbf{k}_i, \mathbf{r}') g_{\mu\delta}(\mathbf{k}_s, \mathbf{r}') d_{i,\nu}^* d_{s,\mu}^*.$$

Now we introduce the differential SFG efficiency as

$$d\Xi^{SFG}(-\mathbf{k}_i, \mathbf{e}_i^*; -\mathbf{k}_s, \mathbf{e}_s^* \rightarrow -\mathbf{k}_p, \mathbf{e}_p^*) = r_p^2 d\Omega_p \frac{\Phi_p(-\mathbf{k}_p, \mathbf{e}_p^*)}{\Phi_i(-\mathbf{k}_i, \mathbf{e}_i^*) \Phi_s(-\mathbf{k}_s, \mathbf{e}_s^*)},$$

(5.57)

where $e_{s,i} = d_{s,i}/|d_{s,i}|$. The quantity in Eq. (5.57) represents the ratio of the power of SFG photons propagating inside the solid angle $d\Omega_p$ in the direction $-k_i$ to the energy fluxes of incoming signal and idler plane waves Φ_i and Φ_s. Eq. (5.57) bears analogies with the scattering cross section in the linear problem. The value of SFG efficiency is found from Eqs. (5.53) to (5.52) as

$$\frac{d\Xi^{SFG}(-k_i,e_i^*;-k_se_s^* \to -k_p,e_p^*)}{d\Omega_p} = \frac{2\pi q_p^4}{c} \times$$

(5.58)

$$\left| \int d^3 r' e_{p,\alpha} g_{\alpha\beta}(k_p,r')\chi^{(2)}_{\gamma\delta,\beta} g_{\nu\gamma}(k_i,r')g_{\mu\delta}(k_s,r')e_{i,\nu}^* e_{s,\mu}^* \right|^2 .$$

Now we proceed to the SPDC process. The complex wavefunction of a photon-pair, generated in a $\chi^{(2)}$-nonlinear structure, has the amplitude [28]

$$T(r_s\mu,r_i\nu \leftarrow r_pe_p) = \int d^3 r' G_{\sigma_s\nu}(r_s,r')G_{\sigma_i\nu}(r_i,r')\chi^{(2)}_{\gamma\delta,\beta}(r')E_{p,\beta}(r',\omega_p), \quad (5.59)$$

where $r_s(r_i)$ and $\mu(\nu)$ are signal (idler) photon coordinates and polarizations, respectively, and E_p is the electric field of the pump with the frequency ω_p. Similar to Eq. (5.48) in the SFG case we now write that the pump wave is generated by the point far-field source,

$$E_p(r) = G(r,r_p)d_p \qquad (5.60)$$

where $d_p \parallel e_p \perp r_p$ is the point dipole amplitude. Plugging in the Green function asymptotic expressions from Eq. (5.54), we rewrite the biphoton wavefunction in the form

$$T(r_s\mu,r_i\nu \leftarrow r_pe_p) = \frac{q_i^2 q_s^2 q_p^2}{r_i r_s r_p} e^{i(q_i r_i + q_s r_s + q_p r_p)}$$

(5.61)

$$\times \int d^3 r' g_{\alpha\beta}(k_p,r')\chi^{(2)}_{\gamma\delta,\beta} g_{\nu\gamma}(k_i,r')g_{\mu\delta}(k_s,r')d_{p,\alpha}.$$

Comparing Eq. (5.61) with Eq. (5.56) we find the identity

$$T(r_s\mu,r_i\nu \leftarrow r_pe_p)d_{s,\mu}^* d_{i,\nu}^* = E_\alpha^{(SFG)}(r_p \leftarrow r_i,d_i^*;r_s,d_s^*)d_{p,\alpha}, \qquad (5.62)$$

which proves the SPDC-SFG correspondence for the localized $\chi^{(2)}$-nonlinear source.

We are also interested in comparing the experimentally accessible quantities, namely, the two photon-pair generation rate and the SFG generation efficiency in Eq. (5.58). In order to determine the photon-pair generation rate, we need to calibrate the photon detection process [28,60]. To this end we explicitly introduce the signal and idler detectors modeled as the two-level systems with the dipole momenta matrix elements $\boldsymbol{d}_i$, $\boldsymbol{d}_s$ and the energies $\hbar\omega_i$, $\hbar\omega_s$. The number of photons absorbed by the detector per unit time is given by

$$\frac{dN_{\text{abs}}}{dt} = \frac{2\pi}{\hbar}\delta(\hbar\omega - \hbar\omega_{i,s})\,|\,\boldsymbol{d}\cdot\boldsymbol{E}_{i,s}\,|^2, \tag{5.63}$$

where $\boldsymbol{E}_{i,s} \propto 1/r_{i,s}$ is the local electric field of emitted signal or idler photon at the corresponding detector. The detector quantum efficiency $dQE_{i,s}/d\Omega_{i,s}$ is the ratio between the number of photons absorbed by the detector and the number of photons $dN_{i,s}/dt$ propagating inside the solid angle $d\Omega_{i,s}$ per unit time,

$$\frac{dN_{i,s}}{dt} = r^2 d\Omega_{i,s}\frac{c}{2\pi\hbar\omega_{i,s}}\,|\,E\,|^2, \tag{5.64}$$

$$\frac{dQE_{i,s}}{d\Omega_{i,s}} = \frac{dN_{\text{abs}}}{dN_{i,s}} = \frac{4\pi\omega\,|\,d_{i,s}\,|^2}{\hbar c}\frac{1}{r_{i,s}^2}. \tag{5.65}$$

The two-photon generation rate per unit of the signal and idler spectra is formally defined as

$$\frac{dN_{\text{pair}}}{dt\,d\omega_i\,d\omega_s\,d\Omega_i\,d\Omega_s} = \frac{W_{is}}{dQE_i\,dQE_s} \tag{5.66}$$

where

$$W_{is} = \frac{2\pi}{\hbar}\delta(\hbar\omega_p - \hbar\omega_i - \hbar\omega_s)\,|\sum_{\nu\mu} d_{i,\nu}^* d_{s,\mu}^* T(r_s\mu, r_i\nu \leftarrow r_p e_p)\,|^2, \tag{5.67}$$

is the uncalibrated rate of two photon counts calculated from the biphoton amplitude Eq. (5.61). Substituting Eqs. (5.61), (5.65), and (5.66) into Eq. (5.67) and comparing with Eq. (5.58) we find a general *absolute* correspondence between the sum-frequency rate and the photon-pair generation rate in the form

$$\frac{1}{\Phi_p}\frac{dN_{\text{pair}}(\boldsymbol{k}_i, \boldsymbol{e}_i; \boldsymbol{k}_s \boldsymbol{e}_s \leftarrow \boldsymbol{k}_p, \boldsymbol{e}_p)}{dt\,d\Omega_i\,d\Omega_s\,d\omega_i\,d\omega_s} =$$

$$\frac{\delta(\omega_i + \omega_s - \omega_p)}{2\pi}\frac{\lambda_p^4}{\lambda_i^3\lambda_s^3}\frac{d\Xi^{SFG}(-\boldsymbol{k}_i, \boldsymbol{e}_i^*; -\boldsymbol{k}_s\boldsymbol{e}_s^* \to -\boldsymbol{k}_p, \boldsymbol{e}_p^*)}{d\Omega_p}. \tag{5.68}$$

Eq. (5.68) is the main result for the SPDC-SFG correspondence, valid for an arbitrary localized $\chi^{(2)}$-nonlinear system. In order to facilitate comparison with actual experimental setup, we first integrate it over the signal and idler frequencies and obtain

$$\frac{1}{\Phi_p} \frac{dN_{\text{pair}}(\boldsymbol{k}_i, \boldsymbol{e}_i; \boldsymbol{k}_s \boldsymbol{e}_s \leftarrow \boldsymbol{k}_p, \boldsymbol{e}_p)}{dt d\Omega_i d\Omega_s} = \frac{\Delta\omega_s}{2\pi} \frac{\lambda_p^4}{\lambda_i^3 \lambda_s^3} \frac{d\Xi^{SFG}(-\boldsymbol{k}_i, \boldsymbol{e}_i^*; -\boldsymbol{k}_s \boldsymbol{e}_s^* \rightarrow -\boldsymbol{k}_p, \boldsymbol{e}_p^*)}{d\Omega_p},$$

$$\tag{5.69}$$

where $\Delta\omega_s \equiv 2\pi c \Delta\lambda_s / \lambda_s^2$ is the signal spectral width.

5.3.3 Experimental Demonstration for a Coupled Waveguide System

Here we overview the experiment on observation of SPDC-SFG correspondence made in Ref. [60] for the system with coupled nonlinear waveguides.

The SFG-SPDC characterization protocol was realized for an array of coupled nonlinear waveguides, representing a practical example of complex multidimensional system. The measurement schemes for SPDC and SFG are shown in Figure 5.8a and b, respectively. The device is made of three evanescently coupled waveguides, fabricated on a z-cut lithium niobate substrate by the use of the reverse proton exchange technique [62,63]. The three waveguides have an inhomogeneous and asymmetric poling pattern along the propagation direction with five defects at different locations of the array introduced by translating the poled domains by half a poling period Λ. This design is based on the recently developed concept of quantum state engineering with specialized poling patterns [58].

The squared amplitudes of the wavefunction elements predicted by SFG measurements are shown in Figure 5.8c and those directly measured through SPDC coincidences are presented in Figure 5.8d. The SFG predictions are obtained by integrating the measured conversion efficiencies over a bandwidth of 6 nm. The two correlation matrices have a fidelity of $F = \sum_{n_s n_i} \sqrt{|\Psi_{n_s n_i}^{SFG}|^2 |\Psi_{n_s n_i}^{SPDC}|^2} = 99.28 \pm 0.3\%$.

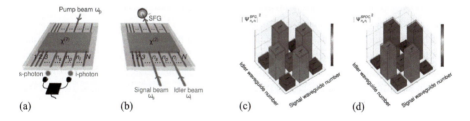

(a)   (b)   (c)   (d)

FIGURE 5.8 Scheme for the characterization of the biphoton state produced by an array of N waveguides with an arbitrary $\chi^{(2)}$-nonlinear process. (a) SPDC: a pump beam is injected into waveguide at the input of the device. Photon-coincidence counting measurements between each pair of waveguides at the output are used to measure photon-pair generation rates and relative absolute squared values of the wavefunction. (b) SFG: Laser light at signal and idler frequencies is injected into waveguides and in the reverse direction of SPDC. Absolute photon-pair generation rates and relative absolute squared values of the wavefunction can be predicted by direct optical power detection of the sum-frequency field emitted from waveguide n_p. (c, d) Normalized biphoton wavefunctions predicted by SFG (c) and measured by SPDC (d). (After Lenzini, F. et al., *Light Sci. Appl.*, 7, 17143, 2018.)

In case of the waveguide geometry, Eq. (5.69) for the correspondence between the absolute photon-pair generation rates for SPDC can be written as

$$\frac{1}{P_p}\frac{dN_{\text{pair}}}{d\omega_s dt} = \frac{\omega_i \omega_s}{2\pi\omega_p^2}\eta_{n_s n_i}^{SFG}(\omega_s, \omega_i). \tag{5.70}$$

Here, P_p is the power of the pump beam during SPDC, $dN_{\text{pair}}/d\omega_s dt$ is the rate of photon-pair coincidence counts per unit signal frequency, and $\eta_{n_s n_i}^{SFG} \equiv P_{SFG}/(P_s P_i)$ is the sum-frequency power conversion efficiency. Using the SFG measurements and Eq. (5.70) a photon-pair generation rate has been found as $N_{SFG} = 2.36 \pm 0.14$ MHz, which is the sum of the rates from all six output combinations. Direct measurement of this rate from the SPDC data gives $N_{SPDC} = 1.67 \pm 0.15$ MHz, showing a good qualitative agreement between the two values.

5.4   EXPERIMENTAL DEMONSTRATION OF PHOTON-PAIR GENERATION IN DIELECTRIC NANOANTENNAS

The experimental demonstration of generation of spontaneous photon pairs in nanoscale photonic structures has been recently demonstrated using an AlGaAs disk nanoantenna exhibiting Mie-type resonances at both pump and biphoton wavelengths [27]. A schematic of the nanoantenna is shown in Figure 5.9a. It is a crystalline AlGaAs cylinder with a diameter $d = 430$ nm and height $h = 400$ nm. A scanning electron microscope (SEM) image of the fabricated structure is shown in Figure 5.9b. The non-centrosymmetric crystalline structure of the (100)-grown AlGaAs offers strong bulk quadratic susceptibility of $d_{14} = 100$ pm/V [64,65]. The AlGaAs also exhibits high transparency in a broad spectral window from 730 nm up to the far infrared, due to its direct electronic bandgap. As such, the one- and two-photon absorption at telecommunication wavelengths is negligible. The distance between neighboring nanocylindors is 10 μm, thereby the response is dominated by the local optical properties of the single antenna [9].

The antenna was excited by a linearly polarized pump beam in the near-infrared spectral range and through the process of SPDC generates signal and idler photons in the telecommunication wavelength range. The dimensions of the nanocylinder are chosen such that it exhibits Mie-type resonances at the pump and signal/idler wavelengths. The simulated linear scattering efficiency is defined as the scattering cross section C_{sca} normalized by the cross area of the nanocylinder πr^2: $Q_{sca} = C_{sca}/\pi r^2$. It is shown in Figure 5.9c along with the two leading multipolar contributions of the scattering. In the infrared region of the spectrum, where the signal and idler photons are generated, the nanocylinder exhibits a magnetic dipolar resonance, which is the lowest order Mie-mode, featuring a Q factor of nine (Figure 5.9c). For the spectral region of the pump 760–790 nm, another strong resonance with a Q factor of 52 is present, represented by a peak in the scattering efficiency spectrum (Figure 5.9c). This is dominated by the electric dipole moment of the antenna, although it also contains higher-order multipolar contributions (not shown). The strong internal fields

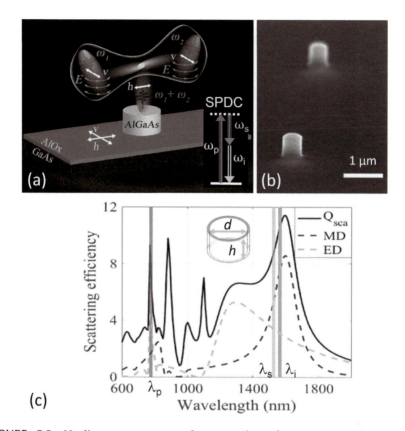

FIGURE 5.9 Nonlinear nanoantenna for generation of spontaneous photon pairs. (a) Schematic representation of the nanoantenna-based source of photon-pairs through the SPDC process. The inset depicts the energy diagram of the SPDC process. The SPDC pump is horizontally polarized along the (100) AlGaAs crystallographic axis. (b) A typical scanning electron microscope (SEM) image of two (100) AlGaAs monolithic nanocylinders, 10 μm apart, such that each cylinder can be excited individually. (c) Simulated scattering efficiency, Q_{sca}, and multipolar decomposition in terms of the two leading electric (ED) and magnetic dipoles (MD) for a nanocylinder with a diameter $d = 430$ nm and height $h = 400$ nm. The vertical lines show the spectral ranges of the pump light and the generated SPDC light (signal and idler), as indicated by labels. The inset shows the geometry of the nanoantenna. (After Marino, G. et al., *Optica*, 6, 1416–1422, 2019.)

at the Mie-type resonances allow for strong enhancement of the nonlinear frequency mixing processes and also imposes a spectral selection for the frequencies of the generated photons.

The SPDC process in the nanocylinder can result in the emission of photon pairs with nontrivial correlations, associated with different angular and polarization components. In order to experimentally determine the optimal conditions for photon-pair generation and ultimately for optimum SPDC efficiency, usually one uses a technique called quantum state tomography [47–50]. However, due to a weak efficiency of the spontaneous processes in the small volume of the nanoantenna, the biphoton rate tends to be low. Thereby long acquisition times of the photon counting statistics are required to obtain sufficient statistical data for significant correlation precision. Therefore, optimizing the experimental parameters directly through SPDC measurements is impractical and an alternative solution is needed.

5.4.1 NONLINEAR CLASSICAL CHARACTERIZATION

We employ the concept of quantum-classical correspondence between SPDC and its reversed process, namely SFG, as described in Section 5.3. We recall that the generated sum-frequency and pump waves should propagate in opposite directions to the SPDC pump, signal and idler [28,60]. The quantum-classical correspondence will allow for the optimization of the excitation parameters, as well as for classical estimation of the SPDC generation biphoton rates. We consider the collection of photons in all directions and accordingly perform the integration of Eq. (5.69) over solid angles within the half-sphere, resulting in the relation:

$$\frac{1}{\Phi_p} \frac{dN_{\text{pair}}}{dt} = 2\pi \, \Xi^{\text{SFG}} \frac{\lambda_p^4}{\lambda_s^3 \lambda_i^3} \frac{c\Delta\lambda_s}{\lambda_s^2}. \tag{5.71}$$

Here, Φ_p is the SPDC pump flux, λ_p, λ_s, and λ_i are the pump, signal, and idler wavelengths, and $\Delta\lambda_s$ is the nonlinear resonance bandwidth at the signal wavelength. The efficiency Ξ^{SFG} is given by the ratio of the sum-frequency photon power to the product of incident energy fluxes at signal and idler frequencies. The number of photon pairs generated through SPDC, in a given optical mode of the nanostructure, is therefore proportional to the SFG amplitude of the classical signal and idler waves, propagating in the opposite direction. In this framework, one can first optimize the SFG efficiency and thus predict the biphoton generation rates, prior to SPDC detection. Importantly, the SFG process can also be characterized for different polarizations, thereby optimizing the parameters for the subsequent SPDC measurements.

The schematic of such SFG experiments is illustrated in Figure 5.10a. Two femtosecond laser pulses at wavelengths 1520 and 1560 nm illuminate the nanoantenna as signal and idler beams. Their spectra are shown in Figure 5.10b. The two beams are focused onto a single AlGaAs nanocylinder by a 0.7 NA objective, with an average power of 10 mW. The beam size of the two incident pulses is 2 μm (diameter) diffraction-limited spots, resulting in 7 GW/cm² peak intensities. The incident linear H polarization is parallel to the AlGaAs nanoresonator's crystallographic axis (100). Figure 5.10c shows the time resolved spectra of H-polarized emission collected in backward direction at different optical delays between the two V-polarized pulses. This polarizations arrangement corresponds to the maximum SFG efficiency, as we discuss below. The two spectral peaks at 760 and 780 nm correspond to the second-harmonic generation (SHG) from the individual signal and idler pulses, and are observed at all delay times. The third peak at 770 nm only occurs at "time zero," when the signal and idler pulses arrive at the nanocylinder simultaneously. This SFG pulse has a full-width at half-maximum (FWHM) of 80 fs, in agreement with the duration of the pump pulses.

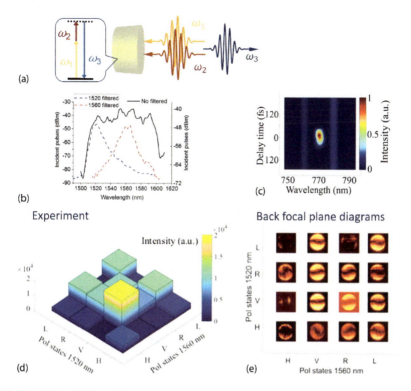

FIGURE 5.10 SFG nonlinear characterization of correlations in the nanoantenna. (a) Schematic of the experimental arrangement and energy conservation diagram of the SFG process in the inset. (b) Spectra of the signal (blue-dashed line) and idler (red-dashed line) spectra filtered from the fs laser source. (c) Spectrum of the nonlinear wave-mixing in the AlGaAs nanocylinder as a function of the time delay between the signal and idler pulses. The SFG signal is only visible when the two pulses overlap onto the nanoantenna. The spectral features at 760 and 780 nm correspond to the second-harmonic generation from the individual signal and idler pulses. (d) Intensity of H-polarized reflected SFG at 770 nm, measured with 16 combinations of horizontal (H), vertical (V), right circular (R), and left circular (L) polarizations of signal and idler beams for the nanocylinder geometry in Figure 5.9. (e) Measured reflected SFG images in *k*-space for the polarization combinations shown in (d) and SFG detected with *NA* = 0.7. (After Marino, G. et al., *Optica*, 6, 1416–1422, 2019.)

By setting different combinations of incident polarizations for the signal and idler pulses, including horizontal (H), vertical (V), right circular (R) and left circular (L), one can measure the SFG for H (or V) polarization. The choice for H-polarized SFG is arbitrary, as for normally incident signal/idler beams V and H are identical due to the cylindrical symmetry of the disk and the isotropy of the material. The resulting SFG signal intensities (normalized to the maximum value) at 770 nm and the corresponding radiation patterns, recorded via a back focal plane (BFP) imaging system, are shown in Figure 5.10d and e, respectively. The maximum signal of H-polarized SFG is obtained when both signal and idler are V-polarized. At the microscopic scale, this corresponds to the excitation of signal, idler and SFG modes whose vectorial components constructively overlap, following the symmetry of AlGaAs second-order susceptibility tensor. The highest SFG conversion efficiency from the nanocylinder is measured to be 1.8×10^{-5}, which is comparable to the SHG efficiency obtained by different groups [8–10]. As shown in the BFP images, the SFG radiation patterns strongly depend on signal and idler polarization combinations, however the general observation is that the SFG signal is emitted under angle, off-axis to the nanocylinder. This is due to the symmetry of the nonlinear tensor, as previously reported for SHG in [8,66].

The experimental results have been compared with finite element simulations under realistic experimental conditions. The simulated SFG intensity is enhanced when the polarizations of both signal and idler beams are VV, or RR, or LL polarized. Lower counts are seen for the mixed polarization cases, and for the case HH. This trend matches the experimental results, particularly for the combinations involving H and V polarizations, while the RR and LL cases appears less bright than VV case. Such discrepancy can be attributed to slight non-uniformity of the fabricated structure, which can deviate from the cylindrical to elliptical shape.

Importantly, knowing the SFG efficiency of 1.8×10^{-5} for the VV → H process, one can estimate the possible biphoton rates for detection of SPDC photon-pair rates from an AlGaAs nanoantenna. The prediction for the photon-pair generation rate, obtained using Eq. (5.71), is about 380 Hz at a pump power of 2 mW. This value is significant and well above the dark count rates for the detectors used in the experiments (estimated at 5 Hz). However, for Eq. (5.71) to exactly predict the following SPDC experiments, one needs to look for SFG emission mainly directed backward. The SFG emission is maximal in the backward direction when the signal and idler beams are incident at oblique angle. Theoretical predictions show that a high normal SHG can be obtained when the signal and idler illuminate the nanoantenna at 45° to the nanocylinder axis. An SFG experiment where the illumination is carried out through a high NA objective, will partially satisfy this criterion. Therefore, under such experimental condition Eq. (5.71) is expected to overestimate the detected SPDC count rate.

5.4.2 SPDC Experiments with Single Nanoantennas

Once knowing the optimal antenna parameters for enhancement of the SFG process, the generation of photon pairs through the SPDC process can be tested. A CW pump laser at 785 nm was incident onto the AlGaAs nanoantenna with a power of 2 mW and a 2 µm (diameter) diffraction-limited spot. The generated photon pairs with frequencies ω_1 and ω_2 are collected by a high-numerical aperture microscope objective and directed through a non-polarizing beam splitter at two single-photon InGaAs detectors. The coincidence of detection of the two photons is recorded by a time-to-digital electronics (Figure 5.11a). The generated SPDC photons are expected to have a large spectral bandwidth of about 150 nm, due to the broad magnetic dipole resonance in the IR spectral range, as shown in Figure 5.9c. This bandwidth is relatively broad with respect to conventional SPDC sources, which typically have sub-nm or few-nm bandwidth. Such broad bandwidth offers a range of advantages, including a short temporal width for timing-critical measurements, such as for temporal

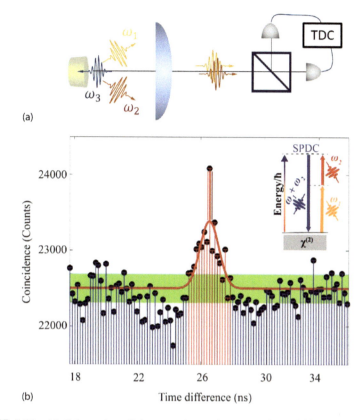

FIGURE 5.11 (a) Schematics of the experiment for generation of biphoton-pairs in an AlGaAs disk nanoantenna. (b) IR coincidence counts integrated over 24 hours on two single-photon detectors after a beam splitter. A significant statistical increase, marked by the red bar, is apparent at a time difference of 26.5 ns, corresponding to the temporal delay between both detectors. Black dots are the measured coincidences, the red-shadowed area indicates correlation due to thermal excitation of the semiconductor materials, while the red solid line is its fitted Gaussian curve. The inset shows a schematic of the SPDC process and energy correlation. The green band depicts the statistical error of the measurement. (After Marino, G. et al., *Optica*, 6, 1416–1422, 2019.)

entanglement [67], or for SPDC spectroscopy [68]. It also dictates a sub-100 fs temporal width of the generated photons, which is much shorter than the coincidence time window τ_c.

The measured coincidences for an H-polarized CW pump are presented in Figure 5.11b, where photon counting statistics is accumulated by integrating over 24 hours. For a time difference of 26.5 ns, corresponding exactly to the time difference between the signal and idler detection arms, a single bin with high coincidence rate is observed. This is consistent with the physics of SPDC generation of signal and idler photons with the estimated temporal correlations of sub-100 fs. Although it only emerges from the background by a limited number of counts, this peak of coincidence rate is statistically relevant and larger than the statistical error (marked with a green band). A second, broader coincidence statistics is also observed underneath the SPDC peak. This Gaussian peak is the indication of correlation due to thermal excitation of the semiconductor materials [69]. It has approximately 2 ns width, as shown with the red-shadowed area in Figure 5.11b. The experimental SPDC rate from the AlGaAs nanocylinder has been analyzed, taking into account the losses in the detection system. The estimate of the total photon-pair generation rate from the nanoantenna is then estimated to $dN_{disk}^{gen} / dt = 35\,Hz$. Normalized to the pump energy stored by the nanoantenna, this rate reaches values of up to 1.4 GHz/Wm, being one order of magnitude higher than conventional on-chip or bulk photon-pair sources [17,26].

Importantly, this rate is significantly higher than the reference measurements of the AlOx/GaAs substrate without the nanocylinder that is of the order of $dN_{sub}^{gen} / dt = 9\,Hz$. The figure of merit calculated for this substrate source results is four-orders of magnitude smaller magnitude than for the nanocylinder. While the AlGaAs nanocylinder is weak in absolute values, such a figure of merit brings to light the nanometer spatial confinement and relatively high Q factors achievable in our AlGaAs nanocylinder. The latter operating in the Mie scattering regime, enables a fine shaping of the spectral and radiation profile, thereby leading to flexible quantum state engineering and possible spatial multiplexing of SPDC sources in a metasurface. Note that the influence of the AlGaAs nanocylinder on the SPDC from the substrate, such as refocusing the pump beam or the photon pairs, was found to be negligible by appropriate numerical modeling [27].

Finally, it is possible to numerically calculate the sub-wavelength mode correlation responsible for the generation of signal and idler photon pairs (Figure 5.12a and b). This can be done again through the quantum-classical analogy presented in Section 5.2.3 by calculating the SFG normal output when exciting the idler and signal fields with combination of ED (p_x or p_y or p_z) and MD (m_x or m_y or m_z) inside the disk at idler or signal wavelengths, respectively. Only orthogonal Cartesian components of magnetic dipole contribute to SPDC process, which leads to coupling of two magnetic dipole moments of the nanoantenna, namely m_x and m_z. The two coupled-modes for an H-polarized pump beam are shown in Figure 5.12c and d. The coupling of these two sub-wavelength modes efficiently generates photon pairs in the far-field via the antenna radiation and underpins the measured photon correlation. As can be seen from Figure 5.12a and b, with $E_H \parallel y$ ($E_V \parallel x$) pump, the generated two photons are dominantly coupled into m_x and m_z (m_y and m_z) modes, respectively. Due to the SFG-SPDC correspondence, the SFG process has the same symmetry properties and can be understood using the group theory [70]. At the signal and idler frequencies, the electric field is controlled by the magnetic dipole resonance, while at the pump/SFG frequency it is determined by the electric dipole mode. In the T_d symmetry group of the AlGaAs crystalline lattice, the magnetic dipole modes transform according to the F_1 irreducible representation, while the electric dipole modes belong to the F_2 representation. The basis functions of the F_2 representation, corresponding to the direct product $F_1 \otimes F_1$, are $p_x \propto m_y m_z, p_y \propto m_x m_z, p_z \propto m_x m_y$ [70] in full agreement with the numerical calculations in Figure 5.12. The states shown in

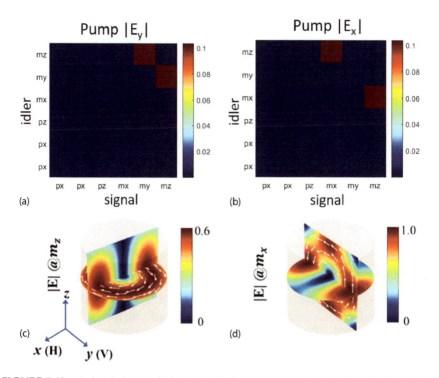

FIGURE 5.12 (a, b) Mode correlation in the AlGaAs nanocylinder for H (left) and V (right) polarized pump excitation. (c, d) Numerically simulated fields inside the nanocylinder when exciting m_x and m_z modes, respectively. The white arrows indicate the electric field vector. Different color bar scales are used, following the different intensities of m_x, m_z inside the nanocylinder, in contrast to the symmetric case of a nanosphere. (After Marino, G. et al., *Optica*, 6, 1416–1422, 2019.)

Figure 5.12a and b have the corresponding Schmidt number of 2, indicating a very strong correlation between the modes. More detailed symmetry analysis of the SFG process in dielectric nanoparticles can be found in [71].

We note that similar SPDC nanoscale sources can be obtained with other nonlinear crystalline nanostructures, including lithium niobate [72]. This would allow one to explore different crystalline symmetries and polarization dependencies. As such, the field of nanoscale sources of two-photon quantum states is expected to grow in the years to come. Therefore, the development of nanoscale quantum sources based on SPDC is likely to be of interest to the wider quantum community.

5.5 OUTLOOK

The experiments on SPDC in a single AlGaAs nanodisk bring confidence that correlated photons can be generated by employing crystalline nanoantennas and arrangement of such nanoantennas in an optical metasurface. Such nanoscale platform can

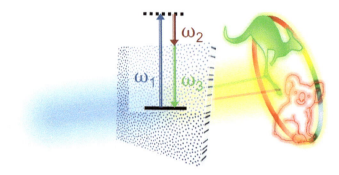

FIGURE 5.13 Vision for a photon-pair generation via down-conversion process in a spatially variant nonlinear metasurface, enabling the generation of correlated images.

open new opportunities for generation, unrestricted by longitudinal phase matching [73], of non-classical spatially entangled states of arbitrary shapes by carefully engineering of the dimensions of different nanoantennas in spatially variant metasurfaces, see artistic image in Figure 5.13. We believe that this opportunity, combined with the capacity of metasurfaces to transform, image, and reconstruct quantum states [74–77], will unleash a potential for ultimate miniaturization of quantum devices suitable for end-user applications [78,79], such us quantum imaging [80], sensing [81], precision spectroscopy [82], free-space communications [83], and cryptography [79].

The authors acknowledge highly productive collaborations, and in particular the results outlined in this chapter were primarily driven by researchers who co-authored [27,28,60].

REFERENCES

1. P. Genevet, F. Capasso, F. Aieta, M. Khorasaninejad, and R. Devlin, Recent advances in planar optics: From plasmonic to dielectric metasurfaces, *Optica* **4**, 139–152 (2017).
2. A. I. Kuznetsov, A. E. Miroshnichenko, M. L. Brongersma, Y. S. Kivshar, and B. Luk'yanchuk, Optically resonant dielectric nanostructures, *Science* **354**, 846–854 (2016).
3. D. Neshev and I. Aharonovich, Optical metasurfaces: New generation building blocks for multi-functional optics, *Light Sci Appl* **7**, 58 (2018).
4. M. R. Shcherbakov, D. N. Neshev, B. Hopkins, A. S. Shorokhov, I. Staude, E. V. Melik-Gaykazyan, M. Decker, et al., Enhanced third-harmonic generation in silicon nanoparticles driven by magnetic response, *Nano Lett* **14**, 6488–6492 (2014).
5. Y. M. Yang, W. Y. Wang, A. Boulesbaa, I. I. Kravchenko, D. P. Briggs, A. Puretzky, D. Geohegan, and J. Valentine, Nonlinear fano-resonant dielectric metasurfaces, *Nano Lett* **15**, 7388–7393 (2015).
6. G. Grinblat, Y. Li, M. P. Nielsen, R. F. Oulton, and S. A. Maier, Efficient third harmonic generation and nonlinear subwavelength imaging at a higher-order anapole mode in a single germanium nanodisk, *ACS Nano* **11**, 953–960 (2017).
7. L. Carletti, A. Locatelli, O. Stepanenko, G. Leo, and C. De Angelis, Enhanced second-harmonic generation from magnetic resonance in AlGaAs nanoantennas, *Opt Express* **23**, 26544–26550 (2015).

8. R. Camacho-Morales, M. Rahmani, S. Kruk, L. Wang, L. Xu, D. A. Smirnova, A. S. Solntsev, et al., Nonlinear generation of vector beams from AlGaAs nanoantennas, *Nano Lett* **16**, 7191–7197 (2016).

9. V. F. Gili, L. Carletti, A. Locatelli, D. Rocco, M. Finazzi, L. Ghirardini, I. Favero, et al., Monolithic AlGaAs second-harmonic nanoantennas, *Opt Express* **24**, 15965–15971 (2016).

10. S. Liu, M. B. Sinclair, S. Saravi, G. A. Keeler, Y. M. Yang, J. Reno, G. M. Peake, et al., Resonantly enhanced second-harmonic generation using III–V semiconductor all-dielectric metasurfaces, *Nano Lett* **16**, 5426–5432 (2016).

11. L. Carletti, D. Rocco, A. Locatelli, C. De Angelis, V. F. Gili, M. Ravaro, I. Favero, et al., Controlling second-harmonic generation at the nanoscale with monolithic AlGaAs-on-AlOx antennas, *Nanotechnology* **28**, 114005 (2017).

12. S. Liu, P. P. Vabishchevich, A. Vaskin, J. L. Reno, G. A. Keeler, M. B. Sinclair, I. Staude, and I. Brener, An all-dielectric metasurface as a broadband optical frequency mixer, *Nat Commun* **9**, 2507 (2018).

13. M. Muller, S. Bounouar, K. D. Jons, M. Glassl, and P. Michler, On-demand generation of indistinguishable polarization-entangled photon pairs, *Nat Photon* **8**, 224–228 (2014).

14. M. A. M. Versteegh, M. E. Reimer, K. D. Jons, D. Dalacu, P. J. Poole, A. Gulinatti, A. Giudice, and V. Zwiller, Observation of strongly entangled photon pairs from a nanowire quantum dot, *Nat Commun* **5**, 5298 (2014).

15. J. L. O'Brien, A. Furusawa, and J. Vučković, Photonic quantum technologies, *Nat Photon* **3**, 687–695 (2009).

16. D. Klyshko, *Photons and Nonlinear Optics* (Gordon and Breach, New York, 1988).

17. P. G. Kwiat, K. Mattle, H. Weinfurter, A. Zeilinger, A. V. Sergienko, and Y. H. Shih, New high-intensity source of polarization-entangled photon pairs, *Phys Rev Lett* **75**, 4337–4341 (1995).

18. A. Sipahigil, R. E. Evans, D. D. Sukachev, M. J. Burek, J. Borregaard, M. K. Bhaskar, C. T. Nguyen, et al., An integrated diamond nanophotonics platform for quantum-optical networks, *Science* **354**, 847–850 (2016).

19. P. Senellart, G. Solomon, and A. White, High-performance semiconductor quantum-dot single-photon sources, *Nat Nanotech* **12**, 1026–1039 (2017).

20. N. Somaschi, V. Giesz, L. De Santis, J. C. Loredo, M. P. Almeida, G. Hornecker, S. L. Portalupi, et al., Near-optimal single-photon sources in the solid state, *Nat Photon* **10**, 340–345 (2016).

21. I. Aharonovich, D. Englund, and M. Toth, Solid-state single-photon emitters, *Nat Photon* **10**, 631–641 (2016).

22. T. T. Tran, K. Bray, M. J. Ford, M. Toth, and I. Aharonovich, Quantum emission from hexagonal boron nitride monolayers, *Nat Nanotech* **11**, 37–42 (2016).

23. M. Fiorentino, S. M. Spillane, R. G. Beausoleil, T. D. Roberts, P. Battle, and M. W. Munro, Spontaneous parametric down-conversion in periodically poled KTP waveguides and bulk crystals, *Opt Express* **15**, 7479–7488 (2007).

24. A. S. Solntsev and A. A. Sukhorukov, Path-entangled photon sources on nonlinear chips, *Rev Phys* **2**, 19–31 (2017).

25. A. Orieux, A. Eckstein, A. Lemaitre, P. Filloux, I. Favero, G. Leo, T. Coudreau, A. Keller, P. Milman, and S. Ducci, Direct Bell states generation on a III–V semiconductor chip at room temperature, *Phys Rev Lett* **110**, 160502 (2013).

26. X. Guo, C. L. Zou, C. Schuck, H. Jung, R. S. Cheng, and H. X. Tang, Parametric down-conversion photon-pair source on a nanophotonic chip, *Light Sci Appl* **6**, e16249 (2017).

27. G. Marino, A. S. Solntsev, L. Xu, V. F. Gili, L. Carletti, A. N. Poddubny, M. Rahmani, et al., Spontaneous photon-pair generation from a dielectric nanoantenna, *Optica* **6**, 1416–1422 (2019).

28. A. N. Poddubny, I. V. Iorsh, and A. A. Sukhorukov, Generation of photon-plasmon quantum states in nonlinear hyperbolic metamaterials, *Phys Rev Lett* **117**, 123901 (2016).

29. P. D. Drummond and M. S. Hillery, *The Quantum Theory of Nonlinear Optics* (Cambridge University Press, Cambridge, 2013).

30. R. W. Boyd, *Nonlinear Optics*, 3rd edition (Academic Press, San Diego, 2008).

31. E. del Valle, A. Gonzalez-Tudela, F. P. Laussy, C. Tejedor, and M. J. Hartmann, Theory of frequency-filtered and time-resolved *n*-photon correlations, *Phys Rev Lett* **109**, 183601 (2012).

32. A. N. Poddubny, P. Ginzburg, P. A. Belov, A. V. Zayats, and Y. S. Kivshar, Tailoring and enhancing spontaneous two-photon emission using resonant plasmonic nanostructures, *Phys Rev A* **86**, 033826 (2012).

33. A. Hayat, P. Ginzburg, and M. Orenstein, Observation of two-photon emission from semiconductors, *Nat Photon* **2**, 238–241 (2008).

34. L. Landau, E. Lifshits, and L. Pitaevski, *Statistical Physics*, Part 2 in Course of theoretical physics (Butterworth-Heinemann, London, 1980).

35. W. Vogel and D. G. Welsch, *Quantum Optics*, 3rd edition (Wiley, Weinheim, Germany, 2006).

36. C. Cohen-Tannoudji, J. Dupont-Roc, and G. Grynberg, *Atom-Photon Interactions: Basic Processes and Applications* (Wiley-VCH, New York, 1998).

37. R. Ghosh and L. Mandel, Observation of nonclassical effects in the interference of 2 photons, *Phys Rev Lett* **59**, 1903–1905 (1987).

38. M. O. Scully and M. S. Zubairy, *Quantum Optics* (Cambridge University Press, Cambridge, UK, 1997).

39. H. Carmichael, *An Open Systems Approach to Quantum Optics* (Springer, New York, 1993).

40. L. Landau and E. Lifshits, *Statistical Physics*, Part 1 in Course of theoretical physics (Butterworth-Heinemann, London, 1980).

41. D. A. Antonosyan, A. S. Solntsev, and A. A. Sukhorukov, Effect of loss on photon-pair generation in nonlinear waveguide arrays, *Phys Rev A* **90**, 043845 (2014).

42. L. G. Helt, M. J. Steel, and J. E. Sipe, Spontaneous parametric downconversion in waveguides: What's loss got to do with it? *New J Phys* **17**, 013055 (2015).

43. M. S. Tomas, Green-function for multilayers—Light-scattering in planar cavities, *Phys Rev A* **51**, 2545–2559 (1995).

44. P. B. Johnson and R. W. Christy, Optical constants of noble metals, *Phys Rev B* **6**, 4370–4379 (1972).

45. G. Di Martino, Y. Sonnefraud, S. Kena-Cohen, M. Tame, S. K. Ozdemir, M. S. Kim, and S. A. Maier, Quantum statistics of surface plasmon polaritons in metallic stripe waveguides, *Nano Lett* **12**, 2504–2508 (2012).

46. B. le Feber, N. Rotenberg, D. M. Beggs, and L. Kuipers, Simultaneous measurement of nanoscale electric and magnetic optical fields, *Nat Photon* **8**, 43–46 (2014).

47. D. F. V. James, P. G. Kwiat, W. J. Munro, and A. G. White, Measurement of qubits, *Phys Rev A* **64**, 052312 (2001).

48. J. B. Altepeter, E. R. Jeffrey, and P. G. Kwiat, Photonic state tomography, *Adv Atom Mol Opt Phys* **52**, 105–159 (2005).

49. A. I. Lvovsky and M. G. Raymer, Continuous-variable optical quantum-state tomography, *Rev Mod Phys* **81**, 299–332 (2009).

50. J. G. Titchener, M. Grafe, R. Heilmann, A. S. Solntsev, A. Szameit, and A. A. Sukhorukov, Scalable on-chip quantum state tomography, *npj Quant Inform* **4**, 19 (2018).

51. M. Liscidini and J. E. Sipe, Stimulated emission tomography, *Phys Rev Lett* **111**, 193602 (2013).

52. A. Eckstein, G. Boucher, A. Lemaitre, P. Filloux, I. Favero, G. Leo, J. E. Sipe, M. Liscidini, and S. Ducci, High-resolution spectral characterization of two photon states via classical measurements, *Laser Photon Rev* **8**, L76–L80 (2014).

53. B. Fang, O. Cohen, M. Liscidini, J. E. Sipe, and V. O. Lorenz, Fast and highly resolved capture of the joint spectral density of photon pairs, *Optica* **1**, 281–284 (2014).

54. I. Jizan, L. G. Helt, C. L. Xiong, M. J. Collins, D. Y. Choi, C. J. Chae, M. Liscidini, M. J. Steel, B. J. Eggleton, and A. S. Clark, Bi-photon spectral correlation measurements from a silicon nanowire in the quantum and classical regimes, *Sci Rep* **5**, 12557 (2015).

55. D. Grassani, A. Simbula, S. Pirotta, M. Galli, M. Menotti, N. C. Harris, T. Baehr-Jones, M. Hochberg, C. Galland, M. Liscidini, and D. Bajoni, Energy correlations of photon pairs generated by a silicon microring resonator probed by stimulated four wave mixing, *Sci Rep* **6**, 23564 (2016).

56. L. A. Rozema, C. Wang, D. H. Mahler, A. Hayat, A. M. Steinberg, J. E. Sipe, and M. Liscidini, Characterizing an entangled-photon source with classical detectors and measurements, *Optica* **2**, 430–433 (2015).

57. B. Fang, M. Liscidini, J. E. Sipe, and V. O. Lorenz, Multidimensional characterization of an entangled photon-pair source via stimulated emission tomography, *Opt Express* **24**, 10013–10019 (2016).

58. J. G. Titchener, A. S. Solntsev, and A. A. Sukhorukov, Generation of photons with all-optically-reconfigurable entanglement in integrated nonlinear waveguides, *Phys Rev A* **92**, 033819 (2015).

59. L. G. Helt and M. J. Steel, Effect of scattering loss on connections between classical and quantum processes in second-order nonlinear waveguides, *Opt Lett* **40**, 1460–1463 (2015).

60. F. Lenzini, A. N. Poddubny, J. Titchener, P. Fisher, A. Boes, S. Kasture, B. Haylock, M. Villa, A. Mitchell, A. S. Solntsev, A. A. Sukhorukov, and M. Lobino, Direct characterization of a nonlinear photonic circuit's wave function with laser light, *Light Sci Appl* **7**, 17143 (2018).

61. M. Born, E. Wolf, and A. Bhatia, *Principles of Optics: Electromagnetic Theory of Propagation, Interference and Diffraction of Light* (Cambridge University Press, Cambridge, 1999).

62. F. Lenzini, S. Kasture, B. Haylock, and M. Lobino, Anisotropic model for the fabrication of annealed and reverse proton exchanged waveguides in congruent lithium niobate, *Opt Express* **23**, 1748–1756 (2015).

63. Y. N. Korkishko, V. A. Fedorov, T. M. Morozova, F. Caccavale, F. Gonella, and F. Segato, Reverse proton exchange for buried waveguides in $LiNbO_3$, *J Opt Soc Am A* **15**, 1838–1842 (1998).

64. I. Shoji, T. Kondo, A. Kitamoto, M. Shirane, and R. Ito, Absolute scale of second-order nonlinear-optical coefficients, *J Opt Soc Am B* **14**, 2268–2294 (1997).

65. M. Ohashi, T. Kondo, R. Ito, S. Fukatsu, Y. Shiraki, K. Kumata, and S. S. Kano, Determination of quadratic nonlinear-optical coefficient of $Al_xGa_{1-x}As$ system by the method of reflected 2nd harmonics, *J Appl Phys* **74**, 596–601 (1993).

66. L. Carletti, A. Locatelli, D. Neshev, and C. De Angelis, Shaping the radiation pattern of second-harmonic generation from AlGaAs dielectric nanoantennas, *ACS Photonics* **3**, 1500–1507 (2016).

67. S. Rogers, D. Mulkey, X. Y. Lu, W. C. Jiang, and Q. Lin, High visibility time-energy entangled photons from a silicon nanophotonic chip, *ACS Photonics* **3**, 1754–1761 (2016).

68. A. S. Solntsev, P. Kumar, T. Pertsch, A. A. Sukhorukov, and F. Setzpfandt, $LiNbO_3$ waveguides for integrated SPDC spectroscopy, *APL Photon* **3**, 021301 (2018).

69. E. B. Flagg, S. V. Polyakov, T. Thomay, and G. S. Solomon, Dynamics of nonclassical light from a single solid-state quantum emitter, *Phys Rev Lett* **109**, 163601 (2012).

70. M. S. Dresselhaus, G. Dresselhaus, and A. Jorio, *Group Theory. Application to the Physics of Condensed Matter* (Springer, New York, 2008).

71. K. Frizyuk, D. Smirnova, I. Volkovskaya, A. Poddubny, and M. Petrov, Second-harmonic generation in Mie-resonant dielectric nanoparticles made of noncentrosymmetric materials, *Phys Rev B* **99**, 075425 (2019).

72. L. Carletti, C. Li, J. Sautter, I. Staude, C. De Angelis, T. Li, and D. N. Neshev, Second harmonic generation in monolithic lithium niobate metasurfaces, *Opt Express* **27**, 33391–33398 (2019).

73. C. Okoth, A. Cavanna, T. Santiago-Cruz, and M. V. Chekhova, Microscale generation of entangled photons without momentum conservation, *Phys. Rev. Lett.* **123**, 263602 (2019).

74. T. Stav, A. Faerman, E. Maguid, D. Oren, V. Kleiner, E. Hasman, and M. Segev, Quantum entanglement of the spin and orbital angular momentum of photons using metamaterials, *Science* **361**, 1101–1103 (2018).

75. K. Wang, J. G. Titchener, S. S. Kruk, L. Xu, H. P. Chung, M. Parry, I. I. Kravchenko, et al., Quantum metasurface for multiphoton interference and state reconstruction, *Science* **361**, 1104–1107 (2018).

76. C. Altuzarra, A. Lyons, G. H. Yuan, C. Simpson, T. Roger, J. S. Ben-Benjamin, and D. Faccio, Imaging of polarization-sensitive metasurfaces with quantum entanglement, *Phys Rev A* **99**, 020101 (2019).

77. P. Georgi, M. Massaro, K. H. Luo, B. Sain, N. Montaut, H. Herrmann, T. Weiss, G. X. Li, C. Silberhorn, and T. Zentgraf, Metasurface interferometry toward quantum sensors, *Light Sci Appl* **8**, 70 (2019).

78. J. W. Silverstone, D. Bonneau, K. Ohira, N. Suzuki, H. Yoshida, N. Iizuka, M. Ezaki, et al., On-chip quantum interference between silicon photon-pair sources, *Nat Photon* **8**, 104–108 (2014).

79. C. Xiong, X. Zhang, Z. Liu, M. J. Collins, A. Mahendra, L. G. Helt, M. J. Steel, et al., Active temporal multiplexing of indistinguishable heralded single photons, *Nat Commun* **7**, 10853 (2016).

80. G. B. Lemos, V. Borish, G. D. Cole, S. Ramelow, R. Lapkiewicz, and A. Zeilinger, Quantum imaging with undetected photons, *Nature* **512**, 409–412 (2014).

81. S. Saravi, A. N. Poddubny, T. Pertsch, F. Setzpfandt, and A. A. Sukhorukov, Atom-mediated spontaneous parametric down-conversion in periodic waveguides, *Opt Lett* **42**, 4724–4727 (2017).

82. D. A. Kalashnikov, A. V. Paterova, S. P. Kulik, and L. A. Krivitsky, Infrared spectroscopy with visible light, *Nat Photon* **10**, 98–102 (2016).

83. A. E. Willner, Y. X. Ren, G. D. Xie, Y. Yan, L. Li, Z. Zhao, J. Wang, M. Tur, A. F. Molisch, and S. Ashrafi, Recent advances in high-capacity free-space optical and radio-frequency communications using orbital angular momentum multiplexing, *Philos Trans R Soc A* **375**, 20150439 (2017).

6 Ultrafast Responses of Mie-Resonant Semiconductor Nanostructures

Maxim R. Shcherbakov, Polina P. Vabishchevich,
Igal Brener, and Gennady Shvets

CONTENTS

6.1 INTRODUCTION

Semiconductor-based metasurfaces are an emergent area of photonics that have surged in the past decade.[1–3] Exciting applications of these novel materials, such as wavefront engineering,[4–7] sensing,[8–11] and nonlinear optics,[12,13] are extensively covered throughout this book. The promising features of semiconductor materials are their transparency below the electronic bandgap, pronounced optical nonlinearities, and charge carrier dynamics that lead to extraordinary opportunities in ultrafast photonics. The goal of this chapter is to provide the reader with an overview of ultrafast semiconductor-based metasurfaces and their applications. The structure of this chapter is as follows: Section 6.2 describes the ultrafast processes taking place after an ultrashort laser pulse excites a semiconducting material; Sections 6.3 and 6.4 review ultrafast phenomena in Mie-resonant nanostructures and metasurfaces that utilize silicon and gallium arsenide, respectively; Section 6.5 is devoted to photon acceleration in semiconductor metasurfaces; Section 6.6 covers the ideas of how semiconductor-based metasurfaces can be utilized for shaping of ultrafast laser pulses; Section 6.7 concludes the chapter and briefly outlines the emerging directions within ultrafast photonics of semiconductor metasurfaces.

6.2 ULTRAFAST PHENOMENA

By convention, ultrafast phenomena are those taking place on a time scale of approximately one picosecond and shorter,[14] allowing characteristic modulation frequencies of above 1 THz. Few electric-current-based devices can provide bandwidths above several hundreds of GHz. Therefore, it is the consensus of the optoelectronics community that an all-optical approach is a major candidate for ultrafast signal processing. Hence, the ultimate challenge of ultrafast photonics is to find materials that enable efficient and practical pathways for photons to interact with each other.

Ultrafast processes in materials are routinely investigated using pump-probe spectroscopy, or time-resolved spectroscopy. A typical pump-probe setup is illustrated in Figure 6.1. A powerful ultrashort laser pulse ("pump") modifies the sample that is then probed by a weaker pulse ("probe"). The probe is monitored as a function of the time delay τ between the pulses, reconstructing the relaxation of the material. The probe and the pump can have different polarizations, frequencies, and temporal profiles, to attack the underlying processes under different angles.

Semiconductors are often the base materials for Mie-resonant nanostructures. The most common semiconductors—silicon, germanium, gallium arsenide, and others—have been well-studied with pump-probe spectroscopy.[15,16] A simplified timeline of the microscopic processes following a femtosecond pulse impinging on a surface of a bulk semiconductor is as follows[17]:

- $\tau \lesssim 100$ fs: Free carriers (FCs) are generated through single- or multi-photon absorption. At this point, the electromagnetic field is typically still within the material, thus subject to coherent frequency-mixing processes.

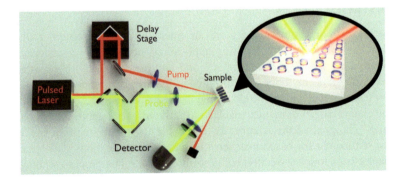

FIGURE 6.1 Pump-probe spectroscopy of semiconductor metasurfaces.

FCs lose coherence via carrier-carrier scattering events. The energy distribution is non-Boltzman, i.e., the electron gas has not yet thermalized.

- $\tau \lesssim 2$ ps: Electrons and holes have thermalized to Boltzman-type energy distributions. The electron-lattice energy exchange is underway through electron-phonon scattering. Light is no longer present with the material.

- $\tau \lesssim 100$ ps (in some cases longer): Electrons and holes have lost their energy to phonons and recombined. Lattice sinks its excess energy to the environment. The end time of this process is highly dependent on many parameters, including the substrate material.

In every step of this process, the transient value of the complex dielectric permittivity of the material $\tilde{\varepsilon}(t) = \varepsilon'(t) + i\varepsilon''(t)$ differs from its equilibrium value $\tilde{\varepsilon}_0 : \tilde{\varepsilon}(t) = \tilde{\varepsilon}_0 + \Delta\tilde{\varepsilon}(t)$. There are three main contributions to $\Delta\tilde{\varepsilon}(t)$: the instantaneous one (often referred to as the Kerr nonlinearity[18]), the one induced by the presence of the free carriers $\Delta\tilde{\varepsilon}_{FC}(t)$ and the one induced by the lattice heating, or the presence of phonons $\Delta\tilde{\varepsilon}_L(t)$. While thermo-optic effects can enable efficient switching of photonic metadevices,[19–21] this chapter will focus on the phenomena that can happen on the timescales around 1 ps or faster.

Optical response of a semiconductor that contains free carriers, such as electrons and holes, is often dominated by the Drude dispersion of the dielectric permittivity[22]:

$$\Delta\tilde{\varepsilon}_{FC} = -\frac{\omega_p^2}{\omega^2 + i\gamma\omega},$$

where $\omega_p = \sqrt{Ne^2/\varepsilon_0 m^*}$ is the plasma frequency, N is the free-carrier concentration, e is the elementary charge, m^* is the effective free-carrier mass, and γ is the damping constant. This approximation holds for most experimental cases, especially for photon energies below the band gap and low free-carrier concentrations.

The rate of FC relaxation to the equilibrium state strongly depends on the material used as the constituent material for a metasurface. Crystalline materials with an indirect bandgap, such as silicon, are poor candidates for ultrafast metasurfaces, as relaxation times in these materials can take hundreds of picoseconds.[15] Slow relaxation can be mitigated by introducing higher-order terms of the rate equation through hard optical pumping. Another approach to shorten the lifetime of FCs in a semiconductor is to increase the probability of monomolecular recombination through inhomogeneities of the crystal structure or nonradiative centers.

The general (simplified) recombination rate equation for FC density $N(t)$ is[23]:

$$\dot{N}(t) = -AN - BN^2 - CN^3, \qquad (6.1)$$

where A, B and C are the monomolecular, bimolecular, and Auger recombination rates. These coefficients are specific to a given semiconductor. For instance, in bulk GaAs, $A < 5\cdot10^7$ s^{-1}, $B = (1.7\pm0.2)\cdot10^{-10}$ cm³/s and $C = (7\pm14)\cdot10^{30}$ cm⁶/s.[23]

At low pump intensities and $N < 10^{18}$ cm^{-3}, the first term in Eq. (6.1) dominates over the second and the third ones, giving a relaxation time of 200 ns. This figure can be improved by either increasing the pump intensity leading to a higher FC density, or decreasing the value of A. The latter can be achieved by introducing impurities (a good example being low-temperature-grown GaAs) or by increasing the surface area by nanostructuring.[24] Both approaches have been utilized in semiconductor metasurfaces, as shown below.

6.3 ULTRAFAST RESPONSE OF SILICON METASURFACES

6.3.1 MANAGING FREE-CARRIER RESPONSE

Since silicon is dominant in micro- and nanoelectronics, for a long time it has been the material of choice for optoelectronic devices. Silicon-based all-optical switches[25–27] showed switching times in the range of several hundreds of picoseconds. This time constant is related to the FC relaxation time that tends to be high in monocrystalline materials. To decrease the relaxation time of FCs and enable ultrafast response times, several approaches are viable, with two of them being in the focus of this chapter.

One of the approaches to decrease the FC relaxation time utilizes amorphous silicon (a-Si). Irregularities in the lattice serve as recombination sites for electrons and holes, modifying the first term in the right-hand side of Eq. (6.1). Typical pump-probe traces of thin a-Si films shown in Figure 6.2a.[28] Characteristic relaxation times related to FC recombination lie on the picosecond scale, which is a two-orders-of-magnitude improvement with respect to crystalline silicon. It is also apparent that the pump intensity and the photoinduced FC density, influences the relaxation time, too. We explore this effect in more detail in Figure 6.2b. Here, FC recombination times decrease significantly as the pump-induced FC density increases, to subpicosecond time scales at FC densities more than $\approx 10^{20}$ cm^{-3}. This fast relaxation due

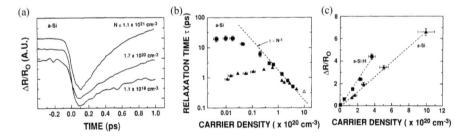

FIGURE 6.2 Ultrafast relaxation of free carriers in thin amorphous silicon (a-Si) films. (a) Typical relaxation traces of reflectance under different free-carrier plasma concentrations. (b) Electronics relaxation time as a function of initial free-carrier density. Squares stand for hydrogenated a-Si, and triangles stand for nonhydrogenated a-Si. (c) Maximum relative reflectance modulation as a function of initial carrier density for amorphous silicon (triangles) and hydrogenated amorphous silicon (squares) films. (Reprinted from Esser, A. et al., *J. Non. Cryst. Solids*, 114, 573–575, 1989. With permission.)

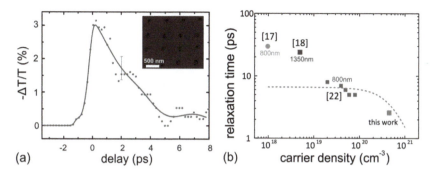

FIGURE 6.3 Ultrafast free-carrier relaxation in Mie-resonant nanoparticles under intense excitations. (a) Transient transmittance of Mie-resonant silicon nanoparticles under a pump fluence of 44 mJ/cm² and estimated FC plasma density of ~5 × 10²⁰ cm⁻³. (b) Comparison of free-carrier relaxation times under various free-carrier densities with data from the references.[17,18,22] (Reprinted from Baranov, D.G. et al., *ACS Photon*, 2016. With permission.)

to the higher-order terms of Eq. (6.1) can be put into use in Mie-resonant nanoparticles. For example, in Baranov et al.,[29] see Figure 6.3a, ultrafast relaxation of dense plasma was demonstrated in silicon nanoparticles populated by dense electron-hole plasma, showing FC relaxation times of down to 2.5 ps. This finding paves the way to FC-driven sub-THz bandwidth for all-optical signal processing.

6.3.2 ALL-OPTICAL MODULATION BY NONLINEAR ABSORPTION

The excited free carriers can take as short as several picoseconds to cool down. However, the so-called instantaneous nonlinearities, such as the Kerr effect and nonlinear absorption, play a significant role in Si metasurfaces. Instantaneous nonlinearities have switching times on the scale of the pump pulse duration and can be shorter than 100 fs, which is much shorter than the typical FC relaxation times. In silicon, the nonlinear figure of merit $F_n = n_2 / (\lambda \beta)$, where n_2 is the nonlinear Kerr effect coefficient, λ is the light wavelength and β is the nonlinear absorption coefficient, is relatively small.[31] Therefore, the prime instantaneous mechanism of all-optical modulation in silicon metasurfaces is two-photon absorption (TPA), whereby two photons get absorbed by a semiconductor simultaneously to produce an electron-hole pair. TPA manifests itself as a sharp dip at near-zero delay on time-resolved transmittance or reflectance traces,[32-34] with a duration of the dip limited to the length of the optical pulses used in the experiment. Specifically, Della Valle et al.[34] modeled their pump-probe spectroscopy traces to break down the response of an evolving amorphous silicon metasurface to three main contributions: TPA, FC, and lattice heating. They revealed a set of operating windows where ultrafast all-optical modulation of transmission exists, with full return to zero in 20 ps. This happens via the distinct dispersive features exhibited by the competing nonlinear processes in transmission and despite the slow (nanosecond) internal lattice dynamics (Figure 6.4).

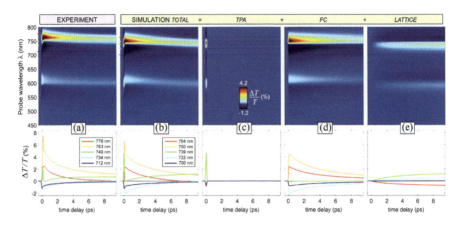

FIGURE 6.4   Transient transmittance of amorphous silicon metasurfaces. The experimental data (a) matches well the numerical results (b), which is broken down into several contributions: two-photon absorption (c), free-carrier dispersion (d) and lattice heating (e). (Reprinted from Della Valle, G. et al., *ACS Photonics*, 4, 2129–2136, 2017. With permission.)

Temporal separation and isolation of the instantaneous and the FC contributions has been implemented in Shcherbakov et al.[32]. Here, the authors used the dispersion of the magnetic Mie-type mode to create conditions that balance away the slow FC response to obtain a purely instantaneous, sub-50-fs response due to TPA. For typical TPA dips in pump-probe traces for metasurfaces, see Figure 6.5a. For one of the samples, the pump-probe trace only contains the sharp, 65-fs-long spike at zero delay (purple curve), whereas the undesirable FC contribution vanishes. Such a sharp separation of responses is explained by the proper spectral positioning of the magnetic Mie-type resonance with respect to the spectrum of the laser pulses. In Figure 6.5b, the red part of the resonance shows a more substantial contribution to transmittance modulation than that at the blue side of the resonance. TPA-induced all-optical modulation has also been observed in Fano-resonant, high-Q metasurfaces, as reported in Yang et al.[33]. Instantaneous phenomena, in sharp contrast with the free-carrier-related contributions, allow for possible switching rates at frequencies of more than 10 THz, paving the way to ultrafast logic gates using photonics.

6.3.3   ACTIVE CONTROL OF SCATTERING

Mie-resonant nanoparticles provide unique opportunities to spatial tailoring of light fields. In Makarov et al.[35], FC plasma was used to dynamically manipulate the scattering diagram of semiconductor nanoparticles. To numerically simulate changes in optical properties of a photoexcited silicon sphere with a diameter of $D=210$ nm at a wavelength of $\lambda = 800$ nm, the authors considered a range of absorbed fluences $F_{eff} < 100$ mJ/cm^2 that generated rather dense free-carrier plasma $\approx 10^{21}$ cm^{-3} for

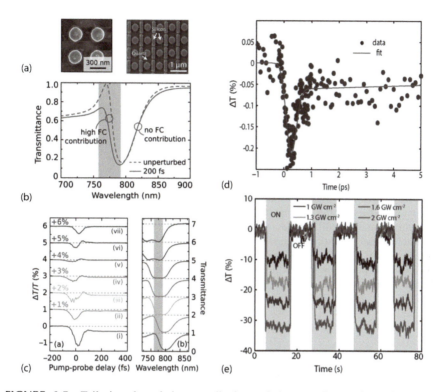

FIGURE 6.5 Tailoring the relative contributions of the two-photon absorption (TPA) and free-carrier (FC) dispersion in silicon metasurfaces. (a) Samples of amorphous silicon metasurfaces. Left: nanodisk metasurface with magnetic-dipole resonance, right: Fano-resonant metasurface with high-Q resonance. (b) Transmittance of the metasurface with a magnetic dipole resonance: unperturbed (solid line) and populated with FCs (dashed line). The left side of the resonance is affected by the FCs while its contribution is small on the right part of the resonance. (c) Transient transmittance of femtosecond pulses through seven different metasurfaces, with the corresponding spectra given on the right. For metasurface (vii), the relative contribution of the slow FC contribution to the transmittance is much smaller than that by the TPA. (d) TPA and FC contributions to the all-optical modulation trace in a Fano-resonant semiconductor metasurface. (e) Thermo-optic contribution to transmittance. (Reproduced from Yang, Y. et al., *Nano Lett.*, 15, 7388–7393, 2015; Shcherbakov, M.R. et al., *Nano Lett.*, 15, 6985–6990, 2015. With permission.)

efficient switching of the nanoparticle optical properties. In Figure 6.6, the scattering diagram of the nanoparticle is almost symmetric at $\lambda = 800$ nm and $F_{eff} \approx 0$, whereas the scattering cross-section has been boosted by a factor of about nine. The latter parameter is changed almost by a factor of three with fluence increasing up to $F_{eff} = 100$ mJ/cm^2 at the fixed wavelength of 800 nm, owing to the strong shift of the peak position of the scattering spectrum (Figure 6.6a). This effect was further explored in Baranov et al.[36]. This concept paves the way to the creation of low-loss, ultrafast, and compact devices for optical signal modulation and beam steering.

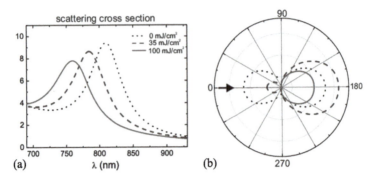

FIGURE 6.6 Dynamic tailoring of the directional scattering by Mie-resonant nanoparticles. (a) Calculated scattering spectra of a single undoped Mie-resonant silicon nanoparticle in thermodynamic equilibrium (dotted line) and under various pump fluences of 35 mJ/cm^2 (dashed line) and 100 mJ/cm^2 (solid line). (b) Corresponding scattering diagrams of light with a wavelength of 800 nm that impinges from the left at an angle of 0°. Highly unidirectional forward scattering is observed at the maximum pump power. The diameter of the particle is 210 nm, the excitation wavelength is 800 nm. (Reprinted from Makarov, S. et al., *Nano Lett.*, 15, 6187–6192, 2015. With permission.)

6.4 ULTRAFAST RESPONSE OF GALLIUM ARSENIDE METASURFACES

Gallium arsenide is a material that has been widely utilized in nanoelectronics. Its high electron mobility[37] for a long time has been enabling the world's fastest field-effect transistors. From the optoelectronics perspective, the bandgap of 1.42 eV makes it attractive to use Ti:Sapphire radiation (photon energy 1.55 eV) to induce interband transitions and efficiently generate FC plasma at low photon flux. This section reviews the attempts to use GaAs as the base material for semiconductor metasurfaces.

6.4.1 RESONANCE POSITION CONTROL

Photonic resonances define the functionalities of metasurfaces. Real-time control of the central resonance wavelength opens opportunities for dynamically reconfigurable metadevices, such as phase-gradient metasurfaces, metalens, and other spatial phase-shaping designs. A GaAs-based metasurface shown in Figure 6.7a, supports a magnetic dipole resonance that manifests itself as a reflectance peak at a wavelength of 1015 nm, see the blue curve in Figure 6.7b. Figure 6.7c shows the dependence of the reflectance spectra of the metasurface as a function of the time delay between the pump and probe. When the metasurface is pumped at 800 nm and free carriers modify the refractive index (pump-probe delay of 1 ps), the resonance wavelength is blue-shifted and gets centered at roughly 985 nm (purple curve in Figure 6.7b). At a delay of 6 ps, the resonance is back at its initial position (yellow curve in Figure 6.7b). Here, a pump fluence of only 310 µJ/cm^2 was used,

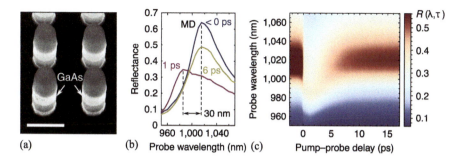

FIGURE 6.7 Active control and ultrafast relaxation of magnetic resonance in GaAs metasurfaces. (a) Scanning electron micrograph of a GaAs-based metasurface. The scale bar is 500 nm. (b) Reflectance of the metasurface under different pump-probe delays. With no pump present, the position of the magnetic dipole (MD) mode of the metasurface is at a wavelength of 1,015 nm (blue curve). At a pump–probe delay of 1 ps, the probe "sees" a blue-shifted resonance, centered at roughly 985 nm (purple curve). At a delay of 6 ps, the resonance is back at its initial position (yellow curve). (c) Reflectance of the metasurface as a function of time delay and wavelength.

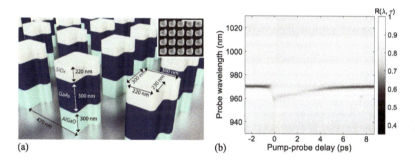

FIGURE 6.8 Resonance position control in high-Q metasurfaces with a bound-state-in-continuum mode. (a) Illustration of a symmetry-broken GaAs-based metasurface with a resonance at 970 nm. The inset shows a scanning electron micrograph of the fabricated sample; the scale bar is 1 μm. (b) Experimental free-carrier-induced shift of the resonance by 10 nm.

which corresponds to 250 fJ per disk.[38] The fast FC relaxation time stems from the large surface area of the nanoparticles, consistent with previous findings reported in photonic crystals.[39]

Even more dramatic tuning of the resonance was achieved in high-quality factor resonances that can be achieved by implementing a broken symmetry design as shown in Figure 6.8a. As shown in the transient spectra in Figure 6.8b, a 10-nm spectral blue-shift of the sharp resonance with a Q factor of 500 was observed with pump fluence of less than 100 μJ cm^{-2}. This corresponds to a relative all-optical reflectance modulation of 45%.

6.4.2 Ultrafast Nonlinear Response

Due to their high nonlinear susceptibilities, GaAs metasurfaces open vast possibilities for frequency conversion. Recently, an optical frequency mixer based on a GaAs-based dielectric metasurface[40] enabled a variety of simultaneous nonlinear optical processes across a broad spectral range. In Figure 6.9a, two collinearly propagating pump beams at frequencies of ω_1 and ω_2 were focused at the same location on the GaAs metasurface sample and generated eleven spectral peaks, ranging from ~380 to ~1,000 nm. Specifically, seven different nonlinear processes (second-harmonic generation (SHG), third-harmonic generation (THG), and fourth-harmonic generation (FHG), sum-frequency generation (SFG), two-photon absorption-induced photoluminescence (PL), four-wave mixing (FWM), and six-wave mixing (SWM)) simultaneously give rise to eleven new frequencies that span the ultraviolet to NIR spectral range. This multifunctional metamixer exploits the combined attributes of resonantly enhanced electromagnetic fields at the metasurface resonant frequencies, large even-order and odd-order optical nonlinearities of non-centrosymmetric GaAs, and significantly relaxed phase-matching conditions due to the subwavelength dimensions of the metasurface.

To investigate the temporal dynamics of the nonlinear generation processes, the signal intensities was measured while varying the optical delay between the two pump pulses. As expected, the harmonic generation signals and PL arising from two-photon absorption are observed regardless of the optical delay. In contrast,

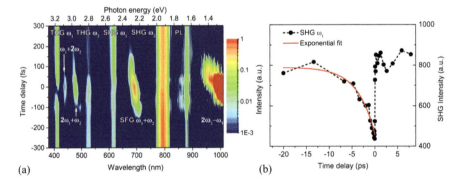

(a)   (b)

FIGURE 6.9 Ultrafast control of nonlinearities and frequency mixing in GaAs metasurfaces. (a) Spectra of the nonlinearly converted signal governed by second-harmonic generation (SHG), third-harmonic generation (THG), sum-frequency generation (SFG), four-wave mixing, and six-wave mixing processes, photoluminescence (PL) in a GaAs-based metasurface. The metasurface is being pumped by two femtosecond laser pulse trains at frequencies of ω_1 and ω_2 with a time delay between them marked in the ordinate axis of the plot. (b) Ultrafast switching of the SHG intensity from the ω_1 beam by the ω_2 beam. (Reprinted from Liu, S. et al., *Nat. Commun.*, 9, 2507, 2018. With permission.)

the frequency-mixing signals such as SFG, FWM, and SWM appear only within a 160 fs window when the two pump pulses temporally overlap at the metasurface. More interestingly, Figure 6.9b shows the dynamics of SHG from the ω_1 beam, where the intensity decreases dramatically at zero delay and then recovers with a time constant of a few picoseconds. The fast recovery of the SHG intensity (~3.7 ps) is due to the relaxation of the free carriers through both nonradiative recombination at surface states[24,39] and higher-order processes such as Auger recombination.[23]

6.5 TIME-VARIANT SEMICONDUCTOR METASURFACES AND PHOTON ACCELERATION

The concept of photon acceleration (PA) was originally introduced in gaseous plasmas[41,42] as a process of frequency conversion that occurs when electromagnetic waves propagate in a medium with a time-dependent refractive index.[43] PA in a solid (e.g., semiconductor) medium can be achieved at much lower laser intensities than in a gas because of the ease of FC generation and can be further enhanced in high-quality factor (high-Q) optical cavities.[44,45]

Recently, PA has been realized in semiconductor metasurfaces.[46] The main idea of photon acceleration in a semiconductor metasurface is given in Figure 6.10a. Briefly, mid-infrared photons interact with, and get trapped by, the metasurface. As FCs are generated by four-photon absorption, the resonant frequency of the metasurface blue-shifts, and the frequency of the trapped photons follows. Accelerated MIR photons then upconvert via the standard $\chi^{(3)}$ nonlinear process, resulting in the observed blue-shifting of the third-harmonic generation. The metasurface, Figure 6.10b, is designed to have a high-Q resonance at $\lambda_R = 3.62$ μm that enables efficient four-photon FC generation at modest pulse intensities. The effect of PA on harmonics generation emerged as upconverted radiation appearing at frequencies of up to $\approx 3.1\,\omega_L$, where ω_L is the central frequency of the MIR pulses. Moreover, anomalous levels of nonlinearly generated signal were detected with frequencies of up to $\approx 3.4\,\omega_L$ in the wings of the THG spectra, corresponding to the spectral density enhancement of $\approx 10^8$ over the projected signal from an unstructured film. In Figure 6.10c, a comparison is given between THG spectra generated by a photon-accelerating metasurface and an unstructured silicon film of the same thickness. In the silicon film, the central peak of the THG spectrum, as well as its width, stays the same for all the fluences of the pump beam. In contrast, THG from the metasurface shows strong blue-shift and broadening of the THG spectrum. An intuitive coupled-mode theory (CMT) model with time-dependent eigenfrequency $\omega_R(t)$ and damping factor $\gamma_R(t)$ accurately captures most features of the experimental data and provides further insights into PA efficiency improvements, thus paving the way to future applications utilizing nonperturbative nonlinear nanophotonics. One of the potential applications of photon-accelerating metasurfaces would be to fill the spectral gap between high optical harmonics to generate satellite-free isolated

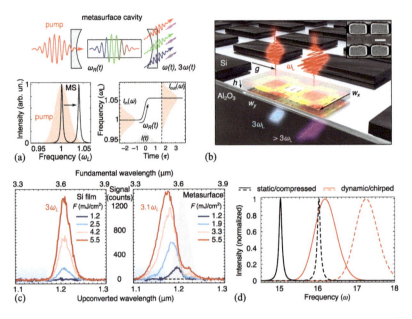

FIGURE 6.10 Photon acceleration in time-variant semiconductor metasurfaces. (a) The concept of blue-shifted harmonics generation. Mid-infrared (MIR) photons are trapped by the metasurface (MS) cavity, blue-shifted by the rapid refractive index variation due to free carrier generation inside the MS, and then nonlinearly upconverted to near-infrared (NIR) photons via the standard THG process. The blue-shifted MIR and NIR photons then leave the MS, and their spectra are detected in transmission. The red shaded areas in the plots illustrate the radiation spectra; the gray area represents the temporal profile of the incoming pulse. Solid and dashed curves illustrate the time-dependent nature of the MS resonance. (b) Schematic of the sample and the MIR beam setup. In experiments, the following dimensions of the sample were used: $w_x = 0.87$ μm, $w_y = 1.54$ μm, $g = 400$ nm, and $h = 600$ nm. Inset: a close-up scanning electron micrograph of the metasurface elements; scale bar: 1 μm. (c) Third harmonic generation (THG) spectra measured for various input fluences for the unpatterned Si film and the metasurface. Shaded gray area: THG from the unstructured film at the highest fluence $F_{max} = 5.5$ mJ cm^{-2}. Shaded blue areas indicate the integration range of $\lambda_{TH} < 1.17$ μm. (d) Overlapping high harmonics from time-variant metasurfaces (TVM). Spectra of the 15th (solid lines) and 16th (dashed lines) optical harmonics generated by a static resonator and a compressed pulse (black) and a TVM and a chirped pulse (red), revealing the spectral overlap of the accelerated harmonics. The spectra are normalized by unity for clarity.

attosecond pulses—something that is currently accomplished using bulky optical components. We propose to utilize the process of high-harmonic generation by a MIR pulse which nonlinearly interacts and gets spectrally broadened by a time-variant metasurface, as shown through CMT calculations in Figure 6.10d.[47] We anticipate that photon-accelerating metasurfaces could play a pivotal role in obtaining resonant solid-state high-harmonic continua for the generation of attosecond pulses in the extreme UV.

6.6 TEMPORAL PULSE SHAPING

In the view of recent advances in ultrashort laser pulse sources, it is essential to seek novel approaches to pulse shape management. Huygens' metasurfaces (Figure 6.11a) have been shown to efficiently manipulate wavefronts via their capability to create a desired discrete phase gradients between 0 and 2π.[48] In Figure 6.11b, the spectrally resolved relative phase delays for a systematic scaling of the nanodisks are given, these nanodisks may be then used to create spatially tailored light fields. However, this kind of metasurface can also support pathways to *temporal* shaping of laser pulses. In Figure 6.11c, a schematic of metasurface-based pulse shaper is shown. It provides a strong group-delay dispersion of more than -2000 fs^2 in combination with unity transmission, resulting in a Fourier-transform limited pulse. The authors of Decker et al.[48] managed to design a metasurface that recompresses a chirped 120-fs laser pulse at $\lambda = 800$ nm wavelength that has passed 20 cm of glass. Huygens' metasurfaces can provide remarkable dispersion control that can be integrated into a linear beam path as an ultrathin transmitting element.

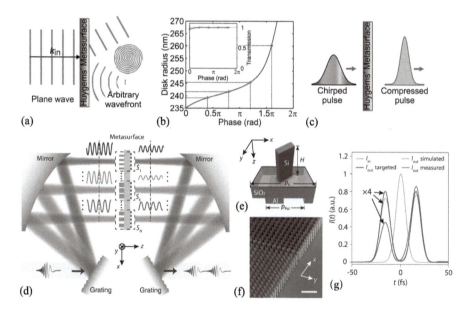

FIGURE 6.11 Shaping ultrashort laser pulses with semiconductor metasurfaces. (a) General concept of spatiotemporal shaping by Huygens' metasurfaces. (b) Engineering Huygens' metasurfaces by silicon nanodisks of various diameters. The phase of light transmitted through the metasurface can be tuned continuously by varying the nanodisk diameter, keeping the transmittance close to unity (see inset). (c) Scheme showing the recompression of a chirped pulse after passing a Huygens' metasurface. (d) Ultrashort pulse shaping setup with metasurfaces as the dispersion control element. (e) The unit cells of the metasurface consist of amorphous silicon cuboids, with their dimensions determining the phase of the output beam, which is then polarization-filtered by an aluminum grid at the bottom. (f) A fragment of the fabricated metasurface. (g) The result of pulse shaping: an incoming ultrashort pulse is split into two.

Another approach to pulse shape management has been recently revealed by Divitt et al.[49] They embedded a silicon-based metasurface in the focal plane of a Fourier-transform (spectral dispersing-recombining) setup shown in Figure 6.11d. Here, the metasurface enabled pulse shaping that tailors the temporal profile of a femtosecond laser pulse. As an example, silicon nanopillars coupled to an aluminum polarizer shown in Figure 6.11e and f were combined to provide a compact, highly dispersive element to replace bulky spatial light modulators in pulse shapers. As an example, Figure 6.11g shows a resulting waveform that contains two isolated pulses, along with the respective calculations. This approach promises novel applications of metasurfaces as two-dimensional wavefront shapers to be translated into the temporal domain, enabling flexible spatiotemporal shaping of light.[50]

6.7   CLOSING REMARKS AND OUTLOOK

Although being a young research field, semiconductor metasurfaces have shown promises for a variety of active and ultrafast photonics applications. We envision, in the foreseeable future, further efforts in several emergent directions. Currently, silicon and gallium arsenide are dominant in semiconductor metasurface photonics. Novel materials, such as GaP,[51] Ge,[21] InSb,[52] BaTiO,[53] LiNbO$_3$[54] chalcogenide glasses,[55] and others have promise for ultrafast photonics and electrooptics, expanding the scope of ultrafast metasurfaces to electromagnetic spectrum beyond the infrared. Integration of semiconductor nanocavities with other materials, such as two-dimensional materials,[56–58] quantum-confined systems,[59] perovskites,[60] may bring qualitative improvements in cooldown times, switching powers and spectral ranges available for ultrafast processes described in this chapter. Since ultrafast phenomena are relevant to various integrated photonics applications,[61,62] we predict more exciting results in this direction. A plethora of ultrafast effects are yet to be explored in semiconductor metasurfaces, including Franz-Keldysh effect, ponderomotive nonlinearities, as well as nonperturbative regimes of light-matter interaction and high-harmonic generation.[63] Finally, the next frontier in ultrafast metasurface research will encompass a wealth of possibilities offered by spatially inhomogeneous metasurfaces, such as phase plates, lens, diffraction gratings, and other flat optics devices. Semiconductor metasurfaces offer unprecedented flexibility in spatiotemporal control of light fields on femto- and nanoscales, and will serve as fundamental building blocks for novel photonic metadevices.

REFERENCES

1. Kuznetsov, A. I., Miroshnichenko, A. E., Brongersma, M. L., Kivshar, Y. S. & Luk'yanchuk, B. Optically resonant dielectric nanostructures. *Science*. **354**, aag2472 (2016).
2. Jahani, S. & Jacob, Z. All-dielectric metamaterials. *Nat. Nanotechnol.* **11**, 23–36 (2016).

3. Staude, I. & Schilling, J. Metamaterial-inspired silicon nanophotonics. *Nat. Photon.* **11**, 274–284 (2017).
4. Khorasaninejad, M. *et al.* Metalenses at visible wavelengths: Diffraction-limited focusing and subwavelength resolution imaging. *Science.* **352**, 1190–1194 (2016).
5. Paniagua-Domínguez, R. *et al.* A Metalens with a near-unity numerical aperture. *Nano Lett.* **18**, 2124–2132 (2018).
6. Chong, K. E. *et al.* Efficient polarization insensitive complex wavefront control using Huygens' metasurfaces based on dielectric resonant meta-atoms. *ACS Photonics* **3**, 514–519 (2016).
7. Shalaev, M. I. *et al.* High-efficiency all-dielectric metasurfaces for ultracompact beam manipulation in transmission mode. *Nano Lett.* **15**, 6261–6266 (2015).
8. Chong, K. E. *et al.* Refractive index sensing with Fano resonances in silicon oligomers. *Phil. Trans. R. Soc. A* **375**, 20160070 (2017).
9. Tittl, A. *et al.* Imaging-based molecular barcoding with pixelated dielectric metasurfaces. *Science.* **360**, 1105–1109 (2018).
10. Yavas, O., Svedendahl, M., Dobosz, P., Sanz, V. & Quidant, R. On-a-chip biosensing based on all-dielectric nanoresonators. *Nano Lett.* **17**, 4421 (2017).
11. Krasnok, A., Caldarola, M., Bonod, N. & Alú, A. Spectroscopy and biosensing with optically resonant dielectric nanostructures. *Adv. Opt. Mat.* **6**, 1701094 (2018).
12. Li, G., Zhang, S. & Zentgraf, T. Nonlinear photonic metasurfaces. *Nat. Rev. Mater.* **2**, 17010 (2017).
13. Krasnok, A., Tymchenko, M. & Alù, A. Nonlinear metasurfaces: A paradigm shift in nonlinear optics. *Mater. Today* **21**, 8–21, 165217 (2018).
14. Weiner, A. *Ultrafast Optics.* (Hoboken, NJ: John Wiley & Sons, 2011).
15. Sabbah, A. & Riffe, D. Femtosecond pump-probe reflectivity study of silicon carrier dynamics. *Phys. Rev. B* **66**, 1–11 (2002).
16. Zu, M. *et al.* Direct and simultaneous observation of ultrafast electron and hole dynamics in germanium. *Nat. Commun.* **8**, 15734 (2017).
17. Shah, J. *Ultrafast Spectroscopy of Semiconductors and Semiconductor Nanostructures.* (New York: Springer, 1999).
18. Boyd, R. W. *Nonlinear Optics.* (Orlando: Academic Press, 2008).
19. Rahmani, M. *et al.* Reversible thermal tuning of all-dielectric metasurfaces. *Adv. Funct. Mater.* **27**, 1–7, 1700580 (2017).
20. Lewi, T., Butakov, N. A. & Schuller, J. A. Thermal tuning capabilities of semiconductor metasurface resonators. *Nanophotonics* **8**, 331–338 (2018).
21. Bosch, M., Shcherbakov, M. R., Fan, Z. & Shvets, G. Polarization states synthesizer based on a thermo-optic dielectric metasurface. *J Appl. Phys.* **126**, 073102 (2019).
22. Ashcroft, N. W. & David Mermin, N. *Solid State Physics.* (Philadelphia: Saunders College Publishing, 1976).
23. Strauss, U., Rühle, W. W. & Köhler, K. Auger recombination in intrinsic GaAs. *Appl. Phys. Lett.* **62**, 55–57 (1993).
24. Bristow, A. D. *et al.* Ultrafast nonlinear response of AlGaAs two-dimensional photonic crystal waveguides. *Appl. Phys. Lett.* **83**, 851–853 (2003).
25. Preble, S. F., Xu, Q., Schmidt, B. S. & Lipson, M. Ultrafast all-optical modulation on a silicon chip. *Opt. Lett.* **30**, 2891–2893 (2005).
26. Almeida, V. R., Barrios, C. A., Panepucci, R. R. & Lipson, M. All-optical control of light on a silicon chip. *Nature* **431**, 1081–1084 (2004).
27. Nozaki, K. *et al.* Sub-femtojoule all-optical switching using a photonic-crystal nanocavity. *Nat. Photon.* **4**, 477–483 (2010).
28. Esser, A. *et al.* Ultrafast recombination and trapping in amorphous silicon. *Phys. Rev. B* **41**, 2879–2884 (1990).

29. Baranov, D. G. *et al.* Nonlinear transient dynamics of photoexcited silicon nanoantenna for ultrafast all-optical signal processing. *ACS Photon.* **3**, 1546–1551 (2016).

30. Esser, A. *et al.* Ultrafast recombination and trapping in amorphous silicon. *J. Non. Cryst. Solids* **114**, 573–575 (1989).

31. Lin, Q., Painter, O. J. & Agrawal, G. P. Nonlinear optical phenomena in silicon waveguides: Modeling and applications. *Opt. Express* **15**, 16604–16644 (2007).

32. Shcherbakov, M. R. *et al.* Ultrafast all-optical switching with magnetic resonances in nonlinear dielectric nanostructures. *Nano Lett.* **15**, 6985–6990 (2015).

33. Yang, Y. *et al.* Nonlinear fano-resonant dielectric metasurfaces. *Nano Lett.* **15**, 7388–7393 (2015).

34. Della Valle, G. *et al.* Nonlinear anisotropic dielectric metasurfaces for ultrafast nanophotonics. *ACS Photonics* **4**, 2129–2136 (2017).

35. Makarov, S. *et al.* Tuning of magnetic optical response in a dielectric nanoparticle by ultrafast photoexcitation of dense electron-hole plasma. *Nano Lett.* **15**, 6187–6192 (2015).

36. Baranov, D. G., Makarov, S. V., Krasnok, A. E., Belov, P. A. & Alù, A. Tuning of near- and far-field properties of all-dielectric dimer nanoantennas via ultrafast electron-hole plasma photoexcitation. *Laser Photon. Rev.* **10**, 1009–1015 (2016).

37. Wolfe, C. M., Stillman, G. E. & Lindley, W. T. Electron mobility in high-purity GaAs. *J. Appl. Phys.* **41**, 3088 (1970).

38. Shcherbakov, M. R. *et al.* Ultrafast all-optical tuning of direct-gap semiconductor metasurfaces. *Nat. Commun.* **8**, 15 (2017).

39. Husko, C. *et al.* Ultrafast all-optical modulation in GaAs photonic crystal cavities. *Appl. Phys. Lett.* **94**, 021111 (2009).

40. Liu, S. *et al.* An all-dielectric metasurface as a broadband optical frequency mixer. *Nat. Commun.* **9**, 2507 (2018).

41. Yablonovitch, E. Self-phase modulation of light in a laser-breakdown plasma. *Phys. Rev. Lett.* **32**, 1101–1104 (1974).

42. Wilks, S. C., Dawson, J. M., Mori, W. B., Katsouleas, T. & Jones, M. E. Photon accelerator. *Phys. Rev. Lett.* **62**, 2600–2603 (1989).

43. Felsen, L. B. & Whitman, G. M. Wave propagation in time-varying media. *IEEE Trans. Antennas Propag.* **AP-18**, 242–253 (1970).

44. Preble, S. F., Xu, Q. & Lipson, M. Changing the colour of light in a silicon resonator. *Nat. Photon.* **1**, 293–296 (2007).

45. Tanabe, T., Notomi, M., Taniyama, H. & Kuramochi, E. Dynamic release of trapped light from an ultrahigh-Q nanocavity via adiabatic frequency tuning. *Phys. Rev. Lett.* **102**, 043907 (2009).

46. Shcherbakov, M. R. *et al.* Photon acceleration and tunable broadband harmonics generation in nonlinear time-dependent metasurfaces. *Nat. Commun.* 10, 1345 (2019).

47. Shcherbakov, M. R. *et al.* Enhancing harmonics generation by time-variant metasurfaces. *Proc. SPIE* **10927**, 109270F (2019).

48. Decker, M. *et al.* High-efficiency dielectric Huygens' surfaces. *Adv. Opt. Mat.* **3**, 813–820 (2015).

49. Divitt, S., Zhu, W., Zhang, C., Lezec, H. J. & Agrawal, A. Ultrafast optical pulse shaping using dielectric metasurfaces. *Science* **894**, 890–894 (2019).

50. Shcherbakov, M., Eilenberger, F. & Staude, I. Interaction of semiconductor metasurfaces with short laser pulses: From nonlinear-optical response towards spatiotemporal shaping. *J. Appl. Phys.* **126**, 085705 (2019).

51. Cambiasso, J. *et al.* Bridging the gap between dielectric nanophotonics and the visible regime with effectively lossless gallium phosphide antennas. *Nano Lett.* **17**, 1219–1225 (2017).

52. Iyer, P. P., Pendharkar, M. & Schuller, J. A. Electrically reconfigurable metasurfaces using heterojunction resonators. *Adv. Opt. Mat.* **4**, 1582–1588 (2016).

53. Timpu, F., Sergeyev, A., Hendricks, N. R. & Grange, R. Second-harmonic enhancement with mie resonances in perovskite nanoparticles. *ACS Photonics* **4**, 76–84 (2017).

54. Sergeyev, A. *et al.* Enhancing waveguided second-harmonic in lithium niobate nanowires. *ACS Photonics* **2**, 687–691 (2015).

55. Xu, Y. *et al.* Reconfiguring structured light beams using nonlinear metasurfaces. *Opt. Express* **26**, 30930 (2018).

56. Wang, K. *et al.* Ultrafast saturable absorption of two-dimensional MoS2 nanosheets. *ACS Nano* **7**, 9260–7 (2013).

57. Tsai, D.-S. *et al.* Few-layer MoS2 with high broadband photogain and fast optical switching for use in harsh environments. *ACS Nano* **7**, 3905–11 (2013).

58. Poellmann, C. *et al.* Resonant internal quantum transitions and femtosecond radiative decay of excitons in monolayer WSe2. *Nat. Mater.* **14**, 889–893 (2015).

59. Müller, K. *et al.* Ultrafast polariton-phonon dynamics of strongly coupled quantum dot-nanocavity systems. *Phys. Rev. X* **5**, 1–7, 031006 (2015).

60. Makarov, S. V. *et al.* Multifold emission enhancement in nanoimprinted hybrid perovskite metasurfaces. *ACS Photonics* **4**, 728–735 (2017).

61. Li, J. *et al.* All-dielectric antenna wavelength router with bidirectional scattering of visible light. *Nano Lett.* **16**, 4396–4403 (2016).

62. Okhlopkov, K. I. *et al.* Optical coupling between resonant dielectric nanoparticles and dielectric nanowires probed by third harmonic generation microscopy. *ACS Photon.* **6**, 189–195 (2019).

63. Liu, H. *et al.* Enhanced high-harmonic generation from an all-dielectric metasurface. *Nat. Phys.* **14**, 1006–1010 (2018).

7 Nonlinear Photonic Crystals

P. Colman, S. Combrié, A. De Rossi,
A. Martin, and G. Moille

CONTENTS

7.1   INTRODUCTION

The variety of nonlinear optical phenomena is amazingly broad, which was quickly realized in the wake of the invention of the laser [7]. Nonlinear optics has been for long time associated with the notion of a large optical field, hence high power laser sources.

As early as the concept of "Photonic Crystals" has been introduced, S. John in 1987 pointed out that: the utilization of localization as a trigger mechanism for non-linear or bistable response, as in the Kerr electro-optic effect, may lead to a number of useful device applications [79]. This suggested that nonlinear effect could be evidenced by achieving large optical fields by localizing energy in tiny volumes instead of using a powerful source.

The practical implementation of this insightful comment has faced formidable technical challenges, yet the goal has been achieved 30 years later. The first challenge was to build such a structure as a "photonic crystal," where geometry has to be controlled at the nanometer scale over several periods. The second challenge was master the large degrees of freedom involved in the design of cavities and waveguides and preventing radiative leakage. The third challenge was to identify suitable materials capable of sustaining the large density of optical energy resulting from the localization of light in tiny volumes. In this chapter, we will provide a perspective on the achievement of strong nonlinear interaction in semiconductor photonic crystals, to which the authors have been contributing for about a decade.

We will first revisit the basic properties of a particular class of Photonic Crystals, namely the two-dimensional periodically patterned semiconductor slabs and discuss the mechanisms leading to enhanced light-matter interaction and large non-linear response. Then we will consider two examples where nonlinear optics in

PhC structures not only has been demonstrated but also reveals peculiar aspects. The first is the nonlinear dynamics of short pulses in PhC waveguides. These nonlinear waves, observed for the first time in fiber in 1981 and named "solitons," result from the interplay of nonlinearity and dispersion. What is remarkable here is that the relevant spatial scale is in the sub-millimeter range, and, therefore, Soliton physics can go "on chip."

Another remarkable achievement is the demonstration of efficient parametric interactions, also within one millimeter, meaning that most of the all-optical signal processing, demonstrated in optical fibers and essentially based on four-wave mixing (FWM), could be integrated on chip. Thus, photonic integrated chips could be able to perform coherent wavelength conversion and amplification, signal regeneration, and many other operations along with established linear functions.

Thus, the research in the field of nonlinear photonic crystals contributes to the convergence of photonics and electronics by providing new functionalities on the chip scale, which, besides signal processing, address application domains ranging from metrology to quantum information and nonclassical computing.

7.2 LOCALIZATION AND PROPAGATION OF OPTICAL FIELDS IN STRUCTURED DIELECTRICS

The localization of light is somehow a balance between different sources of losses, which are in general exacerbated when the size of the mode is reduced and approaches the diffraction limit. In fact, diffraction losses are the main limitation in small dielectric resonators. This appears for instance in micro disks or micro-rings when the radius of curvature is decreased [196]. In microwaves, in contrast, subwavelength high-Q resonator are feasible, as these resonators are made of metals and diffraction loss are avoided. Translating these concepts in the optical domain has not revealed straightforward, however. The Q factor of optical resonators based on a surface plasmon resonance at the surface of a metal tends to be reduced to about 100 or below as the confinement goes below the diffraction limit. This point has been clarified very recently [82] and is summarized hereafter. Subwavelength optical modes are localized within a plasma wavelength, as a consequence, the oscillation of the electromagnetic field involves the periodic exchange between the energy stored in the electric field and the kinetic energy of free carriers (electrons in metal), instead of the magnetic energy. This energy is lost at a rate that is of the same order of magnitude of the electron scattering time in metals, which is below 100 fs. Therefore, the Q factor of subwavelength resonators at optical frequencies are (severely) limited by the material, whatever the geometry.

When the resonant enhancement of the nonlinear interaction matters (e.g., the Q^2/V scaling of the power threshold for optical parametric oscillations or bistable switching, but also propagation loss in waveguides), the severe limitation to Q is a too high price to pay for subwavelength confinement. The interest of PhC in this respect is in the possibility of achieving a close to diffraction limit confinement while preserving a large Q factor, only limited by material. However, for dielectrics, this limit is way more forgiving than metals.

7.2.1 Photonic Crystals

Photonic Crystals (PhC) are optical structures, in general made of transparent dielectrics, with periodic modulation of the refractive index. In one dimensional geometries, such as a Distributed Bragg Reflector (DBR) in an optical fiber, this leads systematically to the onset of a "photonic band-gap," namely a spectral interval where the propagation is forbidden (Figure 7.1a). The width of the gap is proportional to the contrast in the modulation of the refractive index [201]. In two or three dimensional geometries, a complete photonic band-gap only opens if the modulation contrast is large enough [198]. The existence of a complete band-gap, whereby propagation is impossible in any direction, bears very important implications. First, light-matter interaction is strongly modified. In particular, the spontaneous radiative lifetime of electric dipoles placed inside the PhC would either be increased or decreased compared to their value when the dipole is in free space. Applied to semiconductor laser sources, that would improve efficiency by inhibiting spontaneous emission [199]. Second, strong localization [79] would result from disordering the periodic lattice (Figure 7.1a), in analogy with defect and impurity states in semiconductors. S. John noted that strong, wavelength-scale localization, would greatly enhance nonlinear effects.

While the fabrication of truly three-dimensional PhC is still challenging, more interesting two or one dimensional structures have been demonstrated since two decades, and rely on the advanced semiconductor processing technology [87]. In fact, a suspended slab of semiconductor patterned with an hexagonal lattice of holes could provide a complete band-gap in the plane of the PhC, which is enough for the localization of optical modes, while the large index contrast would confine light

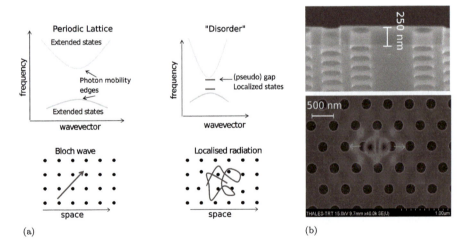

(a)   (b)

FIGURE 7.1 (a) Dispersion of electromagnetic waves in a periodic dielectric lattice. In a disordered lattice localized states appear close to the mobility edges. (From John, S., *Phys. Rev. Lett.*, 58, 2486–2489, 1987.) (b) SEM image representing a suspended PhC membrane waveguide (view of the cleaved edge) and H0 cavity (view from the top) with superimposed the map (linear scale) of the localized electric field.

out of plane, within a minimum in-plane wavevector defining the *light cone* [80]. Figure 7.1b shows the Scanning Electronic Micrograph (SEM) image of a suspended semiconductor slab made of Gallium Arsenide. The PhC pattern is written by electron-beam lithography and transferred to the semiconductor through a two steps etching procedure. The suspended membrane is obtained by removing the underlying sacrificial layer by chemical etching [29].

7.2.2 PHOTONIC CRYSTAL CAVITIES

A resonant cavity is formed by introducing a defect in the periodic lattice [200]. For instance, a missing hole would create the equivalent of a donor level, namely a resonance close to the upper edge of the band-gap, or mobility edge as referred to in Ref. [79]. In two-dimensional PhC slabs, the design of the defect is crucial in order to reduce the radiation leakage and therefore increase the photon lifetime. This was realized long ago [5] and led to a steady improvement of the design such that today the limit is set by the material (residual absorption) and the fabrication (roughness). The current record is 1.1×10^7, achieved in a silicon platform [8]. The design of the cavity can be optimized with respect to localization, instead that for minimal radiation losses. For instance, the design proposed by Zhang and Min Qiu [210] results into a mode volume (see Table 7.1) of $V_m = 2\lambda^3(2n)^{-3}$, namely twice the diffraction limit. An example of this cavity made of GaAs [74] is shown in shown in Figure 7.1b. The map representing the optical mode clearly indicates a very strong confinement within a scale of about 500 nm. This design has been optimized recently [137,128] to achieve a radiation limited Q above 10^7. The large figure of merit Q/V_m is relevant to light-matter interaction and nonlinear optics. It is important to note that the modal volume is relevant to the interaction with a small dipole (e.g., an atom

---

TABLE 7.1
Definitions and Formulas Describing Confinement in Cavities and Waveguides

Definition	Formula	Comment	References
	Cavities		
Normalization	$\int dV \, \varepsilon_r \varepsilon_0 \, |\mathbf{e}|^2 = 2$		
Mode Volume	$V_m^{-1} = \dfrac{1}{2} max(\varepsilon_r \, |\mathbf{e}|^2)$		
NL Volume	$V_\chi^{-1} = \dfrac{\varepsilon_0^2 \varepsilon_r^2}{4} \displaystyle\int_{\chi^{(3)} \neq 0} \dfrac{2|\mathbf{e}|^4 + |\mathbf{e} \cdot \mathbf{e}|^2}{3} \, dV$	int. over NL	[31]
NL detuning	$\dfrac{\Delta\omega}{\omega} = -\dfrac{c_0 n_2}{\varepsilon_r V_\chi} W$		[129]
OPO Threshold	$P_{th} = \dfrac{\varepsilon_r V_{FWM} \omega}{4 c_0 n_2 Q^2}$	overcoupled[a]	[31]

(Continued)

TABLE 7.1 (*Continued*)

Definitions and Formulas Describing Confinement in Cavities and Waveguides

Definition	Formula	Comment	References
	Cavities		
Master Equation	$\dot{a} = (i\Delta\omega - 1/\tau)a - i\dfrac{c_0 n_2}{\varepsilon_r V_\chi}\mid a\mid^2 a$		
Field	$\mathbf{E}(\mathbf{r},t) = \mathbf{e}(\mathbf{r})\exp(i\omega t)a(t)$		
	Waveguides		
Normalization	$\displaystyle\int_S dS\,\hat{z}\cdot(e^*\times h + e\times h^*) = 4$	propagation along z	
Mode Area	$A_m^{-1} = \varepsilon_0 c_0 max(\sqrt{\varepsilon_r}\mid \mathbf{e}\mid^2)/2$	Max Irradiance $= PA_m^{-1}$	[21]
NL Area	$A_\chi^{-1} = \dfrac{\varepsilon_0^2 c_0^2}{4L}\displaystyle\int_{V_{\chi^{(3)}\neq 0}} dV\varepsilon_r\dfrac{2\mid\mathbf{e}\mid^4 + \mid\mathbf{e}\cdot\mathbf{e}\mid^2}{3}$	Bloch period L	[113]
Propagation equation	$\partial_z a = i\dfrac{n_2\omega}{c_0 A_\chi}\mid a\mid^2 a$		
Field	$\mathbf{E}(\mathbf{r},t) = \mathbf{e}(\mathbf{r})\exp[i(kz - \omega t)]a(t)$		
	Parameters and Variables		
Normal Bloch Mode	$\mathbf{e}(\mathbf{r})$	Electric Field	$\mathbf{E}(\mathbf{r},t)$
Third-order susceptibility	$\chi^{(3)}$	Nonlinear index	$n_2\,[m^2/W]$
Energy (cavity)	$W = \mid a\mid^2$	Power (waveguide)	$P = \mid a\mid^2$
Mode amplitude	$a(t)$	Photon lifetime	τ
Irradiance	$I\,(W/m^2)$		

a V_{FWM} denotes the NL volume for the FWM process, which is a generalized definition of V_χ.

or a Quantum Dot). A more appropriate measure of the enhancement of the non-linear material forming the photonic crystal is rather the nonlinear volume V_χ (see Table 7.1). For instance, the power threshold for bistable switching [172] or parametric oscillations depends on V_χ, not on V_m, unless the dominant nonlinear response arises from a point-like dipole as in Ref. [186].

7.2.3 PHOTONIC CRYSTAL WAVEGUIDES

In a translation-invariant dielectric structure such as a waveguide (or an optical fiber), the confinement results from the total internal reflection. Notable exceptions are Photonic Crystal Fibers [157] and antiresonant reflecting optical waveguides (ARROW) [84]. Typically, the effective index n_{eff} increases from the value of the

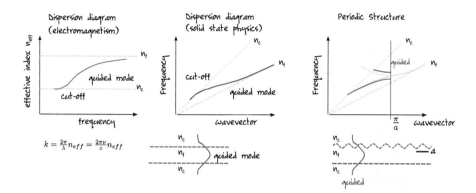

FIGURE 7.2 Typical dispersion (solid line) of a planar waveguide where the confinement results from the total internal reflection. The addition of a periodic modulation results into a modification of the dispersion curve, now represented in the $[-\pi/a, \pi/a]$ reduced Brillouin Zone. Some modes are above the light line and are therefore leaky (thin gray solid line). The gray line represents the unperturbed dispersion and with is folded replica (dashed) in the first BZ.

cladding n_c to that of the guiding layer n_f as frequency is increased. When considering the dispersion, e.g., frequency ω vs. wavevector k (Figure 7.2), then it can be noticed that the group index (proportional to the inverse group velocity) $n_g/c = \partial_\omega k$ first increases then decreases, meaning that the waveguide contribution to the dispersion is negative: $\partial_\omega^2 k < 0$. In tightly confined waveguides such as silicon ribs, this effect can be strong enough to offset the material dispersion, which is positive, therefore allowing dispersion engineering [179]. This result is of considerable practical importance as it opens up the whole field of Soliton physics and parametric interactions [184,207], which will be discussed in the next section.

When the translation invariance is broken, for instance by corrugating the interface periodically, then a forbidden gap opens up in the dispersion of the waveguide mode, the upper branch of which is folded to be represented in the first Brillouin Zone (Figure 7.2). This has two remarkable implications: part of the upper branch will cross the $\omega = k/n_c$ line, meaning that the waveguide mode hybridizes with free space modes in the cladding layer. These *leaky* modes are lossy because they radiate in the cladding. The second property is that if the refractive index is large enough, the opening of the forbidden gap is large enough to induce a very strong change of the dispersion, such that the material dispersion becomes negligible.

A Photonic Crystal Waveguide can be thought as a structure allowing an extended propagating state, for instance obtained by the hybridization of a line of localized states. The most common example is the so-called *missing line-defect* design, consisting, for instance, in a line of missing holes (Figure 7.3). This defines a direction in the reciprocal space of the 2D lattice where translation invariance is maintained for *discrete* displacements. Line-defect PhC waveguides made of 2D slabs have been considered in Ref. [81], the main points of which are summarized below. In the case considered in Figure 7.3, the line is oriented along the K. The reciprocal space is projected on the ΓK axis and the reduced Brillouin Zone (BZ) is defined between Γ and K',

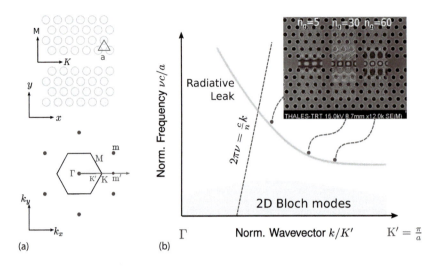

FIGURE 7.3 (a) PhC waveguide formed by a missing line of holes in an hexagonal lattice of period a. The missing line defect is oriented along the K direction. The First Brillouin Zone (and the reduced zone, filled area) is represented in the reciprocal space along with the high symmetry points Γ, K, M. Its projection on the ΓK axis defines the reciprocal space for the waveguide, which is mapped in the interval $[0\ \pi/a]$. (b) Dispersion diagram representing the even gap-guided mode (solid line). The shaded area above the $\omega = kc/n$ line represents leaky modes. The "valence" band (the lower mobility edge in Figure 7.1) is represented by the shading. The color maps superimposed to the SEM image represent the Electric Field corresponding to different modes (dots over the dispersion).

the latter being the projection of M on the axis. The length of the vector $\Gamma - K'$ is π/a, as for a one dimensional lattice with period a.

The most important propagating mode is located inside the band-gap of the background photonic lattice and has even parity relative to the $x - z$ plane (namely: $\mathbf{e}(-y) \cdot \mathbf{y} = \mathbf{e}(y) \cdot \mathbf{y}$). It is therefore *gap-guided* and bears similarity with the upper branch of the mode of a 1D periodic structure shown in Figure 7.2. Part of this branch crosses the light line, defined by the condition $\omega = \frac{c}{n} k$. Within the light cone, delimited by the light line, the mode is hybridized with free-space radiation and it is therefore leaky. For practical applications, it is of the utmost importance to operate the structure outside the light cone.

The strong modulation of the refractive index contrast resulting into an opening of a full (2D) photonic gap also induces a strong modification of the dispersion over a broad spectral range, relative to structures where the periodic modulation is a small perturbation (e.g., DBR fibers). This results in the group velocity being way smaller than the phase velocity [149]. "Slow-light" propagation was first reported in PhC waveguide made of silicon slabs [139]. The decrease of the group velocity is connected to a marked reshaping of the Bloch mode which is apparent in the inset in Figure 7.3b. Moreover, the dispersion is strongly anomalous, which allows the possibility to observe and tailor the nonlinear pulse dynamics on the chip scale. This will be discussed in the next section.

7.2.4 Arrays of PhC Cavities

Resonators can be coupled to form more complex optical structures, similarly to atomic physics where two identical atoms are coupled to form a molecule where energy levels are split proportionally to the coupling strength, which is captured by the tight-binding model. Following the analogy, a PhC "molecule" resonator can be created where the fundamental mode is split into a bonding and an anti-bonding modes (Figure 7.3) and it is conveniently described by the coupled-mode theory (CMT) [110]. This kind of multimode resonator has interesting properties, in particular the bonding and anti-bonding modes are very similar, which maximizes the nonlinear coupling. This has been exploited for all-optical switching [30,129,131]. Moreover, the inherent symmetry of a molecule is the starting point for the demonstration of PT symmetric systems [65] (Figure 7.4).

When more resonators are coupled together, light can "hop" from resonator to resonator with a speed controlled through the hopping time, i.e., the coupling. This was pointed out by Yariv et al. [203], who introduced the concept of Coupled-Resonator Optical Waveguide (CROW), as pictured in Figure 7.3. An infinite chain of resonator has a typical sinusoidal dispersion $\omega = \omega_0[1 + \gamma/2\cos(ka)]$, a being the distance between resonators. N resonators will corresponds to an equal number of states $k_n = n\pi/L$. PhC CROW have been fabricated with up to 1000 cavities [89] and a large group delay observed (≈ 100 ps) corresponding to an estimated photon lifetime $\tau > 400$ ps. This triggers expectations for reconfigurable photonic devices and efficient nonlinear interactions.

7.2.5 "Structural" Slow-Light

In the context of light-matter interaction, e.g., optical resonances with atomic transition, "slow-light" relates to the propagation into a material whose dispersion is such

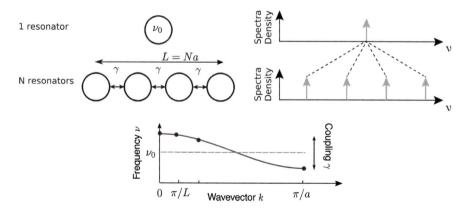

FIGURE 7.4 Schematic of coupled resonators to form a CROW. When a single resonator with a fundamental mode (top part) is coupled to several other identical ones, the fundamental modes is split in as many modes as the structure is made of coupled cavities. The modes are spaced evenly in the wavevector space and the resonances follow a sinusoidal dispersion.

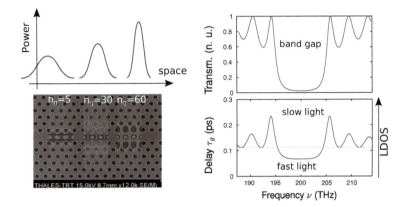

FIGURE 7.5 Slow light in periodic photonic lattices. Left: Enhancement of the electric field regarded as a spatial compression of a pulse, following the slowing down of the group velocity. Right: Distributed Bragg reflector as a model for structural slow light. Modification of the group delay $\tau_d = -\partial_\omega\phi$ relative to free-photons τ_0 in free space with same optical length (dashed line). The field enhancement factor is proportional to τ_g/τ_0.

that the group velocity is very low. Here we consider instead the interaction of waves in a structured yet nondispersive material. When light propagates in a PhC waveguide on a mode with reduced group velocity, the electric field increases proportional to n_g/n, i.e., the ratio between the group and the effective index of the material, also designated as the *slow-down* factor [86]. This can be interpreted as a *spatial compression* of a light pulse [20,134]. This is shown in Figure 7.5, whereby when the group velocity decreases the spatial envelope of the pulse reduces. It has been shown theoretically and experimentally that slow light in structured material is related to an enhancement of the nonlinear response of the material [134,160].

It has to be noted that the enhancement holds for a CW excitation as well, although in this case the definition of the propagation speed is problematic. The confusion arises from interpreting the slow-light enhancement of the light-matter interaction, hence of the nonlinearity, as being due to an increased time of light-matter interaction. A possibly better interpretation is formulated in terms of the Local Density of the Optical States (L-DOS). In cavity electro-dynamics, this effect is known as the Purcell effect. The Purcell formula can be extended to open geometry systems (waveguides) and exhibits then a contribution from the first derivative of the band diagram $\partial_\omega k$, which we have precisely defined as being the group velocity. That is why we differentiate *structured slow light* from *material slow-light* where the L-DOS remain unchanged. As a result, slow-light enhancement of nonlinearity does not exist in latter system. Thus, the slow-down factor $S = n_g/n$ in a structured waveguide can be interpreted to be the counterpart of the quality factor Q in a cavity [107].

This point has been clarified by Boyd, et al. [17]. When entering a "slow-light" medium (dispersive), the energy density increases but not the field strength. This can be formulated as follows: the conservation of the flux of energy links the Irradiance I to the energy density W through the group velocity $I = Wv_g$; therefore, the flux of energy being constant, a decrease of the group velocity leads to an increase of the energy density. In a dispersive material, however, the energy and the field are related through a term depending on the group velocity, that is $W \propto |E|^2 / v_g$. The dispersive contribution entirely accounts for the increase of the energy density, as a result, the relation between I and $|E|^2$ is independent of v_g.

In contrast, when dispersion relation between the wavevector k and the frequency $\omega/2\pi$ is entirely dominated by the *geometry* and therefore the dispersive contribution of the material itself can be ignored, a decrease of the group velocity translates into an increase of the amplitude of the electric field, through the equation $W = \varepsilon |E|^2 / 2 = I v_\varepsilon^{-1}$. The key point is that here it can be shown that in nondissipative materials the energy velocity v_ε, defined by the equation above, coincides with the group velocity [158,204]. Therefore, when slow-light results from the geometry and not from the dispersion of the material, the field strength increases while the energy flux, hence the power is kept constant. This implies that the nonlinear response, defined with respect to the excitation power, is enhanced.

Let us consider a specific case, namely the model of a finite length L Distributed Bragg Reflector, whose typical spectral transmission is calculated and shown in Figure 7.5. Following Winful [193], the corresponding group delay is calculated from the phase of the transmission φ, using the formula $\tau_g = \partial_\omega \varphi$. At the mobility edges, τ_g increases well above Ln/c, meaning that the energy is trapped in the structure. Because of that, the energy density increases. In fact, it is found that the energy density (or, equivalently, the L-DOS) is proportional to the ratio $\tau_g c/(nL)$, hence the slow-down factor. The resonant character of the slow-light enhancement is apparent when considering CROWs, which, obviously, also experience the same enhancement of the electric field, hence the light-matter interaction [203]. A more intriguing case is when considering that inside the band-gap the group delay is shorter than the Ln/c, which might be interpreted as the group velocity were larger than c/n. As Winful pointed out, the correct interpretation is that τ_g relates to a "dwelling time," not propagation, as energy is being reflected back. The shorter the dwelling time, the smaller the fraction of energy entering the structure. This corresponds to a local density of states which is smaller than in free space, thus completing the connection between group velocity, field enhancement, and L-DOS. A direct evidence of the enhancement of light-matter enhancement relates to the linear absorption, which has been demonstrated experimentally to scale as c_0/v_g, which has been experimentally confirmed (see for instance ref. [40]).

7.2.6 IMPACT OF THE FABRICATION IMPERFECTIONS

The experimental characterization of the slow wave propagation in PhC waveguides has revealed a strong deviation from the ideal behavior, mainly in the form of strong backscattering. In fact, the connection between backscattering (and, more generally, attenuation due to elastic scattering) and group velocity has been established theoretically [57,71]. The proposed models consider the fabrication imperfections, namely roughness and fluctuations from the nominal geometry. As an example, let us consider the semiconductor PhC waveguide made of GaAs shown in Figure 7.6. The high-resolution SEM image reveals the deviation of the holes from their ideal

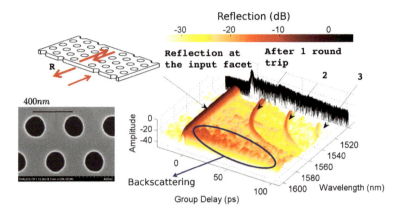

FIGURE 7.6 High-resolution SEM image of a semiconductor PhC. Time-frequency spectrogram obtained from a coherent optical measurement revealing dispersive propagation and backscattering.

rounded shape. Moreover, the sidewalls are rough, to the scale of the nanometer. As both effects appear as random, the translation invariance of the structure is broken, which results into elastic scattering of the Bloch modes, similar to electrons in crystals with impurities or phonons. The two-dimensional band-gap prevents in-plane scattering, except toward the back-propagating mode or to other guided modes. Out of plane scattering is however possible. Figure 7.6 shows the reflection map of a PhC waveguide as a function of the time and the wavelength, measured using a coherent optical technique. The map reveals the reflection at the input facet and several replica of the signal having traveled back and forth in the waveguide [147]. This allows the measurement of the propagation delay, which is observed to increase with the wavelength as the edge of the waveguide dispersion is approached. As the group delay becomes large, backscattering increases sharply, as apparent in the figure. This is consistent with the predicted n_g^2 scaling of backscattering [71], which has been intensely investigated experimentally [90,126]. On a more fundamental side, the disorder in PhC has been investigated theoretically [109,162,178] and experimentally, and, specifically, in the context of the Anderson localization [79] in disordered waveguides [49,168,181,197]. On a more practical standpoint, fabrication disorder has to be tackled when considering applications such as the control of the optical delay or nonlinear frequency conversion, which will be discussed in the next sections.

In photonic crystal waveguides, disorder leads to increased radiative loss due to out of plane scattering and to random fluctuations of the resonance [49,127,176,178,197]. These effects have to be considered in systems of coupled resonators [119], and resonators lattices. Recently, the active control of the disorder, inspired from studies in scattering media, has been introduced in order to counteract fabrication disorder in arrays of PhC cavities [101,102,169–171]. The main idea is to achieve the individual control of the resonances of each of cavities through a spatially shaped optical pump, generating a suitable temperature distribution (Figure 7.7). A permanent reconfiguration of the resonators could be achieved through photo-induced chemical changes, for instance oxidation, which has been investigated in Silicon PhC cavities [25].

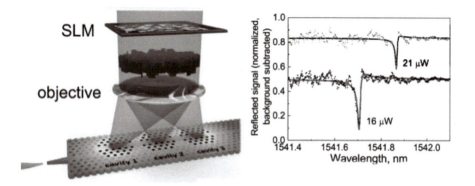

FIGURE 7.7   Thermo-optic tuning using a holographic pattern. (Adapted with permission from [170] © The Optical Society Sokolov, S. et al., *Opt. Express*, 25, 4598–4606, 2017.)

A different perspective comes from the new paradigm of *topological photonics* [62,104]. Inspired by solid state physics, the aim here is to create "topologically protected" states existing at the boundaries of photonic crystals with different topological charge. These states will be robust against disorder and hence expected to result into much better devices and pave the way to breakthrough applications.

7.2.7   RELEVANT NONLINEAR INTERACTIONS

7.2.7.1   Thermo-Optic

A fraction of the carriers injected in the conduction band by absorption of the pump relax nonradiatively by emission of phonons. The dissipation of a fraction α_{abs} of the power leads to an increase of the temperature:

$$\Delta T = R_{th} P_{th} = R_{th} \alpha_{abs} P_{in} \tag{7.1}$$

which, in the steady state, defines its thermal resistance R_{th} ($K \cdot W^{-1}$). This parameter depends on the material through its thermal conductance ($K m W^{-1}$) and on the geometry, in analogy to the electric resistance. The rise of the temperature leads to a change of the refractive index which in semiconductors is particularly large when the photon energy is close to the band-gap. For instance, in GaAs in the Telecom spectral band, the thermo-optic coefficient is $\partial n/\partial T = 2.5 \times 10^{-4} K^{-1}$.

The thermal resistance of suspended PhC structures is about $10^5 W^{-1}K$, meaning that μW level dissipation already induces a temperature shift of about 1 K. In high-Q resonators the resonance shift induced by the thermo-optic effect is enough for inducing bistable switching. This is shown in Figure 7.8. An interesting aspect of thermal nonlinearity in PhC is that the relaxation time scale is below 1 μs. This is because of the very small thermal capacitance, defined in analogy to its electric counterpart and relating the temperature rate of change to the heat flow. It is related to the specific heat, the density of the material and the volume of the cavity $C_{th} = c_v \rho V$ is about $10^{-12} K^{-1} J$,

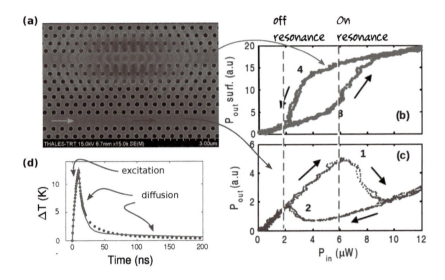

FIGURE 7.8 Thermo-optic bistable switching in a high-Q cavity. (a) SEM image with map of the electric field of the fundamental order mode. Experimental bistable cycle considering the energy in the cavity (b) and the transmission (c). (Adapted with permission from Weidner, E. et al., *Appl. Phys. Lett.*, 90, 101118, 2007.) (d) modeled thermal relaxation in a suspended membrane evidencing fast thermal diffusion out of the cavity. (From de Rossi, A. et al., *Physical Review A*, 79, 043818, 2009.)

hence the thermal time constant $\tau_{th} = R_{th}C_{th} \approx 10^{-7}$s. Dynamical measurements [37] have revealed an even faster thermal effect occurring when the size of the optical mode is much smaller than the thermal diffusion length (typically many μm, depending on the thermal conductivity [171]). In this case, the resonator experiences a decrease of the temperature which is much faster that the thermal relaxation time (Figure 7.8), typically in the ns scale. These fast thermal effects may compete with other noninstantaneous nonlinear effects such as the index change induced by free carriers [37].

7.2.7.2 Free Carriers

The excitation of free carriers in semiconductors induces a change in the refractive index which is proportional to their density. This is a strongly dispersive effect consisting of several contributions [14], one of those is described by the Drude theory of free carriers. As carriers accumulate, the change of index is related to the optical excitation through an integral interaction $\Delta n \propto -\int_{-inf}^{t} \alpha[I(t')]exp[(t'-t)/\tau]dt'$, where $\alpha(I)$ denotes the absorption, e.g., $\alpha_2 I^2$ for two-photon absorption (TPA). In the limit of "slow" dynamics, relative to the carrier lifetime τ, the nonlinear response is equivalent to a Kerr effect with negative coefficient (inducing blue shift), namely $\Delta n \propto -\alpha\tau I$ in case of linear absorption, and fifth order susceptibility in case TPA is the dominant absorption mechanism: $\Delta n \propto -\alpha_2\tau I^2$.

A particularity of PhC is that surface recombination reduces the carrier lifetime drastically. For instance, in GaAs, direct pump-probe measurements on PhC suspended membranes concluded that the lifetime there is reduced to less than 5 ps, namely orders of magnitude smaller than in semiconductor heterostructures protected from the interaction with the surface [74]. Unless τ is in the sub-picosecond range, this nonlinear response dominates the electronic Kerr effect by orders of magnitude, and it is therefore a preferred mechanism for implementing energy-efficient all-optical switching. In photonic crystals, this has been first demonstrated in silicon [177], although there the carrier lifetime is much longer and prevents fast switching. Ultra-efficient switching [140] and memory [141] have been shown owing to an optimized III-V material. A fast ($\tau \approx 10$ ps) III-V PhC nonlinear switch has been integrated heterogeneously in a silicon photonic circuit [13]. A crucial aspect here is the control of the surface recombination. An optimal carrier lifetime of about 10 ps has been achieved in suspended GaAs PhCs using atomic layer deposition (ALD)[131]. The applications of this technology for all-optical signal processing have been discussed in Ref. [130].

7.2.7.3 Second-Order Polarization

III-V semiconductor alloys such as GaAs, InP, etc. have a non-centrosymmetric crystalline structure, which enables a nonvanishing second-order susceptibility. In contrast to nonlinear crystals such as LiNbO$_3$, the large nonlinear response of GaAs is difficult to exploit because of the lack of a convenient phase-matching mechanism. To overcome this issue, artificial birefringence has been proposed [54]. The nonlinear enhancement associated to the large density of optical states in PhC has inspired research aiming at harnessing the second-order susceptibility for second-harmonic generation (SHG), parametric downconversion, etc. The main issue is that the second-order susceptibility involves interacting frequencies spanning over an octave or more. For instance, efficient SHG requires the interaction of confined modes separated exactly by one octave. Designing a suitable photonic crystal is still challenging as it requires to engineer multiple photonic bands simultaneously. Furthermore, when considering two-dimensional PhC, higher order bands are usually associated with strongly leaky modes, which hinders the achievement of an efficient interaction. Nevertheless, there exists hints that enhanced second-order nonlinear interaction might be possible with a suitable design [19,161]. Moreover, resonant [155] or slow-light enhancement [95] of the fundamental mode have shown to compensate in part the lack of phase matching. Considering that large band-gap materials (GaInP, GaP) are transparent at the second harmonic (in the visible), in contrast to silicon, the achievement of fully enhancement SHG in PhC would result in very efficient sources.

7.2.7.4 Third-Order Polarization

Third-order susceptibility relates to a higher order nonlinear process, which is in principle weaker than the effects discussed above. However, it is ubiquitous and is more tolerant relative to phase matching. This review will focus on the particular case where the nonlinear interaction takes the term $P_l^{NL}E_l^* \propto \chi^{(3)}E_iE_jE_k^*E_l^*$, meaning a process where two photons $\hbar\omega_{i,j}$ are annihilated to create a pair $\hbar\omega_{l,k}$ such that the energy is conserved. This implies that nonlinear interactions involving modes within a relatively narrow spectral range are possible. Therefore, a PhC structure can be engineered to allow nonlinear interactions entailing a large density of optical states, and, therefore, a large enhancement.

A particular case is the fully degenerate interaction, $i = j = l = k$, resulting into Self-Phase Modulation (SPM). This is described by the nonlinear coefficient γ, defined as:

$$\Delta\Phi = \gamma PL, \tag{7.2}$$

where $\Delta\Phi$ is the nonlinear phase shift which an optical mode carrying the power P accumulates over a length L. It is shown that [3]:

$$\gamma = \frac{n_2\omega}{c_0 A\chi} \propto \frac{n_2}{v_g^2 A_g} \tag{7.3}$$

where A_g is a geometric cross section which is related to the more rigorous definition of the nonlinear cross section A_χ through the relation $A_g = (n_g/n)^2 A_\chi$, which makes the dependence on the group velocity explicit [15]. This relation defines the length scale $L_{NL}^{-1} = \gamma P$ over which the nonlinear interaction is substantial.

Compared to the core of a standard single-mode fiber, the cross section A_{eff} of a photonic wire/crystal waveguide is more than 2 orders of magnitude smaller (Figure 7.9), which implies that a much larger density of the optical energy corresponds to the same power, hence a much larger nonlinear contribution to the polarizability. Moreover, compared to Silica, the third-order susceptibility $\chi^{(3)}$ is about 3 orders of magnitude larger in semiconductors such as GaAs, InGaP, silicon. The slow-down enhancement also contributes substantially, since the group velocity in PhC waveguides is typically at least one order of magnitude smaller than the phase velocity, yet keeping propagation losses under control [63,159,163].

7.2.8 LIGHT-MATTER INTERACTION IN PhC

The nonlinear interaction of guided waves is conveniently described in terms of the normal modes of waveguides and cavities. The basic assumption is that the

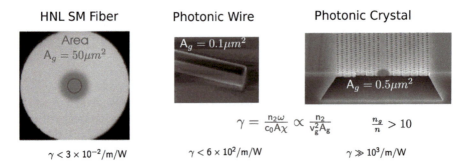

FIGURE 7.9 Influence of geometry and slow light on the nonlinear interaction.

nonlinearity does not affect the confinement itself, yet it merely induces a phase modulation and an exchange of energy. The formalism for nonlinear PhC waveguides and cavities has been developed in the literature (for instance in Refs. [15,103,160], and many others). The main results are summarized and rewritten in a consistent formalism in Table 7.1.

Some figures of merit can be calculated, based on the theory summarized here, for a variety of structures made of semiconductor material, hence based on a large index contrast. The Table 7.1 provides insight on the specific properties of PhC relative to then nonlinear interaction. The photon lifetime can be calculated in cavities and waveguides (assuming it is limited by attenuation). The largest inter-action time has been achieved in Silicon PhC cavities. Surprisingly, this is larger than in semiconductor wires even assuming low propagation loss $\alpha = 0.5$ dB/cm. Only in ultra-low-loss, weakly confining dielectrics such as doped glass, Silicon Nitride, Lithium Niobate, and polished crystals (CaF, MgF) much larger lifetimes have been achieved.

The modal volume V_m and the nonlinear volume V_χ are both representative of the ability to confine light in the resonator. The comparison with other resonators with comparable Q factor shows that confinement in PhC is at least 2 orders of magnitude better. The consequence of the confinement is quantified in the ratio between the effective irradiance in the resonator (or waveguide) relative to the injected power. In resonators, this also accounts for the "photon recycling" as radiation is kept inside the cavity. For instance, in the high-Q cavity demonstrated in silicon [8], an input power of 1 μW corresponds to spatial maximum of the electric field in the cavity corresponding to an irradiance of about 0.6 GW/cm^2. The consequence of a very small mode volume is also seen in terms of irradiance corresponding to the presence of a single photon, which is relevant to experiments in the quantum regime. For instance, in the "H0" cavity, this corresponds to 40 kW/cm^2, which is a sizable field. Finally, the small nonlinear volume combine with large Q factor in multimode cavities could decrease the power threshold for optical parametric oscillation (OPO) to the micro-watt range [31].

7.3 NONLINEAR LOCALIZATION IN PhC WAVEGUIDES

The propagation optical pulses in strongly nonlinear and dispersive structures such as PhC waveguides corresponds to a fairly complex dynamical system which sits into an unique parameter space. Moreover, PhC waveguides offer a remarkable flexibility in engineering these parameters, thereby allowing the full control of this dynami-cal system, in principle. Here we summarize the very recent achievements, such as the Soliton-effect compression, with perspectives of application in compact pulsed sources, and, more broadly, Soliton physics on-chip scale.

7.3.1 SPATIAL AND TEMPORAL LOCALIZATION: THE SOLITON

A nonlinear pulse that propagates without changing its shape despite the pres-ence of chromatic dispersion is called Soliton and it is ubiquitous in physics.

In the context of optical fibers, since their prediction in the early 1970s [1], the properties of solitons in fiber raised expectations for applications in long haul communication as an effective mean to use Kerr nonlinearity in order to counterbalance the detrimental effect of the chromatic dispersion. Of course, it is out of question here to use Photonic Crystal waveguides for km long propagation. However the dispersion in metamaterial can be so steep that it already plays an important role even at the *mm* propagation scale! Having a way to mitigate it is then important.

Moreover, the interplay of nonlinearity and dispersion grants access to a larger variety of optical phenomena than for Kerr nonlinearity alone where the solely stable solutions are constants, namely CW waves. This is of importance regarding applications because ultra-fast optical phenomena, which cannot be generated by electronics, can now be obtained. Finally, at the fundamental physics level, we will see that nonlinear metamaterials constitutes a set of useful test systems for the study of nonlinear oscillators.

Interplay between Kerr nonlinearity and group velocity dispersion can be modeled using the Nonlinear Schrödinger Equation (NLSE) [4,67,85]:

$$\partial_z E(z,t) = \frac{-i\beta_2}{2} \partial_{tt} E(z,t) + i\gamma \mid E(z,t)\mid^2 E(z,t) \tag{7.4}$$

where γ and β_2 stand for the effective Kerr nonlinearity and the group velocity dispersion ($\partial^2_\omega \beta$), respectively. The power flowing through the waveguide is $P(t) = |E(t)|^2$. Before going further, it is important to assess *a priori* the relative importance of the different effects because this provides a hint regarding the phenomena that could be eventually observed. We can associate any effect with an effective length; that is the propagation length required so the effect can develop fully. Here the Nonlinear and the Dispersion Lengths are defined respectively as $L_{NL} = 1/\gamma P_0$ and $L_d = T_0^2/\beta_2$, P and T_0 are the pulse peak power and pulse temporal width. Eq. 7.4 can be normalized in to a dimensionless equation through the transformation: $\xi = z/L_{GVD}$, $\tau = t/T_0$. This yields:

$$\partial_\xi E(\xi,\tau) = \frac{-i}{2} \partial_{\tau\tau} E(\xi,\tau) + iN^2 \mid E(\xi,\tau)\mid^2 E(\xi,\tau) \tag{7.5}$$

Here $N^2 = L_d/L_{NL}$ is called the soliton number and corresponds to relative strength of the nonlinearity compared to the dispersion. If the dispersion in anomalous ($\beta_2 < 0$), nonlinearity and dispersion play against each other; and in case $N = 1$ they compensate exactly so Eq. 7.5 supports a stationary solution (the Soliton) expressed as $s(t) = \sqrt{P_0} \, sech(t/T_0) \exp[iz/(2L_d)]$. According to the shape of the soliton, the full width at half maximum of the pulse is $T_{FWHM} = 1.76 T_0$. It is usually considered that it takes a propagation length of about $L_{soliton} = \sqrt{L_d L_{NL}}$ for the input pulse to turn into a soliton. This sets the characteristic length the metamaterial waveguide must have (if shorter, no soliton effect would be observed).

7.3.1.1   On Demand Properties

Parameters for typical line-defect PhC waveguides are shown in Figure 7.10b–d. The effective Kerr parameter benefits from the slow-light enhancement[1] $\gamma = (n_g/n_0)^2 \omega n_{2I}/cA_g$ so its variations follow closely those of the group index n_g. The chromatic dispersion (GVD) is $\beta_2 = \partial_\omega n_g/c$. The effective parameters of PhC waveguides: $\gamma \approx 1000 \text{ W}^{-1}\text{m}^{-1}$, $\beta_2 \approx -2\times10^3 \text{ps}^2\text{m}^{-1}$ must be compared to either structured macroscopic systems with strong dispersion but no light confinement like Fiber Bragg Grating: $\gamma \approx 1.10^{-3} \text{ W}^{-1}\text{m}^{-1}, \beta_2 \approx -2\times10^3 \text{ps}^2\text{m}^{-1}$ [47], or to integrated silicon nanowires, strong confinement but weak slow-light enhancement and dispersion: $\gamma \approx 300 \text{ W}^{-1}\text{m}^{-1}$, $\beta_2 \approx -2 \text{ ps}^2\text{m}^{-1}$ [41,206]. As shown graphically on Figure 7.10a, only in PhC waveguides both L_d and L_{NL} can fit in a chip, namely

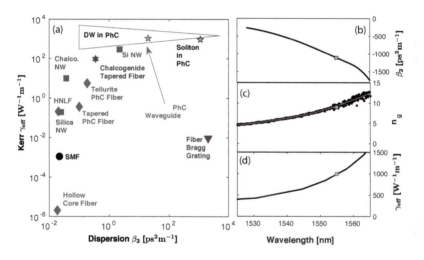

FIGURE 7.10   (a) Nonlinear and dispersive properties of various selected optical systems. squares: Integrated nanowires, diamond: fibers systems, cyan stars: waveguides parameters for the two propagation effects discussed in this chapter. The triangle represents the parameter space for the PhC. (b–d) Typical evolution of PhC waveguide parameters with the wavelength. cyan squares indicates the parameters for the soliton and Dispersive Waves (DW) demonstration. (With kind permission from Springer Science+Business Media: *Nature Photonics*, Temporal solitons and pulse compression in photonic crystal waveguides, 4, 2010, 862–868, Colman, P. et al.)

---

[1] In the definition of γ used here, the contribution of the slow light is made explicit, which is reasonable considering that $(n_g/n_0)^2$ can be in practice as large as 100. Even if not set explicitly, this effect also exists, albeit much weaker as $(n_g/n_0) \approx 1$, in other non-slow light systems such as nanowires, micro-ring resonators, and even fibers! In fact it is implicitly taken into account in the definition of the effective Area (A_χ) through a specific normalization of the photonic mode. That mean ones must be very careful to the latter so the slow-light effect is not counted twice. Finally, we consider here that the geometrical dispersion exceeds by far the material one (dispersion-less material assumption). If that is not the case, a proper procedure to normalize the electromagnetic field can be for example found in Ref. [144].

TABLE 7.2

Parameters and Figures of Merit for the Confinement of the Field and Light-Matter Interaction in Cavities and Waveguides Made of Semiconductor Materials (Silicon, AlGaAs and InGaP) in the Telecom Spectral Range. Definitions in Table 7.1

		Cavities		
FOM	Symbol (Unit)	PhC H0	PhC Heter.	Nanobeam
Photon Lifetime (exp.)[a]	τ (ns)	0.3 [39]	9 [8]	0.8 [24]
Modal Volume	V_m ($\times 10^{-18}$ m^3)	0.034	≈ 0.1	0.08[d]
Field Enhancement[b]	I_{max}/P (m^{-2})	0.9×10^{18}	$\approx 6 \times 10^{18}$	$\approx 1 \times 10^{18}$
Field for 1 photon[c]	$I_{max}|_{1\hbar\omega}$ (kWcm^{-2})	40	12	16
NL Volume	V_χ ($\times 10^{-18}$ m^3)	0.2	0.4	0.4
OPO Threshold[e]	P_{th} (W)	n.a.	3×10^{-5}, †	

		Waveguides		
FOM	Symbol (Unit)	PhC WG	Phot. Wire	μRing
Photon Lifetime (exp.)[a]	τ (ns)	0.2	1	0.08 [152]
NL Area	A_χ ($\times 10^{-12}$ m^3)	0.005	0.2	0.2
Field Enhancement[b]	I_{max}/P	2×10^{14}	5×10^{12}	$\approx 5 \times 10^{14}$

Source: Combrié, S. *Laser & Photonics Rev.*, 11, 1700099, 2017.

[a] cavities: $\tau = Q/\omega$, waveguides $\tau = n_g/(c_0\alpha)$ (where α denotes propagation losses and n_g is the group index)

[b] Spatial maximum of I for unity input power $= \frac{c_0\tau}{V_{mn}}P$ (cavity), $= \frac{P}{A_\chi}$

[c] Spatial maximum of I with 1 photon in the cavity $\frac{c_0\hbar\omega}{V_{mn}}P$

[d] FWM mode width 4 µm.

[e] Assuming $n_2 = 0.6 \times 10^{-17}$ m^2/W.

† Equispaced resonances with $Q = 4 \times 10^5$

within a few mm, still generated by an on-chip laser diode delivering pulses with peak power level about a few Watt. As a rules of thumb, 10 W peak power injected inside the waveguide would correspond to an energy density of about 20 GW/cm^2, considering the slow-light enhancement (Table 7.2). Consequently only PhC waveguides would support a soliton at the mm propagation scale, and allows its evolution. Another interesting feature of metamaterial waveguides lies in the prevalence of the geometry on the optical properties. Thus it is possible, within the same technology -and even on the same photonic chip- to tune by one order of magnitude nonlinearity and dispersion (Section 3.2).

7.3.1.2 Soliton Propagation and Compression

Let consider at first the propagation of a pulse with parameters: $T_{FWHM} = 2.3$ ps, $P_0 = [0.5, 8]$ W through a $L_{wg} = 1.5$ mm-long PhC waveguide at a group index of $n_g = 9.3$, corresponding to $\gamma = 920$ W^{-1}m^{-1}, $\beta_2 = -1.1 \times 10^3$ ps^{2}m^{-1} [28]. The corresponding length scales are $L_d = 1.6$ mm and $L_{NL} = [2.5, 0.15]$ mm. At the output of the waveguide, the pulse spectra and temporal autocorrelation trace are recorded in Figure 7.11a and b for various input powers.[2] At low power ($E_{pulse} < 6$ pJ, hence $P_0 < 2$ W), the pulse widens during the propagation due to the effect of dispersion (Figure 7.11c). For larger power, the Kerr nonlinearity comes into play and counterbalances the GVD. At the same time, the spectra exhibit the emergence of sidebands characteristic of the self-phase modulation (Figure 7.11d). As the soliton is formed the output pulse has about same power and duration as the input pulse. The absence

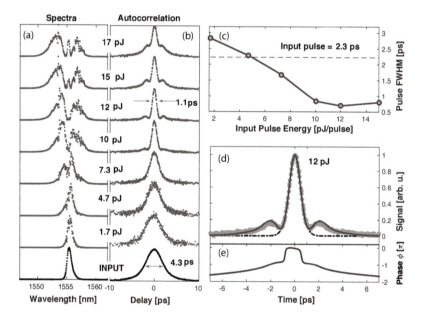

FIGURE 7.11 (a and b) Evolution of output spectra and autocorrelation traces with input power. (c) Deconvoluted pulse duration (Full Width at Half Maximum) corresponding to b). (d) Comparison between experimental and modeling. Black dot-dashed line corresponds to the autocorrelation trace of an ideal soliton (note the absence of any shoulders). (e) Optical phase inferred from simulation, corresponding to (d): phase is almost constant over the soliton duration. (With kind permission from Springer Science+Business Media: *Nature Photonics*, Temporal solitons and pulse compression in photonic crystal waveguides, 4, 2010, 862–868, Colman, P. et al., by permission from Springer Nature.)

---

[2] In an idealized lossless system, because the NLSE is normalized and shows a scaling invariance, it is strictly equivalent to either look at the evolution of a pulse during its propagation (cut-back method) or to scan the input power and characterize only the output of the waveguide. Latter method is usually simpler to implement and gives satisfactory results although integrated waveguides are not ideal.

of pulse distortion during the solitons is characterized in the time domain by a constant phase across the pulse (Figure 7.11e). Indeed, the pulse remains localized in time and space, does not disperse so that all its spectral components stay in phase with each other. Consequently, there is a direct correspondence between the soliton amplitude (since the phase is flat) spectrum and its temporal shape: the time × bandwidth product of the soliton equals $\delta v \delta \tau = 0.315$, corresponding to the fact that its duration is minimal considering its spectral bandwidth.

If the pulse energy is increased further, we see that the pulse gets compressed by a factor up to 4. This compression effect is characteristic of soliton with $N > 1$ [132]. During the compression $\delta v \delta \tau$ stayed within 0.30 to 0.36. This indicates that the compressed pulse remains close to Fourier limited; and consequently any spectral broadening necessarily results in a temporal compression.

7.3.1.3 Other Effects and Limitations

The demonstration of soliton demonstration is an important achievement in nonlinear optics: it certifies the capacity of the metamaterial to support nonlinear pulses. Indeed, the ideal NLSE propagation equation is based on several assumptions and neglects other effects that co-exist inside a PhC waveguide. The presence of deleterious effects, such as higher order chromatic dispersion (TOD), nonlinear absorption, and multiple scattering, could disrupt the interplay between dispersion and Kerr effect, hence preventing any soliton-like pulse propagation.

In particular the two-photon absorption (TPA) effect scales with power exactly as the Kerr effect. The relative strength of the two effect is expressed by a figure of merit $T_\chi = n_2/\alpha_2 \lambda$. If T_χ is small, then the Kerr effect is spoiled by the nonlinear absorption α_2 and the compression of a soliton is not possible. Moreover a side product of the nonlinear absorption is the generation of a free-carrier plasma that further alter the ideal soliton dynamics. In the present demonstration, the use of a large band-gap material GaInP ($E_g = 1.89$ eV) instead of silicon ($E_g = 1.1$ eV) suppressed the two-photon nonlinear absorption ($2E_{photon} = 1.6$ eV). Thus the PhC waveguide can sustain power up to 20 W, and is only limited by thee photon absorption, which scales differently than Kerr effect and does not pose any problem for the range of power considered here [74].

The reader may also have noticed that the nominal NLSE assumes a z-invariant waveguide, while PhCs support Bloch modes which show a structuration along the propagation direction. As a consequence we defined the effective nonlinear parameters as an average over a unit cell: but up to which power level would that description hold? PhC waveguides differ from Fiber Bragg Gratings because the latter are operated very close to the Photonic Band-Gap so the soliton feels its presence, and may eventually have enough energy to distort the band diagram itself (hence the chromatic dispersion). Such pulses are referred as Gap Soliton [26,36,46]. Here in PhC, a rough estimate indicates that the Bloch mode is robust enough to support power up to 100 W [106] before being affected by the nonlinearity. This must be compared with the ≈ 10 W required to get a strong soliton compression.

While the issues that have just been raised may give the impression that the conditions allowing an ideal NLSE-like nonlinear propagation in structured waveguides are hard to meet, they are at the same time a fantastic opportunity to investigate new optical phenomena; and to improve the models and the understanding we have regarding nonlinear optics. For instance, the presence of free-carriers plasma can result in an interesting pulse breaking phenomenon where the soliton splits up into a bunch of smaller pulses [77].

7.3.2 PERTURBATION OF SOLITON: RADIATION OF DISPERSIVE WAVES

We can also investigate the impact of a nonideal dispersion and see how a soliton (stable attractor) adapts to this new situation [48,188]. Note that the nonlinear effects observed in metamaterial, as soon as they are being described by effective parameters, are analog to effects already observed, hence studied, in fibers. That said, the energy level (W, picoJoule/pulse) and pulse duration (picoseconds) inside a PhC greatly differ from the conditions of operation in fibers (kW power, femtoseconds). Consequently nonlinear optics in metamaterial bears some differences, and complementarity, with fibers.

Soliton only exists in anomalous dispersion ($\beta_2 < 0$), still if it is close enough to a point where dispersion changes sign, namely called a zero group velocity dispersion point, it would feel the presence of a forbidden normal dispersion zone ($\beta_2 > 0$). Figure 7.12a shows the spectral evolution of a 1.8 ps pulses launched at

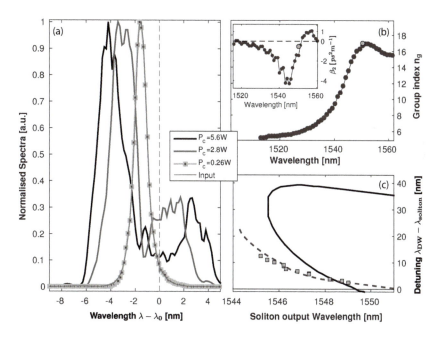

FIGURE 7.12 (a) Spectral evolution of pulse (linear scale). Note the shift of the pulse center and the generation of a Dispersive Wave in the normal dispersion region. λ_0 indicates the frontier between anomalous and normal dispersion regions. (b) Experimental conditions corresponding to (a). (c) Detuning between the soliton and its Dispersive Wave for various input pulse power and wavelength. Black: Phase-matching according to Eq. 7.6. Gray-dashed: Group velocity matching condition. Cyan squares: experimental data. (Reprinted with permission from Colman, P., *Opti. Express*, 3934, 3930–3934, 2012. Copyright 2012 by the American Physical Society.)

$n_g = 15.2$ ($\gamma \approx 10^3$ W^{-1}m^{-1}, $\beta_2 \approx -200$ ps^{2}m^{-1}), only about 1.5 nm away from a zero dispersion point (see inset in Figure 7.12b) [27].

As the power is increased, a side lobe, also referred to as dispersive wave (DW), appears in the forbidden normal dispersion region and builds up. Concomitantly, the main soliton pulse shifts toward shorter wavelength (i.e., higher frequency). In total, the dispersive wave contains an important fraction of the soliton energy, about 30% here.

7.3.2.1   Asymmetric Phase-Matching Condition

At first glance, the generation of a dispersive wave on only one side of the pulse may look surprising. Indeed, parametric gain caused by the Kerr effect results usually in four-wave mixing (FWM), which is a fully symmetric effect. The main difference here is that we are dealing with pulses localized in time, so there is a contribution from the Doppler effect, a moving reference frame resulting into a frequency shift, that modifies the phase-matching condition, which now reads [187]:

$$K(\omega) - K(\omega_0) - \frac{\omega - \omega_0}{v_g(\omega_0)} = K_{soliton} \qquad (7.6)$$

where $K_{soliton} = 1/(2L_d) = \gamma P_0 / 2$ is the nonlinear phase shift induced by the soliton. In PhC waveguides, this term can be neglected in practice. Eq. 7.6 tells that an initial linear dephasing rate $K(\omega) - K(\omega_0)$ could be eventually compensated by the Doppler effect. This optical radiation effect is the exact analog to the Cherenkov radiation in solid. It happens when the group velocity exceeds locally the phase velocity $V_\phi \approx (\omega - \omega_0)/(K(\omega) - K(\omega_0))$. More generally, any phase-matching conditions for the generation of dispersive waves can be retrieved by linearizing the Generalized NLSE propagation equation around the soliton solution. In brief, one adds a small perturbation $\delta E exp[-i\delta\omega t + iK(\omega + \delta\omega)z]$; and search for the conditions so that δE grows up during the propagation. This approach assumes the soliton remains unperturbed and that the perturbation δE remains small.

7.3.2.2   Soliton Recoil

The blue-frequency recoil undergone by the soliton can be understood as a momentum and energy conservation principle. In FWM the generation of waves at high frequency is compensated by the generation of a symmetric wave at lower frequency, so that the total energy and momentum are conserved. Because the dispersive wave generation is here asymmetric, the conservation of energy and momentum must then be satisfied by the whole pulse, hence the spectral recoil in opposite direction to the emission of the dispersive wave. One must note that in fibers, this recoil phenomenon is canceled by the Raman scattering that cause a general red-shift of any nonlinear pulse [61]. The limited bandwidth of the PhC waveguide that has been used here (8 THz) is too small to allow the Stimulated Raman Scattering effect (12 THz) to take place, so a blue shift can actually be observed. The possibility to suppress unwanted effects thought bandwidth engineering is a unique feature of metamaterial.

7.3.2.3 Phase-Matching vs Group Velocity Matching

According to Eq. 7.6 phase-matching conditions only involve the current soliton spectral position, neither its energy (if we assume $K_{soliton} \approx 0$), nor its input spectral position. The detuning between the soliton and its dispersive wave is shown in Figure 7.12c for various soliton input wavelengths and powers; and indeed the position of the dispersive wave (DW) only depends on the output soliton state. That said, the DW-soliton detuning does not obey any phase-matching, but rather a group velocity matching relationship. The phase-matching condition just indicates where would be generated the DW; but does not tell anything about any subsequent DW-soliton interaction. In fiber, this interaction is weak so the DW position does not change after its creation [45]. Indeed the DW is generated at the front of the soliton and immediately propagates away from it. However if the soliton can blue shift, it then accelerates (for anomalous dispersion, a blue shift means lower n_g) and eventually catches up with the DW and traps it [60]. This close interaction might explain the efficient transfer of energy from the soliton to the DW. The trapping mechanism holds as long as the soliton has enough energy (see Figure 7.13).

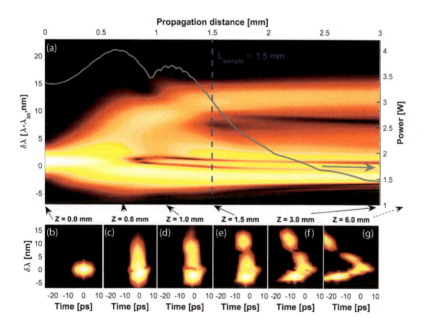

FIGURE 7.13 (a) Simulation: pulse spectrum evolution with propagation length. Overlay: Peak power evolution showing two successive soliton compression cycles. (b–g) Frequency-time map showing the time localization of the spectral features, at specific positions along the waveguide. (b): Input soliton. (c): Generation of the DW. (d–e): DW and Soliton travel locked to each other. (f–g): The pulse loses its soliton properties (not enough energy); the DW is not trapped anymore and travels away from the pulse, which progressively disperses temporally under the action of the chromatic dispersion.

7.3.2.4 Step-Like Evolution of Supercontinuum Generation

The emission of the Dispersive Wave (DW) is directly fueled by the soliton peaks power. Consequently, when a higher order ($N \gg 1$) soliton is launched inside the PhC, it first compresses until it reaches the threshold power that triggers the efficient generation of the dispersive wave. The soliton recoils happened at the same time. After this first event, if the soliton has still enough power, it can undergo a second soliton compression cycle that will result in the emission of a second DW, etc. [182].

The phenomena we just described are the very first steps of the generation of a coherent supercontinuum in the anomalous regime. In other systems, the Raman scattering blurs this step-like evolution which is only clearly visible in PhC. In particular, the blue shift and the subsequent trapping of the DW by the soliton render the overall process very efficient here. The phenomenon we are seeing here is highly coherent in the sense that it does not build up from noise or any instability. This contrast with fibers systems pumped with *ps* pulses. Compared to other effects like the modulation instability or cascaded FWM, the spectral broadening caused by the emission of DW and the subsequent soliton self-frequency shift (SSFS) is much larger.

7.3.3 Perspectives and Applications

7.3.3.1 Issues and Opportunity for PhC

The main hindrances to applications are the propagation loss and the difficulty to fabricate high quality PhC, specifically stitching errors which are critical in long PhC waveguides. This prevents the observation of any cascaded nonlinear effects (e.g., fully developed supercontinuum, cascaded FWM), requiring a longer propagation length. Still, this limitation is not intrinsic to PhC; that is solely a limitation of the technology; and any future improvements, either in nano-fabrication process or in the development of new low nonlinear loss materials like TiO_2 [91] or SiN [135] will directly benefit to the PhC technology as well. In addition, the gradual shift of telecom and signal processing applications toward the $\lambda = 2\,\mu m$ band [70], where silicon does not exhibit any TPA, will also relax some constraints on the fabrications: PhC working at $\lambda = 2\,\mu m$ will have larger lattice constant and hole radii than PhC working at $\lambda = 1.55\,\mu m$, hence the same absolute fabrication errors will have a lesser impact.

7.3.3.2 Nonlinear Filtering and Signal Processing

Taking current limitations into account, the interest in PhC waveguide resides in its capacity to transform efficiently, and with a very low power threshold level, nonlinear pulses. Although PhC waveguide does not have enough bandwidth to support a fully developed supercontinuum, it can still be used in specific signal processing applications where energy-efficient nonlinear pulse transfer functions are required, for instance as nonlinear filter to perform signal generation and regeneration.

Signal regeneration is embraced by the 3R problematic: Reamplification, Reshaping, Retiming. Soliton formation and compression of low energy pulse is a good candidate for reshaping functions, while the spectral recoil can also serve as

Mamyshevs type filter, namely a filter with a bell shaped response to input power, to equalize the bit-to-bit signal power variations [138,151].

Regarding the signal generation, at the heart of ultra-fast lasers lies a mode-locking mechanism. Alternatively to the use of saturable absorber (SESAM [146]) or nonlinear polarization rotation [51], one could also use a nonlinear loop mirror (NOLM) [42]. A good NOLM must be short–for high repetition rate–, with low threshold power, and because it set up most of the final laser characteristics (operating power, spectral bandwidth, temporal shape, etc.), possibility for dispersion, and nonlinearity engineering is important. PhC waveguides possess all these features.

7.3.3.3   New Directions in (integrated) Nonlinear Optics

Most of the new directions taken in nonlinear fibers optics can also be shared by PhC waveguides, which fundamentally remains a nonlinear waveguide, with extreme parameters though. Since about 2012, focus has been set in particular on two specific points we would like to present here. First it has been noticed that optical shock waves could be generated in the normal dispersion ($\beta_2 > 0$). Optical shocks happen when the nonlinearity exceeds the dispersion effect and consist in the formation of a steep front at the head (and/or at the tail) of the pulse, exactly the same way a water wave evolves and breaks-up when arriving to the shore. The breaking of an optical pulse is accompanied by the generation of a shock fan and the creation of multiple Dispersive Shock Waves (DSWs) [32,145,156].

The formation of the optical shock is very efficient in term of spectral broadening and generation of new frequencies; and so it can have strong implication in the generation of frequency comb and supercontinuum. Moreover, both normal and anomalous dispersion regime can now be exploited in nonlinear optics. Optical shock can form either when the waveguide exhibit a frequency dependent nonlinearity [35,38,73], called dispersive nonlinearity, or by using tailor made optical input.

This leads to the second point. Thus far, we only presented the evolution of the nonlinear pulse as being entirely controlled by the waveguide properties. This is true when the waveguides only supports one stable attractor (fundamental soliton); any input will then tend to converge to it. But in case the system exhibits several stable solutions (or not at all), then the trajectory and the optical output state depend highly on the optical input conditions [50,195]. It is thus now possible to promote the emergence of a specific optical phenomenon that would have been otherwise completely dominated, or that even could not have been obtained starting from trivial input conditions (e.g., CW, Soliton, Gaussian pulse). In particular optical shocks can be generated more efficiently by using as input a sine wave on constant background instead of a Gaussian pulse. More generally custom optical input can serve as an efficient way to mitigate for some part the lack of control in the waveguide properties.

7.4   FOUR WAVE MIXING IN PhC

One of the most fascinating and rich nonlinear phenomena, which is also highly relevant to applications, is the optical parametric amplification, involving the nonlinear coherent interaction of three waves resulting into a transfer of energy from the "pump" wave to the "signal." This process only requires the conservation of the momentum ($\vec{k}_1 = \vec{k}_2 + \vec{k}_3$) and the energy of the photons ($\omega_1 = \omega_2 + \omega_3$) and there-fore provides a more flexible mean to amplify photons than stimulated emission. An important extension of optical parametric amplification is optical parametric oscil-lation, whereby the combination of an optical cavity and a nonlinear crystal allows the generation of tuneable laser light as achieved by [58].

The development of low-loss optical fibers by Corning in 1970, after the funda-mental breakthrough by C. K. Kao in 1966, immediately suggested the possibility of nonlinear interactions of confined light over very long distances, thereby compensat-ing for the weak nonlinear response of the Silica glass. Since the second-order non-linear term is zero in symmetric materials such as glasses, parametric interactions involve four waves through the $\chi^{(3)}$ term, hence named four-wave mixing (FWM). Optical fibers have thus become an excellent nonlinear material and this contributed to the interest of FWM and lead to countless schemes to harness this coherent process to perform a variety of functions. The main motivation for slow-light subwavelength waveguides is the perspective of scaling the interaction length of FWM to the chip scale. Thus integrated photonic circuits are endowed with the all-optical process-ing capabilities demonstrated in fibers and semiconductor optical amplifiers [68]. This implies reducing the nonlinear interaction length from hundreds of meters to be able to exploit areas of one mm^2 or less. Besides having large nonlinear response, low absorption, and scattering losses, suitable materials need to be compatible with an integrated photonic technology, based either on Silicon or Silicon Nitride. Here we consider PhC waveguide made of III-V semiconductor alloys.

7.4.1 BACKGROUND

The strength of the parametric interaction involving four waves is quantified by the *Conversion Efficiency*. It is defined as the ratio of the output idler power over the input signal power [3]:

$$\eta_{FWM} = \frac{\bar{P}_{idler}(out)}{P_{signal}(in)} = \left(\frac{\gamma P_0}{g} sinh(gL_{eff}) \right)^2 e^{-\alpha L} \tag{7.7}$$

This defines the parametric gain, $g = \sqrt{(\gamma P_0 r)^2 - (\kappa/2)^2}$, which is related to the phase mismatch, $\kappa = \Delta k + \Delta k_{NL}$, to the effective interaction length L_{eff}, to the linear attenu-ation α and to the nonlinear parameter γ. The parameter $r = 2(P_1P_2)^{1/2}/P_0 < 1$ with $P_0 = (P_1 + P_2)$ characterizes the power penalty in case the FWM is generated by two pumps beam whose power P_1 and P_2 are unbalanced. The phase mismatch consist in a linear term Δk relating to the dispersion and a nonlinear term originating from cross and phase modulation [3]. The parametric gain g is related to the conversion effi-ciency $g = 1 + \eta$. In the limit of small gain, negligible linear attenuation and phase-matching $\eta = \gamma^2 P_0^2 L_{eff}^2$. Efficiency decreases if mismatch is not negligible, which implies that controlling the dispersion is crucial.

The correct evaluation of γ must account for the slow-down enhancement, dis-cussed above. The rigorous calculation in the context of the coupled wave theory has been carried out in Ref. [160]. The main result is that $\eta \propto \prod_{i=0}^{3} n_g(v_i)$, namely the dependence on the fourth power of the group index of the efficiency. More pre-cisely, the product of the group indexes of the interacting waves. The exact scaling accounts for a correction originating from the frequency dependence of spatial over-lap of the Bloch modes of the PhC waveguide. The exact result can be rewritten as follows [113]:

$$\gamma_i = \frac{\omega_i n_2 \epsilon_0^2 c_0}{4A} \int_{\chi^{(3)} \neq 0} dV \epsilon_{r,i} \frac{2\mathbf{e}_i^* \mathbf{e}_j \mathbf{e}_k^* \mathbf{e}_l + (\mathbf{e}_i^* \cdot \mathbf{e}_k^*)(\mathbf{e}_j \cdot \mathbf{e}_l)}{3} \tag{7.8}$$

with i, j, l, k forming a cyclic permutation over $0, 1, 2, 3$, denoting the four Bloch modes $\mathbf{e}_i$. The strong dependence on the group index has been observed experi-mentally [125,133,175] in a variety of materials, with reported values of γ ranging between $10^3 \, \mathrm{W}^{-1}\mathrm{m}^{-1}$ and $10^4 \, \mathrm{W}^{-1}\mathrm{m}^{-1}$ [23,96] hence between six to seven orders of magnitude larger than in single-mode fibers. Consequently, the nonlinear interaction length with pump power levels compatible with an integrated photonic circuit has been reduced to the sub-millimeter scale.

7.4.2 DISPERSION ENGINEERING

One of the benefits of the inherent complexity of Photonic Crystal waveguides is that they can be designed to tailor the dispersion of the (quasi-)normal modes.[3] It was realized long ago that a suitable change of the waveguide width could result into a band with almost zero group velocity [149]. Dispersion engineer-ing has been actively pursued because it provides a very powerful tool to control the parametric interactions in waveguides [27,56,64,88,98]. Figure 7.14 summa-rizes some of the design being proposed to achieve a variety of dispersion profile. Typically, besides the so-called "flat-band" design, which is of particular relevance for broadband FWM [133], a combination of parameters might lead to a situa-tion where the dispersion changes the sign within the transmission band of the waveguide once or even twice (Figure 7.14d). This situation allows a more subtle interaction, where FWM is phase matched only within a very narrow spectral region [192]. A more exotic approach relies on infiltrating the holes with a liq-uid, which proved effective in controlling the group velocity over a broad range (c/20 to c/110) [44]. Also, coupled waveguides [11] has been considered. A very important point is whether a suitable design could also reduce the impact of fabri-cation imperfections. This has been explicitly considered with particular empha-sis on the backscattering [108,142,150,189]. There is experimental evidence that design has helped in reducing propagation attenuation α_{dis} in slow-light wave-guides. The typical figure is a few dB per mm at a moderately slow group velocity $(15 < v_g < c/30)$ [63,159,163]. This leads to a figure of merit $\tau_{dis}^{-1} = \alpha_{dis} v_g$ which can be interpreted as the interaction time allowed in the waveguide. The order of mag-nitude of τ_{dis} is 0.1 ns. We note that this is well below the cavity lifetime achieved in silicon high-Q cavities (9 ns).

---

[3] This denotes complex eigen-solutions in electromagnetic structures allowing radiation [93].

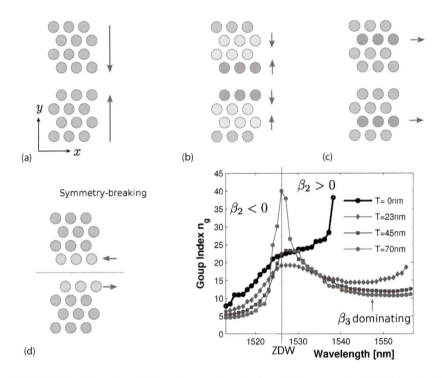

FIGURE 7.14 Examples of dispersion-engineering in a PhC waveguide. (a) "dislocation" defect as two sides of the PhC are moved to change the width of the waveguide. (From Petrov, A.Y., and Eich, M. *Appl. Phys. Lett.*, 85, 4866–4868, 2004.) (b) Independent displacement of the first rows of holes [98], (c) in-axis parallel displacement. (From Hamachi, Y., et al., *Opti. Letter*, 34, 1072–1074, 2009.) and (d) Anti-symmetric in-axis displacement breaking the even-odd symmetry and corresponding experimental dispersion. (Panel d and from Colman, P. et al., *Opt. Express*, 3934, 3930–3934, 2012. With permission of Optical Society of America.)

7.4.3 Parametric Amplification and Coherent Processes

The achievement of on-chip parametric gain, namely when $\eta > 1$, is obviously of great importance because it provides a mean to raise the signal level inside the photonic circuit. The particularity of parametric amplification such as with FWM is that the amplification preserves the coherent nature of the signal. On-chip parametric gain has been achieved in a PhC waveguide, however in pulsed regime [23]. The coherent nature of the parametric amplification appears when both signal and idler are presented at the input. The output will either be amplified or "de-amplified" depending on the mutual phase between signal, idler, and the pump. Unlike phase insensitive processes, where the phase of the generated idler corresponds to a degree of freedom that fulfills the phase-matching condition, Phase-Sensitive Amplification (PSA) generates new photons through FWM *interfering* either constructively or destructively with the input ones. The achievable parametric gain thus depends on the relative phase difference between the waves. The term *interfering* relates to the

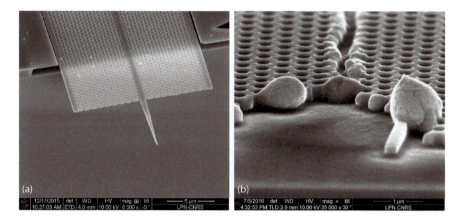

FIGURE 7.15 SEM pictures of the input tapers from top (a) and of a destroyed input tapers of a suspended photonic crystal (b). If the coupled optical field is large enough and the thermal sinking not sufficient, the input tapers may melt or even explode and stick on the bottom.

linear interference of coherent waves with the relative phase between waves determining which photons are annihilated or created [66] (Figure 7.15).

PSA processes have been demonstrated on chip in silicon [208] photonic crystal waveguides but were strongly limited by two-photon absorption (TPA) with an extinction ratio of 11 dB for 2.3 W of input power. Efficient processes were also demonstrated in chalcogenide ridges [209] and using p-i-n junctions on silicon waveguides [34]. PSA experiments in GaInP PhCWG enabled to overcome the TPA limitations and unveil the interplay of competing FWM interactions that leads to a maximum extinction ratio [115]. Interactions between FWM products could hence be tailored through dispersion engineering and lead to enhanced PSA interactions on-chip.

7.4.4 HANDLING LARGE POWER DENSITIES

Most of the all-optical signal processing functions discussed above demonstrated in Highly Nonlinear Fibers were based on continuous wave (CW) pumps. It is indeed a requirement for some applications such as amplification for optical communication. However, pulsed pumps are also of interest for other applications, i.e signal sampling or optical switching for demultiplexing operations. On-chip, even though the involved optical intensities are orders of magnitude smaller compared to fibers, the issue of large average power densities needs to be addressed. Indeed nonlinear enhancement in the highly nonlinear semiconductor structures also comes with increased and undesirable effects such as nonlinear absorption, generation of free carriers inducing dispersion, further absorption, and very intense thermal effect. These effects tend to spoil the benefits of enhanced nonlinearity as they limit the density of the optical field, hence the nonlinear effect itself and might even cause the damage of the structures.

Building on a well-mastered fabrication process and owing to a potential compatibility to CMOS, Photonic crystal waveguides were first developed using silicon. However the electronic band-gap of silicon being smaller than twice the photon energy ($E_g = 1.12$ eV) at Telecom wavelengths ($\hbar\omega = 0.8$ eV), two-photon absorption and the generated free carriers impose an upper limit to the efficiency of the nonlinear processes and generate phonons and consequently heat. Nonlinear performances may be assessed in the pulsed regime as $\chi^{(3)}$ processes are quasi-instantaneous ($\sim fs$) [55] but for practical demultiplexing and optical sampling applications, a higher repetition rate (GHz/THz to CW) still has to be attained. Although Silicon on Oxide provides a very good thermal management, because of free-carriers, a maximum of 14 dBm continuous power has been coupled into a silicon suspended membrane [125] leading to a conversion efficiency of −36 dB.

These limitations have motivated the development of platforms based on other semiconductors, i.e., III-V semiconductors with larger electronic band-gap such as Gallium Indium Phosphide material ($E_g = 1.9$ eV). Demonstration of on-chip soliton generation and compression [28,76] as well as on-chip gain of 11 dB with pulsed pump [23] have raised interest for this material. Unfortunately, in spite of a large nonlinear interaction and absence of two-photon absorption, the residual absorption probably due to surface absorption overwhelms the thermal sinking properties of the material which are 20-fold weaker than silicon.

A way of mitigating the phonon generation is to decrease the thermal resistance by bonding or encapsulating the III-V semiconductor structures in materials with larger thermal conductivity than air or vacuum. The 3-D temperature distribution of a photonic crystal was calculated by solving the heat diffusion equation using the Finite Element Method (FEM) from Comsol Multiphysics. The linear resistivity is defined as the ratio between the local increase of the temperature where heat is generated relative to the background and the dissipated power P_{th}, normalized to the length L:

$$R_{th}^{lin} = L\frac{T_{MAX} - T_{MIN}}{P_{th}}. \tag{7.9}$$

R_{th}^{lin} was also extracted for a suspended, a bonded and an encapsulated membrane (Figure 7.16). As much as a 20-fold decrease of the resistivity between the suspended and bonded or encapsulated membranes is observed because the heat flow is not properly evacuated in a suspended membrane. This result also indicates that the encapsulation does not bring substantial advantage over the structure simply lying on a solid cladding. In fact, the improvement of the thermal conductivity is marginal while the increase of propagation losses is severe [117]. However bonded structures exhibit increased losses due to lower confinement and the increased number of defects caused by the additional process steps.

A material with a better thermal conductivity like Gallium Phosphide ($\times$ 22), while still exhibiting large energy gap ($E_g = 2.26$ eV), could solve this issue. In Table 7.3, conversion efficiencies η of photonic crystals made of Silicon, GaInP, and GaP materials are gathered along with their linear losses α_{lin}, which strongly depends on the group index and mean pump power, $P_{Pump}(in)$. In a low repetition

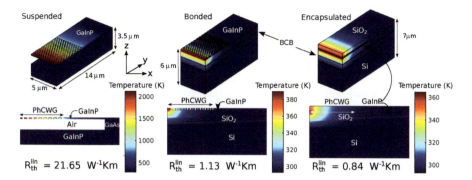

FIGURE 7.16 3D (top) and 2D (bottom) FEM calculations using Comsol of the temperature shift for a given thermal input power of 430 μW emulated with a Gaussian pulse. Only half of a section of the PhC is calculated with appropriate boundary conditions to emulate the whole structure. R_{th} is calculated with the modeled temperature shifts of the suspended (left), bonded (middle) and encapsulated (right) membranes.

TABLE 7.3
FWM in Photonic Crystal Waveguides

Parameters	[97]	[117]	[23]	[114]
Material	Silicon on Oxide	GaInP on Oxide	GaInP Suspended	GaP Suspended
α_{lin} (dB/cm)	?	≈70	≈20	≈60
$P_{Pump}(in)$ (mW)	90	100	4	135
Repetition rate	CW	CW	MHz	GHz
η† (dB)	−24	−18	5	0.8

$$\dagger \ \eta = \frac{P_{idler}(out)}{P_{probe}(out)}$$

rate regime, gain was reached [23] in a GaInP suspended PhCWG. The efficiency is much more challenging to attain with continuous wave interactions and could only be demonstrated with bonded membranes [97,117]. Larger optical densities could be input to a suspended membrane of GaP and efficient nonlinear interactions with GHz regime was demonstrated [114].

7.4.5 RESONANT FOUR WAVE MIXING

In semiconductor materials commonly used in photonics, and considering an interaction length in the millimeter scale, the FWM process becomes effective when the optical power density is about $1\,\mathrm{GWcm}^{-2}$. These levels are reached with power levels well below 1 Watt, owing to the slow-light enhancement; however, this is still large enough when considering miniaturization and large scale integration of nonlinear devices to perform classical and quantum signal processing on chip. Moreover, although nonlinear absorption is in many cases no longer a concern owing to the use of large band-gap material, the residual absorption, and the resulting thermal instabilities need to be tackled, specifically in the CW regime.

Therefore, when harnessing parametric interactions for narrowband signals in optical signal processing, metrology, or quantum optics, the resonant enhancement of the field in optical cavities offers a superior advantage over waveguides, as the conversion efficiency of the parametric process (*stimulated* FWM) scales as $\eta_{sp} \propto Q^4 V^{-2}$ and $\eta_{st} \propto Q^3 V^{-1}$ for its spontaneous counterpart [9]. In crystalline toroidal microcavities, for instance, the Q factor is about 10^9, meaning interaction time in the microsecond range, comparable to the propagation delay through kilometers of optical fibers. In PhC structures, the photon lifetime reached in a PhC cavity is well above the maximum propagation delay observed in PhC waveguides (see Table 7.2). This obvious advantage comes with a severe constraint, namely the resonator must offer equispaced resonances,[4] with a tolerance of their frequency mismatch $\delta v = v_p + v_p - v_s - v_i$ (subscripts denoting pump, signal and idler) smaller than their linewidth, $|\delta v| < v/Q$.

Optical resonators such as toroids, or rings, are highly multimodal, meaning $V_m \gg (\lambda/n)^3$. They are however particularly suited for parametric interactions because equispaced resonances can be obtained over a very broad spectral range owing to dispersion engineering [118]. Importantly, this enabled the emergence of microcavity combs [83] and solitons [69]. Small ring resonators with high-Q are however not straightforward. Considering for instance Silicon Photonics, bending losses set a severe limit [196], for instance, $Q < 10^4$ for $R = 1.5 \times \mu m$.

7.4.5.1 Coupled Resonators

In photonic crystals, in contrast, the volume of the mode can be close to the diffraction limit still allowing $Q \gg 10^6$. However, controlling the frequency spacing of the cavity modes is extremely challenging. The main reason is that in PhC cavities each mode has a specific field distribution and therefore has an unique overlap with the perturbation of the structure due to the fabrication imperfections. Thus, each resonance fluctuates relative to the others, and this is more pronounced if their field distributions are more different. On the other hand, the modal volume of PhC cavities is orders of magnitude smaller than in rings (Table 7.2), which implies increased efficiency of the parametric conversion, as $\eta_{st} \propto V_m^{-2}$. The very first demonstration of FWM in PhC has been reported in three coupled cavities [10]. However, the yield of stimulated FWM was limited to $\eta \approx -60$ dB due to the low Q factor of the modes ($Q \approx 5000$).

The main idea of coupled cavities is that, according to tight binding, the resonances of the supermodes split symmetrically, which would therefore lead to triplets of evenly spaced resonances. This is however prevented by two issues. First, the fabrication disorder is such that the resonances of nominally identical high-Q PhC cavities almost never overlap, and therefore symmetry is lost. To fix the ideas, the

---

[4] More precisely, this includes the nonlinear spectral shift due to thermal effects, self and phase modulation.

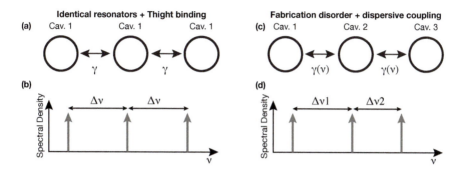

FIGURE 7.17 Creation of a triplet by coupling three cavities. A cartoon representing the 3 cavities and the coupling, γ between them is shown in (a) and (c). The related resonance spectrum is shown in (b) and (d). In the (a)-(b) case, the three cavities are identical and the tight binding approximation is considered, leading to a spectrally equispaced set of three modes. Considering the dispersive coupling, $\gamma(v)$, and the fabrication disorder leading to coupling three different cavities, the spacing between the triplet is no longer the same. (From Lian, J. et al., *Opt. Lett.*, 40, 4488, 2015.)

fluctuation of the resonances of high-Q PhC cavities fabricated on silicon, arguably the best technology available, has been reported to be about 40 GHz [176], hence three orders of magnitude larger than the narrowest linewidth measured (20 MHz) there [8]. The second issue relates to the limitation itself of the tight-binding model, which assumes a constant coupling strength. This is not correct in coupled PhC cavities. In these strongly dispersive structures, the level splitting is enough to modify the field distribution depending on the frequency of the supermode and, therefore, the coupling. As a consequence, the level splitting of the supermodes is never symmetric [101]. This situation is illustrated in Figure 7.17.

A straightforward way to overcome such difficulty is to increase the number of resonators. The availability of large number of supermodes increases the probability of the occurrence of evenly spaced resonances. This was shown by Matsuda et al. with 200 coupled cavities [122], with supermodes having $Q > 10^5$. As the optical mode volume of such structure corresponds, in a first order approximation, to the number of coupled cavities times the volume of a single uncoupled resonator, the much increased V_m implies reduced localization, hence reduced nonlinear interaction. Also, the phase-matching condition in translation-invariant waveguides translates here into the spatial overlap of the interacting supermodes, to which the scaling with the number of resonators only sets an upper bound. In other words, depending on the selected resonances, the interaction might be very weak. More recently, it has been shown that a moderate number of cavities might offer a sensible trade-off between achieving frequency matching and mode volume. Matsuda et al. have demonstrated much more efficient FWM in a chain of 10 coupled PhC cavities [121] made of silicon. We also demonstrated efficient FWM using a similar design based on 10 coupled cavities, made of GaInP [116]. Specifically, the conversion efficiency of −17 dB was reached using about 100 μW of pump, which is orders of magnitude lower than in semiconductor waveguides.

TABLE 7.4

Stimulated FWM in Resonant Cavities. Normalized Efficient $\eta_{norm} = \dfrac{P_i}{P_s P_p^2}$

Technology	Material	Q_{avg}	P_{pmp} (μW)	η_{norm} (W^{-2})	References
Ring res.	Si	2×10^4	1000	10^3	[183]
Ring res.	Doped SiO$_2$	6×10^4	700	4.5×10^4	[52]
Ring res.	Si	1.2×10^6	8000	2	[174]
PhC cavities	Si	4×10^3	60	0.9×10^3	[10]
PhC CROW (10)	Silicon	$> 5 \times 10^5$	100	4×10^4	[121]
PhC CROW (10)	GaInP	7×10^4	36	3×10^6	[116]

It is interesting to compare with recent experiment in resonators, and considering different geometries as well as different materials. This is shown in Table 7.4.

7.4.5.2 Multimode Resonators

A radically different approach aims at the creation of equispaced resonances in a single PhC resonator. The idea is to implement an effective potential for the photons which is harmonic, i.e., parabolic. More precisely, let us consider Bloch modes of a PhC near the band edge. Then, to the first approximation, their dispersion is $\omega = \omega_0 + \partial_{kk}\omega \,|_{CB}\, (k - k_{CB})^2$, with the subscript CB referring to the edge of the conduction band, as an example. Under this conditions, it has been shown [165] that the envelope ϕ of wavepackets follows a Schrödinger equation, namely:

$$i\partial_t\phi = -\frac{1}{2}\partial_{kk}\omega\,|_{CB}\,\partial_{xx}\phi + V(x) \tag{7.10}$$

where the term $V(x)$ takes the role of an effective photonic potential. A suitable spatial perturbation of the PhC structure might turn $V(x)$ into an harmonic potential. As a consequence, the localized photonic modes will be endowed with the properties of the quantum harmonic oscillator, hence they are evenly spaced with envelopes corresponding to the Hermite-Gauss functions.

This has been successfully achieved using a cavity design involving a "bichromatic" lattice, hence involving two close but different periods [6]. It has been shown [31] that, applied to a suspended semiconductor membrane, the localized resonances of this cavity are evenly spaced and, moreover, that the field envelope follows the Hermite-Gauss distribution (Figure 7.18).

Resonators fabricated on InGaP (see Section 2.1.3) have shown up to eight high-Q resonances (Figure 7.18). A statistical analysis performed over 68 resonators demonstrated that the free spectral range is fluctuating within about 20 GHz from the perfect equal spacing, and, in many cases, triplets or quadruplets of resonances are aligned within a few GHz. The fluctuations are ascribed to the fabrication disorder. We note that the fluctuation of the frequency spacing is expected to be smaller than the fluctuation absolute frequency (about 100 GHz), which is consistent with other

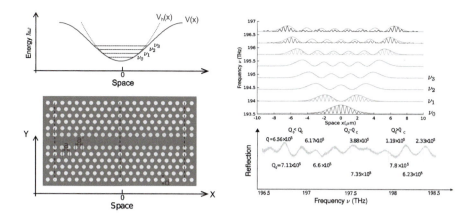

FIGURE 7.18 Effective Harmonic Potential localizing light in a bichromatic PhC cavity. Left, design of the 2D lattice and effective potential. Right, calculated normal modes of the cavity along the $y = z = 0$ axis, compared with the Gaussian-Hermite modes corresponding to the effective potential represented as a dashed line. Below the experimental reflection spectrum revealing 5 high-Q resonances. (From Combrié, S. *Laser & Photonics Rev.*, 11, 1700099, 2017.)

statistical results [164]. Hence, the residual frequency mismatch is small enough to be compensated by thermal tuning, for instance, as demonstrated in ref. [170]. Finally, the statistics of the intrinsic Q factor of the first 4 resonances (hence about 300 in total), reveal a log-normal distribution with most frequent value equal to 0.7 million.

The power threshold for optical parametric oscillation in this cavity is given by the formula (See ref. [31]):

$$P_{th} \approx \frac{\varepsilon_r V_{FWM} \omega}{4 c_0 n_2 Q^2} \qquad (7.11)$$

which is rigorously valid for over-coupled pump where $Q = \prod_{m=1}^{4} Q_m^{1/4}$ is the average of the loaded Q factors of the interacting modes. Considering for instance the first triplet of resonances and an average loaded $Q = 4 \times 10^5$, hence about half the most frequent value of the intrinsic Q, this gives about $P_{th} = 30$ μW, which is remarkably low. This shows that the approach of a multimode cavity might lead to ultra-low parametric resonances in an extremely small resonator owing to the strong localization of the interacting modes.

7.5 NONLINEAR PhCs AND THEIR APPLICATIONS

7.5.1 ALL-OPTICAL SIGNAL PROCESSING WITH FWM

Parametric amplification in optical fibers has been investigated since the 1970s (see [173]) and has opened a wide range of new applications for all-optical signal processing. One of the most successful is perhaps the supercontinuum generation, where the nonlinear dynamics of an intense picosecond pulse is exploited in octave spanning coherent light sources [43]. In the previous section, we have shown that pulse broadening, although limited by the PhC bandwidth, is achieved with moderate power and in a very small structures. More generally, we will discuss the perspective of integration of other known functions owing to PhCs.

7.5.1.1 Coherent Wavelength Conversion, Phase Conjugation, and All-Optical Sampling

As Four Wave Mixing (FWM) is a coherent process, the complex field $E(t)\exp(i\omega_s)$ near a frequency ω_i can be transferred to a new frequency ω_i, preserving the phase information. As FWM based on the electronic third-order nonlinearity (we refer to it as electronic Kerr FWM), which is very fast, the process is broadband, virtually only limited by the waveguide dispersion. Thus, high-speed signals, carrying arbitrarily complex data encoding, from the simple differential phase shift keying to Quadrature and Amplitude Modulation (QAM), can be converted seamlessly. Importantly, the converted signal is phase-conjugated. This property is the cornerstone of many optical processing techniques, for instance dispersion compensation, proposed long ago [202].

As the Kerr FWM is very fast, wavelength conversion can be reconfigured to perform high-speed optical sampling up to the Terahertz range [143], as the time profile of the gating function is limited by the pump pulse and the device dispersion. Moreover, the broad bandwidth can be used to ensure the simultaneous conversion/ gating of multiple channels [191].

Wavelength conversion has been demonstrated in about 1 mm-long PhC waveguides and at sub-Watt pump power levels [33,94]. While waveguide bandwidth, or more precisely the phase-matching bandwidth are limited to about 1 THz, this prevents using PhC for really ultra-fast signal processing. However, the compactness and the potential improvement of the nonlinear enhancement through a better design and reduced losses is likely to make this approach extremely attractive in the context of integrated all-optical signal processing.

7.5.1.2 All-Optical Logic, Phase-Sensitive Amplification, and Signal (re)generation

Ultra-fast all-optical logic has been demonstrated in the context of optical fibers. AND/OR/XOR-logical gates have been implemented using nonlinear optical loop mirrors (NLOM) [16]. In PhC waveguides, FWM has been used to implement these functions by manipulating the phase of optical signals and achieving a data rate of 40 Gb/s [72]. PSA enables the discretization [148] and the regeneration [166] of phase-encoded signals, owing to the selective processes of the two complex quadratures of the optical field used to encode the information [123]. PSA also enables the regeneration of data packets in fiber loops with robustness against the noise to enable optical storage (see [12]). PSA not only re-amplifies a signal but also performs reshaping and re-timing that are crucial for the regeneration of a signal (see [153]). Besides being a phase-sensitive process, FWM also enables the regeneration of the phase and thereby opens the way to phase-encoded modulation formats as in [154] or amplification approaching 0 dB ideal noise figure predicted for parametric amplifiers in general [22]. Remarkably, PSA based on Highly Nonlinear

Fibers (HNLF) [112] has been exploited to achieve an almost ideal noiseless amplification ($F = 1.1$ dB) [180].

PSA in the CW regime is still missing in PhC devices. This achievement would contribute to the integration of these all-optical processing schemes.

7.5.2 QUANTUM INFORMATION AND NONCLASSICAL COMPUTING

An important property of parametric interactions is that they lead to the creation of nonclassical states of the light [167]. In the so-called spontaneous FWM, in analogy with the emission process from excited dipoles, the "pump" radiation interacts with the vacuum fluctuations of the signal and idler modes which are populated with half a photon thus contain an energy of $E_{vac,i} = 0.5\hbar\omega$. This process used to manipulate the nonclassical states of light because of their coherent nature. As highlighted by [194], linear optics alone is indeed not sufficient for entangling two states or generating deterministic gates. Some applications of the Kerr effect for quantum information are described below.

7.5.2.1 Quantum Frequency Conversion and Squeezing

FWM encompasses a variety of parametric mixing processes [124]. This is summarized in Figure 7.19. The annihilation of two pump photons into a signal and idler is also referred to as Modulation Interaction (MI), which is similar to phase conjugation (PC). Bragg Scattering FWM (BS-FWM) is different, because one pump and one signal photons annihilate into another pump and idler photons. BS does not add noise from the pump in contrast with MI or PC, as vacuum fluctuations are not amplified. Thus, BS allows the conversion of the optical frequency of a quantum state while preserving momentum and phase, and it

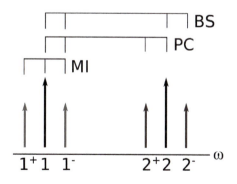

FIGURE 7.19 Parametric amplification (PA) involves two pumps (1,2) and four products ($1^+, 1^-, 2^+, 2^-$). PA is based on three different processes: The modulation interaction (MI), the phase conjugation (PC) and the Bragg Scattering (BS). In MI and PC, the pump photons are annihilated to generate the idler and signal photons whereas in the case of BS, the signal photon is destroyed to generate the idler and vacuum fluctuations are not amplified. (From McKinstrie, C.J., *Opt. Express*, 13, 9131–9142, 2005.)

is therefore referred to as Quantum Frequency conversion. This was realized in Highly Nonlinear Fibers (HLNF) [59,111,124], SiN mm-sized waveguides [2] and micro-rings [100]. The main application of such conversion would be to build a tunable source of squeezed light for interferometric purposes. Squeezed states [167] are generated using nonlinear effects, for instance, phase-sensitive amplification. They are particularly praised for improving the precision of inter-ferometric measurements or atomic clocks. By squeezing states and phases, phase-sensitive amplification exhibits for one of the quadratures a noise fig-ure below the quantum limit (3 dB) [22,136,205] that can be applied to ultra-low noise amplification as well.

These states differ from classical states that fulfill the Heisenberg relation $\Delta E_P \Delta E_Q \geq \frac{\hbar L^3}{2\varepsilon_0}$, with ΔE_P and ΔE_Q the variances of both quadratures of the field. In this description, an optical field is fully described with its amplitude and phase. A quasi-classical state, whose uncertainty in the measurement of the field comes only from quantum fluctuations, corresponds to a situation where both variances of the quadratures are minimal. If, however the noise of one of the quadratures is mini-mized below those minimal states at the price of increasing the noise of the other quadrature, we call it a squeezed state.

PhC have been used to demonstrate the manipulation of quantum optical states [75,122]. An interesting perspective is offered by multimode resonators, as ultra-low power pump could be used for the generation of correlated quantum states [31,116,120].

7.5.2.2 Comb Generation

Kerr effects even broaden their span of application to the metrology, spectros-copy, and sensing with the generation of combs, first developed in HLNF [185] and later transposed to microresonators such as microdisks and micro-rings [83,105]. This periodic train of equally spaced pulses are generated directly from a mode locked laser or from a continuous wave laser using parametric frequency conversion in a nonlinear medium. Interestingly, spontaneous FWM is one of the major contri-bution of modulation instability frequency comb [69,99]. The creation of combs with PhC is however limited by the transmission bandwidth.

7.5.2.3 Reservoir Computing

Reservoir Computing [78] is a concept introduced recently in the context of neural networks and non-Von-Neumann computing. The main idea is that a nonlinear sys-tem with a rich enough internal dynamics does not need to be modified in the pro-cess of training. This makes the training process much faster as it does not require any nonlinear optimization procedure. It has been shown that laser with delayed feedback behave as "photonic reservoirs," therefore suggesting a novel approach where photonics could perform some specialized high-performance computing [18]. In particular, multimode PhC structures have been proposed for reservoir comput-ing [92], in particular, a photonic reservoir based on nonlinear PhC cavities has been proposed [53].

7.6 CONCLUSION

About 30 years after the prediction made by S. John, Photonic Crystals have provided a route for drastically enhanced nonlinear interactions and, thereby, their integration in an integrated photonic chip. This comes timely with the emergence of integrated photonics, where the footprint and the power budget set a severe challenge to the integration of novel functionalities. Yet, it is fair to say that photonic crystals have delivered. In fact, nonlinear effects such as optical bistability and parametric conversion are triggered with power levels as low as microwatts, optical memories with second storage capability have been demonstrated. Moreover, pulse compression due to soliton effect is possible over length scales below one millimeter, hence orders of magnitude shorter than in optical fibers. Combined with on-chip amplifiers and lasers, and, of course the complex photonic circuitry, this technology would allow the implementation of complex devices for optical signal processing and quantum information, following a pattern similar to that of more popular resonators such as micro-rings and disks.

The optical bandwidth of PhCs is intrinsically limited to a few tens of THz as it is related to the photonic band-gap, which clearly restrains the application domain, compromising for instance some application in spectroscopy or metrology. Moreover, PhC appear still difficult to fabricate and way more complicated to design and master as rings and other simpler yet powerful structures. Still, their complexity provides degrees of freedom for optimizing devices for very specific applications. For instance, maximizing the nonlinear interaction under some constraints of size and bandwidth might be possible using PhC resonators. Also, fabrication and design have progressed remarkably, such that, for instance, fabricating high-Q resonators in a variety of material other than silicon is no longer limited to a few groups.

Ongoing research should address manufacturability and the integration on silicon integrated circuits. This route has been opened by some groups, for instance the group of Prof. T. Baba reported one of the earliest demonstration of PhC fabricated in a photonic foundry, while the heterogeneous integration of PhC on a circuit has been pioneered by C2N (F. Raineri). Heterogeneous integration paves the way to the choice of the most suitable materials for any specific function and enables the coexistence of passive, nonlinear, and active devices on the same platform. In this context, PhC cavities and waveguides have a role to play.

REFERENCES

1. F. Tappert, and A. Hasegawa. Transmission of stationary nonlinear optical pulses in dispersive dielectric fibers. I. anomalous dispersion. *Applied Physics Letters*, 23:142–144, 1973.
2. I. Agha, M. Davanço, B. Thurston, and K. Srinivasan. Low-noise chip-based frequency conversion by four-wave-mixing Bragg scattering in SiN_x waveguides. *Optics Letters*, 37(14):2997, 2012.
3. G.P. Agrawal. *Nonlinear Fiber Optics*. Optics and photonics. Academic Press, 3. ed., [nachdr.] edition, 2007.
4. G.P. Agrawal. *Nonlinear fiber optics*. Academic press, Amsterdam, the Netherlands, 2007.

5. Y. Akahane, T. Asano, B.-S. Song, and S. Noda. High-Q photonic nanocavity in a two-dimensional photonic crystal. *Nature*, 425(6961):944–947, 2003.

6. F. Alpeggiani, L.C. Andreani, and D. Gerace. Effective bichromatic potential for ultra-high Q-factor photonic crystal slab cavities. *Applied Physics Letter*, 107(26):261110, 2015.

7. J.A. Armstrong, N. Bloembergen, J. Ducuing, and P.S. Pershan. Interactions between light waves in a nonlinear dielectric. *Physical Review*, 127(6):1918, 1962.

8. T. Asano, Y. Ochi, Y. Takahashi, K. Kishimoto, and S. Noda. Photonic crystal nano-cavity with a Q factor exceeding eleven million. *Optical Express*, 25(3):1769, 2017.

9. S. Azzini, D. Grassani, M. Galli, L.C. Andreani, M. Sorel, M.J. Strain, L.G. Helt, J.E. Sipe, M. Liscidini, and D. Bajoni. From classical four-wave mixing to parametric fluorescence in silicon micro-ring resonators. *Optics Letters*, 37(18):3807–3809, 2012.

10. S. Azzini, D. Grassani, M. Galli, D. Gerace, M. Patrini, M. Liscidini, P. Velha, and D. Bajoni. Stimulated and spontaneous four-wave mixing in silicon-on-insulator coupled photonic wire nano-cavities. *Applied Physics Letters*, 103(3):031117, 2013.

11. T. Baba, T. Kawasaki, H. Sasaki, J. Adachi, and D. Mori. Large delay-bandwidth product and tuning of slow-light pulse in photonic crystal coupled waveguide. *Optical Express*, 16(12):9245–9253, 2008.

12. G.D. Bartolini, D.K. Serkland, P. Kumar, and W.L. Kath. All-optical storage of a picosecond-pulse packet using parametric amplification. *IEEE Photonics Technology Letters*, 9(7):1020–1022, 1997.

13. A. Bazin, K. Lenglé, M. Gay, P. Monnier, L. Bramerie, R. Braive, G. Beaudoin, I. Sagnes, R. Raj, and F. Raineri. Ultrafast all-optical switching and error-free 10 Gbit/s wavelength conversion in hybrid InP-silicon on insulator nanocavities using surface quantum wells. *Applied Physics Letter*, 104(1):011102, 2014.

14. B.R. Bennett, R.A. Soref, and J.A.D. Alamo. Carrier-induced change in refractive index of inp, gaas and ingaasp. *IEEE Journal of Quantum Electronics*, 26(1):113–122, 1990.

15. N.A.R. Bhat, and J.E. Sipe. Optical pulse propagation in nonlinear photonic crystals. *Physical Review E*, 64(5):056604, 2001.

16. A. Bogoni, L. Poti, R. Proietti, G. Meloni, F. Ponzini, and P. Ghelfi. Regenerative and reconfigurable all-optical logic gates for ultra-fast applications. *Electronics Letters*, 41(7):1, 2005.

17. R.W. Boyd. Material slow light and structural slow light: Similarities and differences for nonlinear optics. *JOSA B*, 28(12):A38–A44, 2011.

18. D. Brunner, M.C. Soriano, C.R. Mirasso, and I. Fischer. Parallel photonic information processing at gigabyte per second data rates using transient states. *Nature Communications*, 4:1364, 2013.

19. S. Buckley, M. Radulaski, J.L. Zhang, J. Petykiewicz, K. Biermann, and J. Vučković. Multimode nanobeam cavities for nonlinear optics: High quality resonances separated by an octave. *Optical Express*, 22(22):26498–26509, 2014.

20. M. de Sterke, C. Monat, and B.J. Eggleton. Slow light enhanced nonlinear optics in periodic structures. *Journal of Optics*, 12:104003, 2010.

21. C. Caër, S. Combrié, X.L. Roux, E. Cassan, and A.D. Rossi. Extreme optical confinement in a slotted photonic crystal waveguide. *Applied Physics Letters*, 105(12):121111, 2014.

22. C.M. Caves. Quantum limits on noise in linear amplifiers. *Physical Review D*, 26(8):1817–1839, 1982.

23. I. Cestier, S. Combrié, S. Xavier, G. Lehoucq, A.D. Rossi, and G. Eisenstein. Chip-scale parametric amplifier with 11 dB gain at 1550 nm based on a slow-light GaInP photonic crystal waveguide. *Optics Letters*, 37(19):3996–3998, 2012.

24. J. Chan, A.H. Safavi-Naeini, J.T. Hill, S. Meenehan, and O. Painter. Optimized opto-mechanical crystal cavity with acoustic radiation shield. *Applied Physics Letters*, 101(8):081115, 2012.

25. C.J. Chen, J. Zheng, T. Gu, J.F. McMillan, M. Yu, G.-Q. Lo, D.-L. Kwong, and C.W. Wong. Selective tuning of high-Q silicon photonic crystal nanocavities via laser-assisted local oxidation. *Optical Express*, 19(13):12480–12489, 2011.

26. W. Chen, and D.L. Mills. Gap solitons and the nonlinear optical response of superlattices. *Physics Review Letter*, 58:160–163, 1987.

27. P. Colman, S. Combrié, and A.D. Rossi. Control of dispersion in photonic crystal waveguides using group symmetry theory. *Optics Express*, 3934:3930–3934, 2012.

28. P. Colman, C. Husko, S. Combrié, I. Sagnes, C.-W. Wong, and A.D. Rossi. Temporal solitons and pulse compression in photonic crystal waveguides. *Nature Photonics*, 4(12):862–868, 2010.

29. S. Combrié, S. Bansropun, M. Lecomte, O. Parillaud, S. Cassette, H. Benisty, and J. Nagle. Optimization of an inductively coupled plasma etching process of GaInP/GaAs based material for photonic band gap applications. *Journal of Vacuum Science & Technology B*, 23(4):1521–1526, 2005.

30. S. Combrié, G. Lehoucq, A. Junay, S. Malaguti, G. Bellanca, S. Trillo, L. Ménager, J.P. Reithmaier, and A. de Rossi. All-optical signal processing at 10 GHz using a photonic crystal molecule. *Applied Physics Letters*, 103(19):193510–5, 2013.

31. S. Combrié, G. Lehoucq, G. Moille, A. Martin, and A. Rossi. Comb of high-Q resonances in a compact photonic cavity. *Laser & Photonics Reviews*, 11(6):1700099, 2017.

32. M. Conforti, F. Baronio, and S. Trillo. Resonant radiation shed by dispersive shock waves. *Physical Review A*, 89:013807, 2014.

33. B. Corcoran, M.D. Pelusi, C. Monat, J. Li, L. O'Faolain, T.F. Krauss, and B.J. Eggleton. Ultracompact 160 Gbaud all-optical demultiplexing exploiting slow light in an engineered silicon photonic crystal waveguide. *Optics Letters*, 36(9):1728–1730, 2011.

34. F. Da Ros, D. Vukovic, A. Gajda, K. Dalgaard, L. Zimmermann, B. Tillack, M. Galili, K. Petermann, and C. Peucheret. Phase regeneration of DPSK signals in a silicon waveguide with reverse-biased p-i-n junction. *Optics Express*, 22(5):5029, 2014.

35. J.R. de Oliveira, and M.A. de Moura. Analytical solution for the modified nonlinear Schrödinger equation describing optical shock formation. *Physics Review E*, 57:4751–4756, 1998.

36. A. De Rossi, C. Conti, and S. Trillo. Stability, multistability, and wobbling of optical gap solitons. *Physics. Review Letter*, 81:85–88, 1998.

37. A. de Rossi, M. Lauritano, S. Combrié, Q. Vy Tran, and C. Husko. Interplay of plasma-induced and fast thermal nonlinearities in a GaAs-based photonic crystal nanocavity. *Physical Review A*, 79(4):043818, 2009.

38. F. DeMartini, C.H. Townes, T.K. Gustafson, and P.L. Kelley. Self-steepening of light pulses. *Physics Review*, 164:312–323, 1967.

39. U.P. Dharanipathy, M. Minkov, M. Tonin, V. Savona, and R. Houdré. High-Q silicon photonic crystal cavity for enhanced optical nonlinearities. *Applied Physics Letter*, 105(10):101101, 2014.

40. I. Dicaire, S. Chin, and L. Thévenaz. Structural slow light can enhance beer-lambert absorption. In *OSA Technical Digest (CD), Slow and Fast Light*, page SLWC2. Optical Society of America, Washington, DC, 2011. doi:10.1364/SL.2011.SLWC2

41. W. Ding, C. Benton, A.V. Gorbach, W.J. Wadsworth, J.C. Knight, D.V. Skryabin, M. Gnan, M. Sorrel, and R.M. De La Rue. Solitons and spectral broadening in long silicon-on-insulator photonic wires. *Optical Express*, 16:3310, 2008.

42. N.J. Doran, and D. Wood. Nonlinear-optical loop mirror. *Optical Letter*, 13(1):56–58, 1988.

43. J.M. Dudley, G. Genty, and S. Coen. Supercontinuum generation in photonic crystal fiber. *Reviews of Modern Physics*, 78(4):1135, 2006.

44. M. Ebnali-Heidari, C. Grillet, C. Monat, and B.J. Eggleton. Dispersion engineering of slow light photonic crystal waveguides using microfluidic infiltration. *Optical Express*, 17(3):1628–1635, 2009.

45. A. Efimov, A.V. Yulin, D.V. Skryabin, J.C. Knight, N. Joly, F.G. Omenetto, A.J. Taylor, and P. Russell. Interaction of an optical soliton with a dispersive wave. *Physical Review Letters*, 95(21):213902, 2005.

46. B.J. Eggleton, C. Martijn de Sterke, and R.E. Slusher. Bragg solitons in the nonlinear schrödinger limit: Experiment and theory. *J. Opt. Soc. Am. B*, 16(4):587–599, Apr 1999.

47. B.J. Eggleton, R.E. Slusher, C.M. de Sterke, P.A. Krug, and J.E. Sipe. Bragg grating solitons. *Physical Review Letters*, 76:1627–1630, 1996.

48. J.N. Elgin, T. Brabec, and S.M.J. Kelly. A perturbative theory of soliton propagation in the presence of third order dispersion. *Optics Communications*, 114(3–4):321–328, 1995.

49. R. Faggiani, A. Baron, X. Zang, L. Lalouat, S.A. Schulz, B. O'Regan, K. Vynck, B. Cluzel, F. de Fornel, T.F. Krauss, and P. Lalanne. Lower bound for the spatial extent of localized modes in photonic-crystal waveguides with small random imperfections. *Scientific Reports*, 6:27037, 2016.

50. J. Fatome, C. Finot, G. Millot, A. Armaroli, and S. Trillo. Observation of optical undular bores in multiple four-wave mixing. *Physical Review X*, 4:021022, 2014.

51. M.E. Fermann, M.J. Andrejco, Y. Silberberg, and M.L. Stock. Passive mode locking by using nonlinear polarization evolution in a polarization-maintaining erbium-doped fiber. *Optical Letter*, 18(11):894–896, 1993.

52. M. Ferrera, D. Duchesne, L. Razzari, M. Peccianti, R. Morandotti, P. Cheben, S. Janz, D.X. Xu, B.E. Little, S. Chu, and D.J. Moss. Low power four wave mixing in an integrated, micro-ring resonator with Q = 1.2 million. *Optical Express*, 17(16):14098, 2009.

53. M.A.A. Fiers, T.V. Vaerenbergh, F. Wyffels, D. Verstraeten, B. Schrauwen, J. Dambre, and P. Bienstman. Nanophotonic reservoir computing with photonic crystal cavities to generate periodic patterns. *IEEE Transactions on Neural Networks and Learning Systems*, 25(2):344–355, 2014.

54. A. Fiore, V. Berger, E. Rosencher, P. Bravetti, and J. Nagle. Phase matching using an isotropic nonlinear optical material. *Nature*, 391(6666):463, 1998.

55. M.A. Foster, A.C. Turner, J.E. Sharping, B.S. Schmidt, M. Lipson, and A.L. Gaeta. Broadband optical parametric gain on a silicon photonic chip. *Nature*, 441(7096):960–963, 2006.

56. L.H. Frandsen, A.V. Lavrinenko, J. Fage-Pedersen, and P.I. Borel. Photonic crystal waveguides with semi-slow light and tailored dispersion properties. *Optics Express*, 14(20):9444, 2006.

57. D. Gerace, and C. Andreani. Effects of disorder on propagation losses and cavity Q-factors in photonic crystal slabs. *Photonics and Nanostructures: Fundamentals and Applications*, 3:120, 2005.

58. J.A. Giordmaine, and R.C. Miller. Tunable coherent parametric oscillation in LiNbO$_3$ at optical frequencies. *Physical Review Letters*, 14(24):973, 1965.

59. A.H. Gnauck, R.M. Jopson, C.J. McKinstrie, J.C. Centanni, and S. Radic. Demonstration of low-noise frequency conversion by Bragg scattering in a fiber. *Optical Express*, 14(20):8989–8994, 2006.

60. A.V. Gorbach, and D.V. Skryabin. Theory of radiation trapping by the accelerating solitons in optical fibers. *Physics Review A*, 76(5):053803, 2007.

61. J.P. Gordon. Theory of the soliton self-frequency shift. *Optical Letter*, 11(10):662–664, 1986.

62. F.D.M. Haldane, and S. Raghu. Possible realization of directional optical waveguides in photonic crystals with broken time-reversal symmetry. *Physical Review Letters*, 100(1), 2008.

63. Y. Hamachi, S. Kubo, and T. Baba. Slow light with low dispersion and nonlinear enhancement in a lattice-shifted photonic crystal waveguide. *Optical Letter*, 34(7):1072–1074, 2009.

64. Y. Hamachi, S. Kubo, and T. Baba. Slow light with low dispersion and nonlinear enhancement in a lattice-shifted photonic crystal waveguide. *Optical Letter*, 34(7):1072–1074, 2009.

65. P. Hamel, S. Haddadi, F. Raineri, P. Monnier, G. Beaudoin, I. Sagnes, A. Levenson, and A.M. Yacomotti. Spontaneous mirror-symmetry breaking in coupled photonic-crystal nanolasers. *Nature Photonics*, 9(5):311, 2015.

66. J. Hansryd, P.A. Andrekson, M. Westlund, J. Li, and P.-O. Hedekvist. Fiber-based optical parametric amplifiers and their applications. *IEEE Journal of Selected Topics in Quantum Electronics*, 8(3):506–520, 2002.

67. A. Hasegawa, and Y. Kodama. Nonlinear pulse propagation in a monomode dielectric guide. *IEEE Journal Quantum Electron*, 23:510–524, 1987.

68. S.M. Hendrickson, A.C. Foster, R.M. Camacho, and B.D. Clader. Integrated nonlinear photonics: Emerging applications and ongoing challenges. *JOSA B*, 31(12):3193–3203, 2014.

69. T. Herr, K. Hartinger, J. Riemensberger, C.Y. Wang, E. Gavartin, R. Holzwarth, M.L. Gorodetsky, and T.J. Kippenberg. Universal formation dynamics and noise of Kerr-frequency combs in microresonators. *Nature Photonics*, 6(7):480–487, 2012.

70. T. Hu, B. Dong, X. Luo, T.-Y. Liow, J. Song, C. Lee, and G.-Q. Lo. Silicon photonic platforms for mid-infrared applications. *Photon. Res.*, 5(5):417–430, 2017.

71. S. Hughes, L. Ramunno, J.F. Young, and J.E. Sipe. Extrinsic optical scattering loss in photonic crystal waveguides: Role of fabrication disorder and photon group velocity. *Physical Review Letters*, 94(3):033903, 2005.

72. C. Husko, T.D. Vo, B. Corcoran, J. Li, T.F. Krauss, and B.J. Eggleton. Ultracompact all-optical XOR logic gate in a slow-light silicon photonic crystal waveguide. *Optical Express*, 19(21):20681–20690, 2011.

73. C. Husko, and P. Colman. Giant anomalous self-steepening in photonic crystal waveguides. *Physical Review A*, 92(1), 2015.

74. C. Husko, S. Combrié, Q.V. Tran, F. Raineri, C.W. Wong, and A. de Rossi. Non-trivial scaling of self-phase modulation and three-photon absorption in III-V photonic crystal waveguides. *Optics express*, 17(25):22442–51, December 2009.

75. C.A. Husko, A.S. Clark, M.J. Collins, A. De Rossi, S. Combrié, G. Lehoucq, I.H. Rey, T.F. Krauss, C. Xiong, and B.J. Eggleton. Multi-photon absorption limits to heralded single photon sources. *Scientific reports*, 3:3087, 2013.

76. C.A. Husko, P. Colman, S. Combrie, J. Zheng, A. De Rossi, and C.W. Wong. Soliton dynamics in the multiphoton plasma regime. In *CLEO*: 2013, page QF1D.5. Optical Society of America, 2013.

77. C.A. Husko, S. Combrié, P. Colman, J. Zheng, A. De Rossi Rossi, and C.W. Wong. Soliton dynamics in the multiphoton plasma regime. *Science Report*, 3:1100, 2013.

78. H. Jaeger, and H. Haas. Harnessing nonlinearity: Predicting chaotic systems and saving energy in wireless communication. *Science*, 304(5667):78–80, 2004.

79. S. John. Strong localization of photons in certain disordered dielectric superlattices. *Physical Review Letter*, 58(23):2486–2489, 1987.

80. S.G. Johnson, S. Fan, P.R. Villeneuve, J.D. Joannopoulos, and L.A. Kolodziejski. Guided modes in photonic crystal slabs. *Physical Review B*, 60(8):5751, 1999.

81. S.G. Johnson, P.R. Villeneuve, S. Fan, and J.D. Joannopoulos. Linear waveguides in photonic-crystal slabs. *Physical Review B*, 62(12):8212, 2000.

82. J.B. Khurgin. How to deal with the loss in plasmonics and metamaterials. *Nature Nanotechnology*, 10(1):2–6, 2015.

83. T.J. Kippenberg, R. Holzwarth, and S.A. Diddams. Microresonator-based optical frequency combs. *Science*, 332(6029):555–559, 2011.

84. Y. Kokubun, T. Baba, T. Sakaki, and K. Iga. Low-loss antiresonant reflecting optical waveguide on si substrate in visible-wavelength region. *Electronic Letter*, 22(17):892, 1986.

85. M. Kolesik, and J.V. Moloney. Nonlinear optical pulse propagation simulation: From maxwell's to unidirectional equations. *Physical Review E*, 70:036604, 2004.

86. F. Krauss. Slow light in photonic crystal waveguides. *Journal of Physics D: Applied Physics*, 40:2666–2670, 2007.

87. T.F. Krauss, R.M. De La Rue, and S. Brand. Two-dimensional photonic-bandgap structures operating at near-infrared wavelengths. *Nature*, 383(6602):699–702, 1996.

88. S. Kubo, D. Mori, and T. Baba. Low-group-velocity and low-dispersion slow light in photonic crystal waveguides. *Optics Letters*, 32(20):2981, 2007.

89. E. Kuramochi, N. Matsuda, K. Nozaki, A.H.K. Park, H. Takesue, and M. Notomi. Wideband slow short-pulse propagation in one-thousand slantingly coupled l3 photonic crystal nanocavities. *Optical Express*, 26(8):9552–9564, 2018.

90. E. Kuramochi, M. Notomi, S. Hughes, A. Shinya, T. Watanabe, and L. Ramunno. Disorder-induced scattering loss of line-defect waveguides in photonic crystal slabs. *Physical Review B*, 72(16):161318, 2005.

91. M. Lamy, K. Hammani, J. Arocas, J. Fatome, J.C. Weeber, and C. Finot. Titanium dioxide waveguides for data transmissions at 1.55 µm and 1.98 µm. In *2017 19th International Conference on Transparent Optical Networks (ICTON)*, pp. 1–4, 2017.

92. F. Laporte, A. Katumba, J. Dambre, and P. Bienstman. Numerical demonstration of neuromorphic computing with photonic crystal cavities. *Optical Express*, 26(7):7955–7964, 2018.

93. G. Lecamp, J.-P. Hugonin, and Philippe Lalanne. Theoretical and computational concepts for periodic optical waveguides. *Optical Express*, 15(18):11042–11060, 2007.

94. K. Lenglé, L. Bramerie, M. Gay, M. Costa e Silva, S. Lobo, J.-C. Simon, P. Colman, S. Combrié, and A. De Rossi. Investigation of fwm in dispersion-engineered gainp photonic crystal waveguides. *Optics Express*, 20(15):16154–16165, 2012.

95. K. Lenglé, L. Bramerie, M. Gay, J.-C. Simon, S. Combrié, G. Lehoucq, and A. De Rossi. Efficient second harmonic generation in nanophotonic waveguides for optical signal processing. *Applied Physics Letters*, 102(15):151114, 2013.

96. J. Li, L. O'Faolain, I.H. Rey, and T.F. Krauss. Four-wave mixing in photonic crystal waveguides: Slow light enhancement and limitations. *Optical Express*, 19(5):4458–4463, 2011.

97. J. Li, L. O'Faolain, I.H. Rey, and T.F. Krauss. Four-wave mixing in photonic crystal waveguides: Slow light enhancement and limitations. *Optical Express*, 19(5):4458–4463, 2011.

98. J. Li, T.P. White, L. O'Faolain, A. Gomez-Iglesias, and T.F. Krauss. Systematic design of flat band slow light in photonic crystal waveguides. *Optics Express*, 16(9):6227, 2008.

99. Q. Li, T.C. Briles, D.A. Westly, T.E. Drake, J.R. Stone, B.R. Ilic, S.A. Diddams, S.B. Papp, and K. Srinivasan. Stably accessing octave-spanning microresonator frequency combs in the soliton regime. *Optica*, 4(2):193, 2017.

100. Q. Li, M. Davan Co, and K. Srinivasan. Efficient and low-noise single-photon-level frequency conversion interfaces using silicon nanophotonics. *Nature Photonics*, 10(6):406–414, 2016.

101. J. Lian, S. Sokolov, E. Yüce, S. Combrié, A. De Rossi, and A.P. Mosk. Dispersion of coupled mode-gap cavities. *Optics Letters*, 40(19):4488, 2015.

102. J. Lian, S. Sokolov, E. Yüce, S. Combrié, A. De Rossi, and A.P. Mosk. Measurement of the profiles of disorder-induced localized resonances in photonic crystal waveguides by local tuning. *Optics Express*, 24(19):21939, 2016.

103. Z. Lin, T. Alcorn, M. Loncar, S.G. Johnson, and A.W. Rodriguez. High-efficiency degenerate four-wave mixing in triply resonant nanobeam cavities. *Physics Review A*, 89:053839, 2014.

104. L. Lu, J.D. Joannopoulos, and M. Soljačić. Topological photonics. *Nature Photonics*, 8(11):821–829, 2014.

105. K. Luke, Y. Okawachi, M. Re Lamont, A.L. Gaeta, and M. Lipson. Broadband mid-infrared frequency comb generation in a Si_3N_4 microresonator. *Optics Letters*, 40(21):4823–4826, 2015.

106. S. Malaguti, G. Bellanca, S. Combrié, A. De Rossi, and S. Trillo. Nonlinear propagation below cut-off in line-defect waveguides. In *Advanced Photonics Congress*, page NTu4D.5. Optical Society of America, Washington, DC, 2012. https://doi.org/10.1364/NP.2012.NTu4D.5martin

107. V.S.C. Manga Rao, and S. Hughes. Single quantum-dot Purcell factor and β factor in a photonic crystal waveguide. *Physical Review B*, 75(20), 2007.

108. N. Mann, S. Combrié, P. Colman, M. Patterson, A. De Rossi, and S. Hughes. Reducing disorder-induced losses for slow light photonic crystal waveguides through bloch mode engineering. *Optics Letters*, 38(20):4244, 2013.

109. N. Mann, A. Javadi, P.D. García, P. Lodahl, and S. Hughes. Theory and experiments of disorder-induced resonance shifts and mode-edge broadening in deliberately disordered photonic crystal waveguides. *Physical Review A*, 92(2):023849, 2015.

110. C. Manolatou, M.J. Khan, S. Fan, P.R. Villeneuve, H.A. Haus, and J.D. Joannopoulos. Coupling of modes analysis of resonant channel add-drop filters. *IEEE Journal of Quantum Electronics*, 35(9):1322–1331, 1999.

111. M.E. Marhic, Y. Park, F.S. Yang, and L.G. Kazovsky. Widely tunable spectrum translation and wavelength exchange by four-wave mixing in optical fibers. *Optics Letters*, 21(23):1906, 1996.

112. M.E. Marhic, P.A. Andrekson, P. Petropoulos, S. Radic, C. Peucheret, and M. Jazayerifar. Fiber optical parametric amplifiers in optical communication systems: Fiber OPAs. *Laser & Photonics Reviews*, 9(1):50–74, 2015.

113. A. Martin, S. Combrié, and A. De Rossi. Photonic crystals waveguides based on wide-gap semiconductor alloys. *Journal of Optics*, 19(3):033002, 2016.

114. A. Martin, S. Combrié, A. de Rossi, G. Beaudoin, I. Sagnes, and F. Raineri. Nonlinear gallium phosphide nanoscale photonics. *Photonics Research*, 6(5):B43–B49, 2018.

115. A. Martin, S. Combrié, A. Willinger, G. Eisenstein, and A. de Rossi. Interplay of phase-sensitive amplification and cascaded four-wave mixing in dispersion-controlled waveguides. *Physical Review A*, 94:023817, 2016.

116. A. Martin, G. Moille, S. Combrié, G. Lehoucq, T. Debuisschert, J. Lian, S. Sokolov, A. Mosk, and A. De Rossi. Triply-resonant continuous wave parametric source with a microwatt pump. In *2016 Conference on Lasers and Electro-Optics (CLEO): QELS_Fundamental Science*, pages FM1A–1. San Jose, CA, pp. 1–2, 2016.

117. A. Martin, D. Sanchez, S. Combrié, A. de Rossi, and F. Raineri. GaInP on oxide nonlinear photonic crystal technology. *Optics Letters*, 42(3):599–602, 2017.

118. A.B. Matsko, A.A. Savchenkov, D. Strekalov, V.S. Ilchenko, and L. Maleki. Review of applications of whispering-gallery mode resonators in photonics and nonlinear optics. *IPN Progress Report*, 42(162):1–51, 2005.

119. N. Matsuda, E. Kuramochi, H. Takesue, and M. Notomi. Dispersion and light transport characteristics of large-scale photonic-crystal coupled nanocavity arrays. *Optics Letters*, 39(8):2290–2293, 2014.

120. N. Matsuda, E. Kuramochi, H. Takesue, K. Shimizu, and M. Notomi. Resonant photon pair generation in coupled silicon photonic crystal nanocavities. In *Lasers and Electro-Optics Europe & European Quantum Electronics Conference (CLEO/Europe-EQEC, 2017 Conference on)*, pp. 1–1. IEEE, 2017.

121. N. Matsuda, E. Kuramochi, H. Takesue, K. Shimizu, Y. Tokura, and M. Notomi. Ultra-narrowband nonlinear wavelength conversion using coupled photonic crystal nanocavities. In *Lasers and Electro-Optics Europe (CLEO EUROPE/IQEC), 2013 Conference on and International Quantum Electronics Conference*, pp. 1–1. IEEE, 2013.

122. N. Matsuda, H. Takesue, K. Shimizu, Y. Tokura, E. Kuramochi, and M. Notomi. Slow light enhanced correlated photon pair generation in photonic-crystal coupled-resonator optical waveguides. *Optical Express*, 21(7):8596, 2013.

123. C. McKinstrie, and S. Radic. Phase-sensitive amplification in a fiber. *Optical Express*, 12(20):4973–4979, 2004.

124. C.J. McKinstrie, J.D. Harvey, S. Radic, and M.G. Raymer. Translation of quantum states by four-wave mixing in fibers. *Optical Express*, 13(22):9131–9142, 2005.

125. J.F. McMillan, M. Yu, D.-L. Kwong, and C.W. Wong. Observation of four-wave mixing in slow-light silicon photonic crystal waveguides. *Optics Express*, 18(15):15484–15497, 2010.

126. A. Melloni, A. Canciamilla, C. Ferrari, F. Morichetti, L. O'Faolain, T.F. Krauss, R. De La Rue, A. Samarelli, and M. Sorel. Tunable delay lines in silicon photonics: Coupled resonators and photonic crystals, a comparison. *IEEE Photonics Journal*, 2(2):181–194, 2010.

127. M. Minkov, U.P. Dharanipathy, R. Houdré, and V. Savona. Statistics of the disorder-induced losses of high-Q photonic crystal cavities. *Optical Express*, 21(23):28233, 2013.

128. M. Minkov, V. Savona, and D. Gerace. Photonic crystal slab cavity simultaneously optimized for ultra-high Q/V and vertical radiation coupling. *Applied Physics Letters*, 111(13):131104, 2017.

129. G. Moille, S. Combrié, and A. De Rossi. Modeling of the carrier dynamics in nonlinear semiconductor nanoscale resonators. *Physical Review A*, 94(2):023814, 2016.

130. G. Moille, S. Combrié, and A. De Rossi. Nanophotonic approach to energy-efficient ultra-fast all-optical gates. In G. Eisenstein, and D. Bimberg. *Green Photonics and Electronics*, pp. 107–137. Springer, the Netherlands, Cham Springer International Publishing, 2017.

131. G. Moille, S. Combrié, L. Morgenroth, G. Lehoucq, F. Neuilly, B. Hu, D. Decoster, and A. de Rossi. Integrated all-optical switch with 10 ps time resolution enabled by ALD. *Laser & Photonics Reviews*, 10(3):409–419, 2016.

132. L.F. Mollenauer, R.H. Stolen, and J.P. Gordon. Experimental observation of picosecond pulse narrowing and solitons in optical fibers. *Physical Review Letters*, 45(13):1095–1098, 1980.

133. C. Monat, M. Ebnali-Heidari, C. Grillet, B. Corcoran, BJ Eggleton, TP White, L. O'Faolain, J. Li, and TF Krauss. Four-wave mixing in slow light engineered silicon photonic crystal waveguides. *Optics Express*, 18(22):22915–22927, 2010.

134. C. Monat, B. Corcoran, M. Ebnali-Heidari, C. Grillet, B.J. Eggleton, T.P. White, L. O'Faolain, and T.F. Krauss. Slow light enhancement of nonlinear effects in silicon engineered photonic crystal waveguides. *Optics Express*, 17(4):2944, 2009.

135. D.J. Moss, R. Morandotti, A.L. Gaeta, and M. Lipson. New CMOS-compatible platforms based on silicon nitride and hydex for nonlinear optics. *Nature Photonics*, 7:597, 2013.

136. Y. Mu, and C.M. Savage. Parametric amplifiers in phase-noise-limited optical communications. *Journal of the Optical Society of America B*, 9(1):65, 1992.

137. T. Nakamura, Y. Takahashi, Y. Tanaka, T. Asano, and S. Noda. Improvement in the quality factors for photonic crystal nanocavities via visualization of the leaky components. *Optical Express*, 24(9):9541, 2016.

138. T.N. Nguyen, M. Gay, L. Bramerie, T. Chartier, J.-C. Simon, and M. Joindot. Noise reduction in 2r-regeneration technique utilizing self-phase modulation and filtering. *Optical Express*, 14(5):1737–1747, 2006.

139. M. Notomi, K. Yamada, A. Shinya, J. Takahashi, C. Takahashi, and I. Yokohama. Extremely large group-velocity dispersion of line-defect waveguides in photonic crystal slabs. *Physical Review Letters*, 87(25), 253902, 2001.

140. K. Nozaki, T. Tanabe, A. Shinya, S. Matsuo, T. Sato, H. Taniyama, and M. Notomi. Sub-femtojoule all-optical switching using a photonic-crystal nanocavity. *Nature Photonics*, 4(7):477–483, 2010.

141. K. Nozaki, A. Shinya, S. Matsuo, Y. Suzaki, T. Segawa, T. Sato, Y. Kawaguchi, R. Takahashi, and M. Notomi. Ultralow-power all-optical RAM based on nanocavities. *Nature Photonics*, 6(4):248, 2012.

142. L. O'Faolain, S.A. Schulz, D.M. Beggs, T.P. White, M. Spasenović, L. Kuipers, F. Morichetti, A. Melloni, S. Mazoyer, J.-P. Hugonin, et al. Loss engineered slow light waveguides. *Optics Express*, 18(26):27627–27638, 2010.

143. L.K. Oxenlowe, Hua Ji, M. Galili, Minhao Pu, Hao Hu, H.C.H. Mulvad, K. Yvind, J.M. Hvam, A.T. Clausen, and P. Jeppesen. Silicon photonics for signal processing of Tbit/s serial data signals. *IEEE Journal of Selected Topics in Quantum Electronics*, 18(2):996–1005, 2012.

144. N.C. Panoi, J.F. McMillan, and C.W. Wong. Theoretical analysis of pulse dynamics in silicon photonic crystal wire waveguides. *Journal of Selected Topics in Quantum Electronics*, 16(1):257–266, 2010.

145. N.C. Panoiu, X. Liu, and R.M. Osgood Jr. Self-steepening of ultrashort pulses in silicon photonic nanowires. *Optics Letters*, 34(7):947–949, 2009.

146. T.L. Paoli. Saturable absorption effects in the self-pulsing (alga)as junction laser. *Applied Physics Letters*, 34(10):652–655, 1979.

147. A. Parini, P. Hamel, A. De Rossi, S. Combrié, Y. Gottesman, R Gabet, A. Talneau, Y. Jaouen, G. Vadala, et al. Time-wavelength reflectance maps of photonic crystal waveguides: A new view on disorder-induced scattering. *Journal of Lightwave Technology*, 26(23):3794–3802, 2008.

148. F. Parmigiani, G.D. Hesketh, R. Slavk, P. Horak, P. Petropoulos, and D.J. Richardson. Optical phase quantizer based on phase sensitive four wave mixing at low nonlinear phase shifts. *IEEE Photonics Technology Letters*, 26(21):2146–2149, 2014.

149. A.Y. Petrov, and M. Eich. Zero dispersion at small group velocities in photonic crystal waveguides. *Applied Physics Letters*, 85(21):4866–4868, 2004.

150. A. Petrov, M. Krause, and M. Eich. Backscattering and disorder limits in slow light photonic crystal waveguides. *Optics Express*, 17(10):8676–8684, 2009.

151. L. Provost, C. Finot, P. Petropoulos, and J. Richardson. A 2R Mamyshev regeneration architecture based on a three-fiber arrangement. *Journal of Lightwave Technology*, 28(9):1373–1379, 2010.

152. M. Pu, L. Ottaviano, E. Semenova, and K. Yvind. Efficient frequency comb generation in AlGaAs-on-insulator. *Optica*, 3(8):823, 2016.

153. M.G. Raymer, and K. Srinivasan. Manipulating the color and shape of single photons. *Physics Today*, 65(11):32–37, 2012.

154. D.J. Richardson, R. Slavk, J. Kakande, F. Parmigiani, A. Bogris, D. Syvridis, and P. Petropoulos. *Phase-Encoded Signal Regeneration Exploiting Phase Sensitive Amplification*. Institute of Electrical and Electronics Engineers, Piscataway, NJ, 2011.

155. K. Rivoire, Z. Lin, F. Hatami, W.T. Masselink, and J. Vučković. Second harmonic generation in gallium phosphide photonic crystal nanocavities with ultralow continuous wave pump power. *Optics Express*, 17(25):22609–22615, 2009.

156. T. Roger, M.F. Saleh, S. Roy, F. Biancalana, C. Li, and D. Faccio. High-energy, shock-front-assisted resonant radiation in the normal dispersion regime. *Physical Review A*, 88:051801, Nov 2013.

157. P. Russell. Photonic crystal fibers. *Science*, 299(5605):358–362, 2003.

158. K. Sakoda. *Optical Properties of Photonic Crystals*. Vol. 80. Springer, Berlin, Germany, 2004.

159. J. Sancho, J. Bourderionnet, J. Lloret, S. Combrié, I. Gasulla, S. Xavier, S. Sales, P. Colman, G. Lehoucq, D. Dolfi, and A. De Rossi. Integrable microwave filter based on a photonic crystal delay line. *Nature Communications*, 3:1075, 2012.

160. M. Santagiustina, C.G. Someda, G. Vadala, S. Combrie, A. De Rossi, et al. Theory of slow light enhanced four-wave mixing in photonic crystal waveguides. *Optics Express*, 18(20):21024–21029, 2010.

161. S. Saravi, S. Diziain, M. Zilk, F. Setzpfandt, and T. Pertsch. Phase-matched second-harmonic generation in slow-light photonic crystal waveguides. *Physical Review A*, 92:063821, 2015.

162. V. Savona. Electromagnetic modes of a disordered photonic crystal. *Physical Review B*, 83(8), 2011.

163. S.A. Schultz, L. O'Faolain, D.M. Beggs, Thomas P. White, A. Melloni, and Thomas F. Krauss. Dispersion engineered slow light in photonic crystals: A comparison. *Journal of Optics*, 12(10):104004, 2010.

164. H. Sekoguchi, Y. Takahashi, T. Asano, and S. Noda. Photonic crystal nanocavity with a Q-factor of 9 million. *Optical Express*, 22(1):916, 2014.

165. J.E. Sipe and H.G. Winful. Nonlinear Schrödinger solitons in a periodic structure. *Optics Letters*, 13(2):132, 1988.

166. R. Slavk, F. Parmigiani, J. Kakande, C. Lundström, M. Sjödin, P.A Andrekson, R. Weerasuriya, S. Sygletos, A.D. Ellis, L. Grüner-Nielsen, et al. All-optical phase and amplitude regenerator for next-generation telecommunications systems. *Nature Photonics*, 4(10):690–695, 2010.

167. R.E. Slusher, L.W. Hollberg, Bernard Yurke, J.C. Mertz, and J.F. Valley. Observation of squeezed states generated by four-wave mixing in an optical cavity. *Physical Review Letters*, 55(22):2409, 1985.

168. S. Smolka, H. Thyrrestrup, L. Sapienza, T.B. Lehmann, K.R. Rix, L.S. Froufe-Pérez, P.D. García, and P. Lodahl. Probing the statistical properties of Anderson localization with quantum emitters. *New Journal of Physics*, 13(6):063044, 2011.

169. S. Sokolov, J. Lian, S. Combrié, A. De Rossi, and A.P. Mosk. Measurement of the linear thermo-optical coefficient of $Ga_{0.51}In_{0.49}P$ using photonic crystal nanocavities. *Applied Optics*, 56(11):3219–3222, 2017.

170. S. Sokolov, J. Lian, E. Yüce, S. Combrié, A. De Rossi, and A.P. Mosk. Tuning out disorder-induced localization in nanophotonic cavity arrays. *Optics Express*, 25(5):4598–4606, 2017.

171. S. Sokolov, J. Lian, E. Yüce, S. Combrié, G. Lehoucq, A. De Rossi, and, A.P. Mosk. Local thermal resonance control of GaInP photonic crystal membrane cavities using ambient gas cooling. *Applied Physics Letter*, 106(17):171113, 2015.

172. M. Soljačić, M. Ibanescu, S.G. Johnson, Y. Fink, and J.D. Joannopoulos. Optimal bistable switching in nonlinear photonic crystals. *Physical Review E*, 66(5), 2002.

173. R. Stolen, and J. Bjorkholm. Parametric amplification and frequency conversion in optical fibers. *IEEE Journal of Quantum Electronics*, 18(7):1062–1072, 1982.

174. M.J. Strain, C. Lacava, L. Meriggi, I. Cristiani, and M. Sorel. Tunable Q-factor silicon microring resonators for ultra-low power parametric processes. *Optics Letters*, 40(7):1274–1277, 2015.

175. K. Suzuki, and T. Baba. Nonlinear light propagation in chalcogenide photonic crystal waveguide. *Optics Express*, 18(25):26675–26685, 2010.

176. Y. Taguchi, Y. Takahashi, Y. Sato, Takashi Asano, and Susumu Noda. Statistical studies of photonic heterostructure nanocavities with an average Q factor of three million. *Optical Express*, 19(12):11916, 2011.

177. T. Tanabe, M. Notomi, S. Mitsugi, Akihiko Shinya, and Eiichi Kuramochi. Fast bistable all-optical switch and memory on a silicon photonic crystal on-chip. *Optics Letters*, 30(19):2575–2577, 2005.

178. H. Thyrrestrup, S. Smolka, L. Sapienza, and P. Lodahl. Statistical theory of a quantum emitter strongly coupled to Anderson-localized modes. *Physical Review Letters*, 108(11), 2012.

179. L. Tong, J. Lou, and E. Mazur. Single-mode guiding properties of subwavelength-diameter silica and silicon wire waveguides. *Optics Express*, 12(6):1025–1035, 2004.

180. Z. Tong, C. Lundström, P.A. Andrekson, C.J. McKinstrie, M. Karlsson, D.J. Blessing, E. Tipsuwannakul, B.J. Puttnam, H. Toda, and L. Grüner-Nielsen. Towards ultrasensitive optical links enabled by low-noise phase-sensitive amplifiers. *Nature Photonics*, 5(7):430–436, 2011.

181. J. Topolancik, B. Ilic, and F. Vollmer. Experimental observation of strong photon localization in disordered photonic crystal waveguides. *Physical Review Letters*, 99(25), 2007.

182. T.X. Tran, and F. Biancalana. Dynamics and control of the early stage of supercontinuum generation in submicron-core optical fibers. *Physical Review A*, 79(6):065802, 2009.

183. A.C. Turner, M.A. Foster, A.L. Gaeta, and M. Lipson. Ultra-low power parametric frequency conversion in a silicon microring resonator. *Optical Express*, 16(7):4881, 2008.

184. A.C. Turner, C. Manolatou, B.S. Schmidt, M. Lipson, M.A. Foster, J.E. Sharping, and A.L. Gaeta. Tailored anomalous group-velocity dispersion in silicon channel waveguides. *Optics Express*, 14(10):4357, 2006.

185. T. Udem, R. Holzwarth, and T.W. Hänsch. Optical frequency metrology. *Nature*, 416(6877):233–237, 2002.

186. T. Volz, A. Reinhard, M. Winger, A. Badolato, K.J. Hennessy, E.L. Hu, and A. Imamoğlu. Ultrafast all-optical switching by single photons. *Nature Photonics*, 6(9):607–611, 2012.

187. P.K.A. Wai, H.H. Chen, and Y.C. Lee. Radiations by "solitons" at the zero group-dispersion wavelength of single-mode optical fibers. *Physical Review A*, 41(1):426–439, 1990.

188. P.K.A. Wai, C.R. Menyuk, Y.C. Lee, and H.H. Chen. Nonlinear pulse propagation in the neighborhood of the zero-dispersion wavelength of monomode optical fibers. *Optics Letters*, 11(7):464–466, 1986.

189. F. Wang, J.S. Jensen, J. Mørk, and O. Sigmund. Systematic design of loss-engineered slow-light waveguides. *Journal of the Optical Society of America A*, 29(12):2657, 2012.

190. E. Weidner, S. Combrié, A. de Rossi, N.-V.-Q. Tran, and S. Cassette. Nonlinear and bistable behavior of an ultrahigh-Q GaAs photonic crystal nanocavity. *Applied Physics Letters*, 90(10):101118, 2007.

191. A.O. Wiberg, C.-S. Bres, B. P.-P. Kuo, J.M. C. Boggio, N. Alic, and S. Radic. Multicast parametric synchronous sampling of 320-Gb/s return-to-zero signal. *IEEE Photonics Technology Letters*, 21(21):1612–1614, 2009.

192. A. Willinger, S. Roy, M. Santagiustina, S. Combrié, A. De Rossi, and G. Eisenstein. Narrowband optical parametric amplification measurements in Ga_05in_05p photonic crystal waveguides. *Optics Express*, 23(14):17751, 2015.

193. H.G. Winful. The meaning of group delay in barrier tunnelling: A re-examination of superluminal group velocities. *New Journal of Physics*, 8(6):101, 2006.

194. L.-A. Wu, P. Walther, and D.A. Lidar. No-go theorem for passive single-rail linear optical quantum computing. *Scientific Reports*, 3, 2013.

195. G. Xu, M. Conforti, A. Kudlinski, A. Mussot, and S. Trillo. Dispersive dam-break flow of a photon fluid. *Arxiv*, page 1703.09019v2, 2017.

196. Q. Xu, D. Fattal, and R.G. Beausoleil. Silicon microring resonators with 1.5-μm radius. *Optical Express*, 16(6):4309–4315, 2008.

197. W. Xue, Y. Yu, L. Ottaviano, Y. Chen, E. Semenova, K. Yvind, and J. Mork. Threshold characteristics of slow-light photonic crystal lasers. *Physical Review Letter*, 116(6), 2016.

198. E. Yablonovitch, T. Gmitter, and K. Leung. Photonic band structure: The face-centered-cubic case employing nonspherical atoms. *Physical Review Letters*, 67(17):2295–2298, 1991.

199. E. Yablonovitch. Inhibited spontaneous emission in solid-state physics and electronics. *Physical Review Letters*, 58(20):2059–2062, 1987.

200. E. Yablonovitch, T.J. Gmitter, R.D. Meade, A.M. Rappe, K.D. Brommer, and J.D. Joannopoulos. Donor and acceptor modes in photonic band structure. *Physical Review Letters*, 67(24):3380, 1991.

201. A. Yariv, and M. Nakamura. Periodic structures for integrated optics. *IEEE Journal of Quantum Electronics*, 13(4):233–253, 1977.

202. A. Yariv, D. Fekete, and D.M. Pepper. Compensation for channel dispersion by nonlinear optical phase conjugation. *Optics Letters*, 4(2):52–54, 1979.

203. A. Yariv, Y. Xu, R.K. Lee, and A. Scherer. Coupled-resonator optical waveguide?: A proposal and analysis. *Optics Letters*, 24(11):711–713, 1999.

204. P. Yeh. Electromagnetic propagation in birefringent layered media. *Journal of the Optical Society of America A*, 69(5):742–756, 1979.

205. H.P. Yuen. Reduction of quantum fluctuation and suppression of the Gordon-Haus effect with phase-sensitive linear amplifiers. *Optics Letters*, 17(1):73, 1992.

206. J. Zhang. Optical solitons in a silicon waveguide. *Optical Express*, 15:7682–7688, 2007.

207. J. Zhang, Q. Lin, G. Piredda, R.W. Boyd, G.P. Agrawal, and P.M. Fauchet. Optical solitons in a silicon waveguide. *Optics Express*, 15(12):7682, 2007.

208. Y. Zhang, C. Husko, J. Schröder, S. Lefrancois, I.H. Rey, T.F. Krauss, and B.J. Eggleton. Phase sensitive amplification in silicon photonic crystal waveguides. *Optics Letters*, 39(2):363–366, 2014.

209. Y. Zhang, J. Schröder, C. Husko, S. Lefrancois, D.-Y. Choi, S. Madden, B. Luther-Davies, and B.J. Eggleton. Pump-degenerate phase-sensitive amplification in chalcogenide waveguides. *Journal of the Optical Society of America B*, 31(4):780, 2014.

210. Z. Zhang, and M. Qiu. Small-volume waveguide-section high Q microcavities in 2D photonic crystal slabs. *Optical Express*, 12(17):3988–3995, 2004.

8 Nonlinear and Quantum Effects in Integrated Microcavities

B. Fischer, P. Roztocki, C. Reimer, S. Sciara,
Y. Zhang, M. Islam, L. Romero Cortés,
S. Bharadwaj, D. J. Moss, J. Azaña,
L. Caspani, M. Kues, and R. Morandotti

CONTENTS

8.1 INTEGRATED MICROCAVITIES

Cavities describe resonant structures where optical fields experience constructive enhancement at resonant frequencies. These structures are among the most fundamental and prevalent devices in optics. Typically of macroscopic size, they have served as tools in fundamental investigations of physics (e.g., for nonlinear mode investigations (Haken and Sauermann 1963), mode-locking dynamics (Kelly 1989), and quantum optics (Pysher et al. 2011)), as well as the basis for countless photonics applications. Among those, the most outstanding has been the development of the laser (Schawlow and Townes 1958), but also include telecommunications (Kaminov et al. 2013), spectroscopy (Parson 2015), and life sciences (Schindl et al. 2000) to name a few.

In recent decades, photonic integrated circuit fabrication techniques have enabled the realization of a multitude of chip-based optical device types (e.g., diode lasers, photodiodes) and have elevated integrated photonics to one of the leading optics platforms. Integrated photonics has also enabled the miniaturization of cavities, transitioning them from bulky setups down to the micron scale (Vahala 2003; Moss et al. 2013). Microcavity-based photonic systems offer increased versatility, operational stability, ease-of-use, and compatibility with existing infrastructures (e.g., electronics, fibers). While these characteristics make them particularly attractive for the miniaturization of existing cavity-based technologies, the transition to novel time and length scales, as well as cavity media, will open the door to unique physics and innovative applications.

The exploitation of microcavities for linear, nonlinear, and quantum applications has thus been gaining increasing interest. Microcavities have found use in a vast variety of different applications—ranging from their exploitation as filters in photonic integrated circuits (Chu et al. 1999), optical signal processing (Ferrera et al. 2010, 2011; Pasquazi et al. 2010), to the generation of frequency combs for optical metrology, spectroscopy, and coherent telecommunications (Del'Haye et al. 2007). More recently, the use of microcavities has also been extended to quantum science (Caspani et al. 2016, 2017).

Now, different microcavity approaches have been extensively investigated[1], where some examples are shown in Figure 8.1. A simple yet reliable way to exploit microcavities is the use of microring resonators (MRRs). These consist of a closed

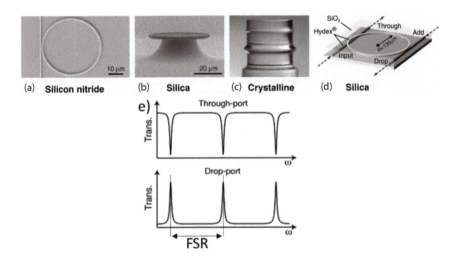

FIGURE 8.1 Different types of microcavities: (a) silicon nitride (SiN) based two-port microring resonator; (b) silica based toroidal microcavity (also known as microdisk); (c) millimeter scale high-Q crystalline ring resonator; (d) silica based four-port microring resonator. (From Kippenberg, T.J. et al., *Science* 332, 555–559, 2011. Reprinted with permission of AAAS.) (reordered); (e) typical output spectra of a four-port resonator at the "through" and "drop" port of the resonator for a broadband signal at the "input" port.

---

[1] For a recent comprehensive review see (Pasquazi et al. 2018).

ring waveguide structure evanescently coupled to one or two bus waveguides (Figures 8.1a, d respectively). An alternative type of microcavity is a microdisk-based resonator (Figure 8.1b). Microdisks can achieve ultrahigh quality factors (Q-factor, which is a measure for the photon lifetime in the resonator multiplied by the angular frequency) (Schwelb 2008). Crystalline microresonators (see Figure 8.1c), made of, e.g., calcium fluoride (CaF_2), are characterized by a millimeter scale and ultrahigh quality factors exceeding those of microdisks (Savchenkov et al. 2008).

Despite the different manufacturing processes used in fabricating microcavities, is still challenging to achieve full integration. Microring structures offer the most stable option to couple light into the resonators, since for such schemes the required bus waveguides (see Figure 8.1(a,d)) can be directly implemented on the chip. This provides a good integrability to other components as well as improved scalability. For microdisks, tapered fibers are usually used to couple light evanescently in and out (Schwelb 2008). Lastly, for crystalline microring resonators light is typically injected into the device via free-space optical paths using prisms or tapered fibers (Schwelb 2008; Strekalov et al. 2016). Although this method can achieve coupling efficiencies of 35%, it poses significant limitations for realizing fully integrated solutions. Thus, although microdisks and crystalline resonators offer high performance with excellent optical properties, they often require elaborate fabrication methods as well as coupling concepts. Therefore, these structures are not very favorable for applications requiring mass production, scalability, or connectivity to additional components in integrated photonic circuit solutions.

In the following, a mathematical description of the key parameters of a micro resonator is given. The spectral separation between two resonances (as depicted in Figure 8.1e) is called the free spectral range (FSR) and is determined by the optical length of the resonator:

$$FSR = \left(\lambda_0\right)^2 / n_g 2\pi R \qquad (8.1)$$

Here, λ_0 is the central resonance wavelength, while the denominator $n_g 2\pi R$ indicates the optical length of the resonator. Furthermore, n_g is the group refractive index and R the radius of the microring. Another important characteristic of microring resonators is the quality factor (Q-factor), which is proportional to the number of optical cycles before the stored energy within the cavity decays to 1/e of its original value. The Q-factor is defined as the ratio between the central wavelength of a given resonance to the linewidth of the same resonance:

$$Q = \lambda_0/\Delta\lambda = \omega_0/\Delta\omega \qquad (8.2)$$

Here, λ_0 and ω_0 describe the central resonance wavelength and frequency, respectively, while $\Delta\lambda$ and $\Delta\omega$ relate to the linewidth of the resonance.

The last important parameter of a microring resonator, closely related to the Q-factor, is the finesse, which is proportional to the number of field round trips before the stored energy decays to 1/e of its original value. It is defined as the ratio between the FSR and spectral bandwidth of the resonance and can act as a measure of the resonator quality:

$$F = FSR / \Delta\omega = \left(\lambda_0\right)^2 / n_g \, 2\pi R \Delta\omega \qquad (8.3)$$

In this chapter, we first discuss the fundamentals of nonlinear effects in microcavities, focusing on microring resonators and their use in classical applications such as mode-locked lasers, optical parametric amplifiers (OPAs), and frequency comb sources. Finally, we will discuss the generation and manipulation of optical quantum states by exploiting integrated microring resonators.

8.2 NONLINEAR OPTICAL EFFECTS IN MICROCAVITIES

Nonlinear processes such as four-wave mixing (FWM) have been reported in various kinds of microcavities and materials, but their efficient use for practical applications is still challenging since integration comes with issues related to short nonlinear interaction lengths, times, and other limitations.

Generally, the nonlinear generation process is based on parametric optical interactions in nonlinear optical crystals or glasses (Boyd 2008). Among the many different nonlinear interactions that can occur in second- and third-order nonlinear media, two of the most common for on-chip applications are parametric down-conversion (PDC) and four-wave mixing (FWM). Which of these two processes dominate depends on the properties of the host crystal, particularly its symmetry. Non-centrosymmetric crystals (e.g., AlN, LiNb) have a second-order nonlinearity, while centrosymmetric materials (e.g., silicon) and noncrystalline materials (e.g., amorphous glass) exhibit third-order nonlinear effects which are typically significantly smaller than second-order nonlinear responses (Boyd 2008; Leuthold et al. 2010).

However, the most crucial part for the efficient exploitation of nonlinearities is to achieve the phase/mode matching for the different involved optical fields in the nonlinear medium (Boyd 2008). If, for example, phase matching is not perfectly attained, the generation efficiency is significantly reduced. There are many different approaches to achieve this ideal condition. Besides the "classic" phase-matching methods (e.g., critical and noncritical phase matching (Boyd 2008)), mainly used in second-order bulk materials, phase matching in third-order nonlinear materials can be obtained through waveguide dispersion design. As well, the dependence of the refractive index on the optical power provides another route for slight mismatch compensation, by choosing an appropriate optical pump power. For on-chip microcavities, third-order nonlinear materials have found great interest due to their use in CMOS-compatible platforms (Leuthold et al. 2010; D. Liang et al. 2010). Additionally, third-order nonlinear materials allow easier pump control since the processes typically involve the mixing of optical signals that are in approximately the same wavelength region and thus it requires less effort to achieve, e.g., spatial mode matching. Nevertheless, in recent years, interest in second-order materials has grown as chip integration (such as crystal growth) techniques have improved. Microcavities in different second-order nonlinear materials such as AlN and LiNb (also known as LN) have recently been reported (Jung et al. 2014; M. Zhang et al. 2017).

The downside of using silicon and silicon-based glasses is the absence of second-order nonlinearities. In order to achieve practical efficiencies with third-order nonlinearities, higher optical pump powers or longer interaction times and/or lengths are required. For this, long (~ few cm) on-chip waveguides have been exploited (Y. Zhang et al. 2017) but also field enhancement within resonant structures can be distinctly useful, in particular optical waveguide-based cavities such as resonators (Ferrera et al. 2008) or photonic crystal waveguides (Corcoran et al. 2009) have achieved significant success in this regard.

For the practical realization of on-chip FWM, many highly nonlinear materials such as AlN (Jung et al. 2014), AlGaAs (Orieux et al. 2017), LiNb (M. Zhang et al. 2017), SiC (Lu et al. 2014), silicon-on-insulator (SOI) (Soltani et al. 2007), and SiN (Moss et al. 2013) have been investigated. Silicon, in the form of SOI, has attracted significant interest since it combines a high nonlinearity, advanced (i.e., CMOS) fabrication methods, low linear loss, and very high refractive index contrast which enables very powerful dispersion engineering methods. However, it has been known (T. K. Liang and Tsang 2004) that two-photon absorption (TPA) in silicon generates the free carriers responsible for high optical nonlinear losses in the telecom wavelength regime near 1550 nm (Rong et al. 2005). An extremely effective way to at least minimize the effect of the TPA generated carriers, even though it cannot eliminate the intrinsic TPA itself, is to use P-I-N structures (Rong et al. 2005; T. K. Liang and Tsang 2004; Engin et al. 2013) to sweep the free carriers out of the region of interest. Despite the presence of TPA in silicon in the telecom band there is still significant interest in this platform for telecom applications because of the previously mentioned features. Indeed, for applications at wavelengths beyond 2000 nm (Miller et al. 2017), extending well into the mid infrared region near 6 µm (Singh et al. 2015) the search has been on for alternative platforms that exhibit even better linear and nonlinear optical properties than silicon in the telecom band. In this regard, silicon nitride (Moss et al. 2013) and more recently amorphous silicon (Leo et al. 2014) as well as silicon rich SiN (Wang et al. 2015; Tan et al. 2016) have been the subject of extensive research.

The first CMOS-compatible platforms studied for integrated nonlinear optical applications, other than SOI itself, were silicon nitride and a closely related high refractive index glass called "Hydex" (Okawachi et al. 2011; Moss et al. 2013; Wang et al. 2015). Silicon nitride (and related materials) has been researched as a platform for linear optics for many years (Henry et al. 1987) but only very recently (Ikeda et al. 2008, Ferrera et al. 2008) was proposed as a platform for nonlinear optics. Its low linear and negligible nonlinear loss in particular make it extremely amenable to realizing high Q-factor ring resonator structures for nonlinear optical applications. With these platforms, on-chip ring resonators succeeded in demonstrating optical parametric and hyperparametric oscillators (Razzari et al. 2010)—enabled by their very low linear loss and negligible nonlinear loss (i.e., TPA) in the telecom wavelength range and by a relatively high nonlinearity compared to that of optical fibers. Of course both platforms offer the advantages of a high compatibility degree to CMOS fabrication processes (Moss et al. 2013). However, as successful as these platforms have been, they are not without their challenges. The ability to grow the thick (>700 nm) SiN layers needed to achieve anomalous dispersion for effective phase-matching in the telecom band has represented a significant challenge, since these thick films introduce significant stress, resulting in film cracking. However, recently these challenges have been overcome using innovative deposition techniques

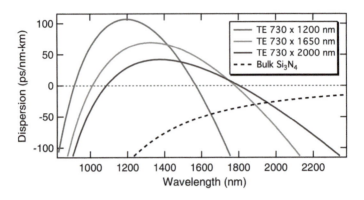

FIGURE 8.2 Anomalous dispersion for fundamental TE modes in SiN waveguides for different waveguide dimensions and the bulk material, for comparison. Here, each of the waveguide structures shows two zero dispersion wavelengths (zero dispersion represented by the thin dashed line) in the telecom window below 2000 nm, whereas the bulk material does not provide these. (From Y. Okawachi, et al., *Opt. Lett.*, 36, 3398, 2011. With permission of Optical Society of America.)

(Ji et al. 2016; Pfeiffer et al. 2016), which have resulted in extraordinarily low loss SiN waveguides—below 1 dB/m. The index contrast of SiN with respect to silica, while not as large as that of silicon, is nonetheless large enough to allow versatile dispersion engineering. In turn, this allows to set the zero dispersion wavelength within the telecom band (see Figure 8.2). This very low and anomalous dispersion enables ideal phase matching for FWM over a broad frequency band, resulting in efficient operation at very low optical pump powers.

8.2.1 MODE-LOCKED LASERS

Mode-locked lasers are characterized by having the frequency mode fields circulating within the cavity phase-locked with respect to each other. This locking results in the generation of laser pulses with an extremely well-defined repetition rate, forming a frequency-domain structure known as a coherent optical frequency comb (see next section). When the cavity and gain support a large number of modes over a broad frequency range, ultrashort optical pulses (typically in the ps- and fs-regimes) are generated (Weiner 2009). However, mode-locked lasers are not only interesting for investigating fundamental physics, such as the generation of attosecond pulses (Paul et al. 2001), but also for practical applications such as optical metrology and spectroscopy, which benefit from the unique dynamics of tightly timed ultrashort pulses (Mandon et al. 2009). In this context, the use of microcavities provides significant advantages over conventional bulk schemes, due to the unique features of the resonator—in particular the resonator acting as both a filter and nonlinear medium at the same time—while providing high robustness against external perturbations such as temperature fluctuations (Peccianti et al. 2012; Pasquazi et al. 2012, 2013; Kues, Reimer, Wetzel, et al. 2017; Roztocki et al. 2017; MacLellan et al. 2018). In order to achieve mode-locking in lasers, several different active (e.g., electro- and acusto-optic modulation) and passive (Kerr nonlinearity, saturable absorbers) mechanisms have been demonstrated (Ippen 1994), with passive mode-locking achieving the shortest pulse durations to date. On the other hand, reaching high repetition rates, while maintaining narrow line widths and low noise, is still very challenging for bulk-based ultrafast pulse mode-locked laser schemes. In order to achieve high pulse repetition rates, harmonic mode-locking is commonly used (Becker et al. 1972). However, this technique is susceptible to generate considerable amplitude noise which results in unstable lasing operation (Becker et al. 1972). To overcome this issue, research groups developed in-fiber Fabry-Perot cavity schemes where the filter is placed inside the resonator cavity in order to filter the generated amplitude noise (Yoshida and Nakazawa 1997). This approach uses the fiber as the nonlinear element and the Fabry-Perot as a filter which allows many modes to oscillate within the bandwidth of the filter. This technique is known as dissipative FWM.

Yet, all of these elaborated schemes struggle with generating nanosecond pulses (featuring ultra-narrow spectral widths) or high repetition rate pulse trains.

The unique characteristics of microring resonators are well-suited to produce ultrashort (sub-ps) pulses with high repetition rates. On the other hand, there is also a need for developing narrow spectral bandwidth mode-locked laser emitting nanosecond pulse durations (Mandon et al. 2009).

For this, a figure-of-eight setup was developed, consisting of two combined loops (Kues, Reimer, Wetzel, et al. 2017). A nonlinear cavity loop (NALM—nonlinear amplification loop mirror) was expanded by an amplifier loop, as depicted in Figure 8.3.

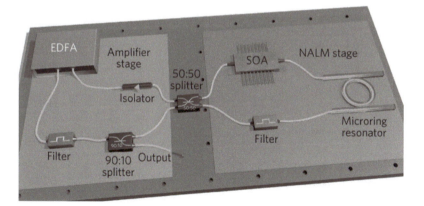

FIGURE 8.3 Figure-8 fiber laser scheme with a nonlinear microring resonator. The setup is composed of two parts, on the left the amplifier stage consisting of an isolator for defining the propagation direction, filter and erbium doped fiber amplifier (EDFA) and on the right, connected by a 50:50 coupler, the nonlinear amplification loop mirror (NALM). Here, the crucial parts are the microring resonator for the nonlinearity and a semiconductor optical amplifier to keep the pulse intensity. (With kind permission from Springer Science+Business Media: Kues, M. et al., *Nature Photonics*, Passively Mode-Locked Laser with an Ultra-Narrow Spectral Width, 11, 2017, 159–162.)

Here, the NALM stage mimics the behavior of a saturable absorber in "classic" semiconductor-based passive mode-locking schemes, but with the advantage of near instantaneous response times. Schemes containing resonant nonlinear elements are superior to previous based on non-resonant nonlinear elements due to an increase in the nonlinear phase-shift required for mode-locking operation. Furthermore, this setup allows wide tunability of the laser frequency with narrow spectral bandwidths (Fourier-transform-limited), sub-ns pulse durations, as well as it enables low power operation due to the use of both microcavities and standard telecommunication components.

In order to generate short transform-limited pulses with high repetition rates, a microcavity nested in one fiber loop is an elegant solution. As in the previous concept, this solution relies mainly on the unique properties of the microcavity to act simultaneously as both filter and nonlinear element. During operation, the multiple oscillating modes in the main cavity are partly suppressed by the microring resonator generating pulse trains with high and adjustable repetition rates. The operation principle of this laser is termed Filter-Driven Four-Wave Mixing (FD-WFM). The first working mode-locked laser based on such a scheme was demonstrated in 2012 (see Figure 8.4) (Peccianti et al. 2012).

This scheme allows the generation of transform-limited sub-ps pulse trains, sub kHz linewidths and repetition rates of over 200 GHz. Furthermore, due to the decreased phase noise limit, the proposed scheme offers negligible amplitude and phase noise for the generated pulses.

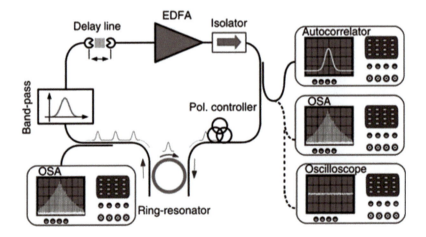

FIGURE 8.4 Nonlinear fiber loop including a microring resonator. The setup consists of an erbium doped fiber amplifier (EDFA), a delay line, bandpass filter, microring resonator, polarization controller, coupler for the output, and an isolator for determining the pulse propagation direction. (With kind permission from Springer Science+Business Media: Peccianti, M. et al., *Nature Communications*, Demonstration of a Stable Ultrafast Laser Based on a Nonlinear Microcavity, 3, 2012, 765.)

8.2.2 OPTICAL PARAMETRIC OSCILLATORS

Since their discovery, optical parametric oscillators (OPO) have formed the basis of state-of-the-art devices for applications such as widely tunable high-power laser sources for, e.g., spectroscopy (Thorpe et al. 2006). Although challenging to achieve, OPOs based on integrated microresonators are highly attractive since integration significantly increases the portability of these devices. In OPOs, the optical gain is provided by nonlinear parametric effects (PDC or FWM) which leads to a parametric gain within the oscillator (Razzari et al. 2010; Ji et al. 2016). OPO in a microcavity not only demands suitable phase-matching (i.e., dispersion) conditions, but also a sufficient input power to reach the threshold required to oscillate (i.e., parametric gain must be higher than the roundtrip loss within the resonator). For microcavities, this threshold is given by (Li et al. 2012)

$$P_{Threshold} = (1+K)^3 n_{eff} \omega A / 8K n_2 \Delta\omega_{FSR} Q^2 \tag{8.4}$$

Here, K is the normalized external coupling rate, n is the effective refractive index of the waveguide, ω is the optical frequency used for excitation, A is the effective mode area, n_2 is the nonlinear index, $\Delta\omega_{FSR}$ is the free spectral range, and Q is the Q-factor of the resonator.

Figure 8.5 shows the output spectra of a silicon nitride based microring resonator excited with different pump powers around 1550 nm.

In order to achieve OPO, the microring resonator was pumped at 1544 nm by a continuous wave (cw) laser, subsequently amplified by an erbium doped fiber amplifier. Before coupling the light into the ring, the polarization of the pump light was adjusted and filtered to only excite a single resonance (here the TM mode). The operation above the OPO threshold led, via nonlinear effects (FWM), to the generation of signal and idler fields as well as optical parametric oscillation, as shown in Figures 8.4a and b. As the power was further increased, cascading of nonlinear effects was observed which resulted in additional frequency modes (Figures 8.4c, d).

8.2.3 COHERENT OPTICAL FREQUENCY COMBS

Frequency combs are characterized by equidistantly spaced spectral lines, resulting in a periodic train of pulses in the time domain, as shown in Figure 8.6.

In order to generate a (coherent) frequency comb, the phases of different modes in a resonator have to be phase-locked. In a classical frequency comb, the locking of the phase can be achieved by active- or passive mode-locking techniques using Q-Switches (active), optical modulators (active), saturable absorbers (passive) (Weiner 2009). However, the generation of stable frequency combs in microcavities (so-called "microcombs") is much more challenging. The evolution of microcombs over most of the last decade has largely been a quest to gain understanding and control of their coherence, or phase stability. In the process, a great deal has been learned of the rich array of physics that microcombs can exhibit, from Turing patterns (Pasquazi et al. 2018) to cavity solitons (Del'Haye et al. 2011), soliton crystals (Cole et al. 2017)

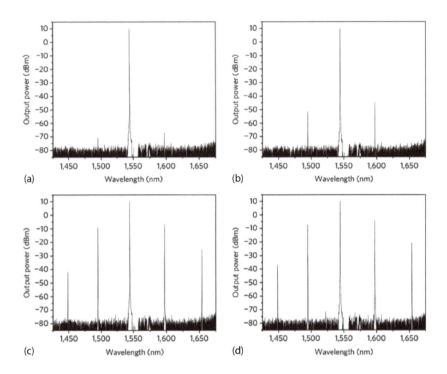

FIGURE 8.5 Output spectra of a silicon nitride microring OPO pumped at 1544.15 nm with increasing (cw-)pump power from a–d. (a) pump power of 50.8 mW; (b) 53.8 mW; (c) 56.8 mW; (d) 59.8 mW. While a and b show only one side peak (signal and idler) symmetric to the pump position, c and d show two side peaks resulting from cascaded nonlinear effects (OPO and FWM). The OPO threshold of the used microring resonator is 54 mW. (With kind permission from Springer Science+Business Media: Razzari, L. et al., *Nature Photonics*, CMOS-Compatible Integrated Optical Hyper-Parametric Oscillator, 4, 2010, 41–45.)

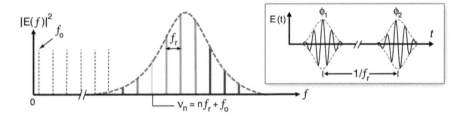

FIGURE 8.6 Typical spectrum of a frequency comb. The comb structure is characterized by the repetition rate f_r and the offset frequency f_0. Inset: Temporal spectra of the frequency comb. The pulse trains are separated by the inverse of the repetition rate. ϕ_1 and ϕ_2 describe the phase difference of the envelope compared to the pulse. (From Kippenberg, T.J. et al., *Science* 332, 555–559, 2011. Reprinted with permission of AAAS.)

and more. In the early days of microcombs, it was difficult to achieve stable and phase-coherent combs (Levy et al. 2010; Razzari et al. 2010). However, while much has been learned of their dynamic behavior, and stable soliton based operation is well understood, it still remains a challenging and complex process to initiate stable and coherent frequency combs in microresonators, particularly if one wants to generate a broad bandwidth comb over a wide wavelength range, needed for many applications—particularly for self-referencing (Telle et al. 1999; The Royal Swedish Academy of Sciences 2005). In order to understand and control the dynamic behavior of microcombs, research groups have put considerable effort into investigating the nonlinear behavior of these devices (Kippenberg et al. 2011; Okawachi et al. 2011; Yi et al. 2016; Herr et al. 2014; Del'Haye et al. 2007). In particular, for self-referencing it is important to generate octave-spanning frequency combs that are coherent (i.e., phase-stable). Such a wide frequency bandwidth is typically needed for stabilizing the carrier-envelope offset (CEO) frequency of the comb, via f-2f interferometry (Reichert et al. 1999; Jones et al. 2000). This concept can generate a comb where the absolute frequency of each comb line does not drift with time, and is stable and related to an absolute reference frequency. CEO-stable frequency combs are the key enabling technology for applications such high-resolution spectroscopy and high-precision radio-astronomy (Lomsadze and Cundiff 2017; Ycas et al. 2012).

In order to generate a high number of modes with a small FSR, a large resonator cavity is required (see Eq. 8.1). In solid state lasers, increasing the cavity length is a challenge due to an accompanying decrease in the peak intensity (Butterworth et al. 1996) associated to each frequency mode as the number of modes scales with the cavity length. However, microcavities can offer both a large number of modes simultaneously with substantial power output. In 2007, Del'Haye et al. opened a new direction by exploiting the periodic resonances of microcavity resonators to generate frequency combs, so-called Kerr-combs (Del'Haye et al. 2007), a more general term than microcombs, which is typically reserved for integrated structures. The focus thereafter has relied on stretching the limits of those combs to achieve long term stable operation, octave-spanning combs, and phase coherence. By using microring resonators with a small FSR, researchers achieved different methods for stabilization such as external-referencing (Del'Haye et al. 2008), feedback controlled (Yi et al. 2016) and coherent soliton (Brasch et al. 2016) frequency combs. External-referencing schemes are a powerful and reliable yet complex and expensive technique which relies on beat measurements of a Kerr comb, generated by a microring resonator, with a reference comb, see Figure 8.7.

Here, the microtoroid resonator was excited by an amplified external cavity diode laser (ECDL) to generate the frequency comb. Afterwards, the 86 GHz beat signal was mixed with the 6th harmonic of a local oscillator (14.3 GHz) in order to achieve a 30 MHz signal, which could be measured and subsequently allowed phase locking of the generated comb by adjusting the output power of the ECDL.

Another way of achieving a stable comb is by measuring the average power of the soliton and implementing a feedback loop for the laser frequency, so-called frequency detuning, see Figure 8.8.

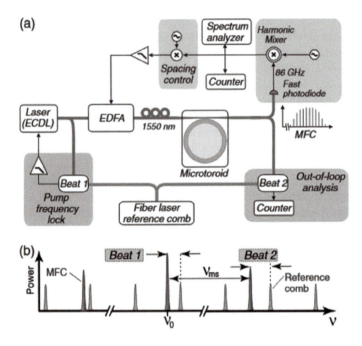

FIGURE 8.7 (a) Experimental setup for a stable comb based on an external reference comb. (b) Output of the stabilized comb including the two beat modes. (From Del'Haye, P. et al., *Phys. Rev. Lett.*, 101, 053903, 2008.)

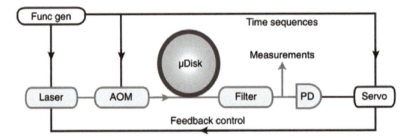

FIGURE 8.8 Schematic of the frequency detuning setup used in the work of Yi et al. The laser power which is coupled into the ring is controlled by an acousto-optic modulator (AOM). Afterwards, the excitation wavelength is filtered by a fiber Bragg grating before the soliton power is measured by means of a photodiode (PD). This PD generates an error signal which is used in a feedback loop to adjust the laser frequency. (From X. Yi. et al., *Opti. Lett.*, 41, 2037, 2016, With permission of Optical Society of America.)

Although microresonators have directly generated octave-spanning frequency combs (Okawachi et al. 2011; Del'Haye et al. 2011), this is difficult to realize, as most conventional laser sources do not provide enough bandwidth to cover one complete octave. Therefore, an additional external setup is often required. The most common way to achieve an octave-spanning frequency comb is through broadening of the pulse by using nonlinear interactions including supercontinuum generation (SCG) in

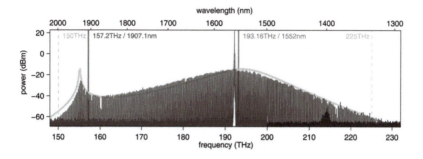

FIGURE 8.9 Spectrum of the comb generated in 2016 by Bratsch et al. using a two-port SiN microresonator. The dashed lines indicate the beginning and end of the comb which results in the spanning of two third of an octave. The solid envelope indicates results obtained from simulations, and the vertical solid lines at 157.2 and 193.16 THz arise from coherence (beat note) measurements of the Cherenkov radiation carried out with a CW laser source emitting at 1552 nm. (From Brasch, V. et al., *Science*, 351, 357–360, 2016. Reprinted with permission of AAAS.)

nonlinear or micro-structured fibers outside of the resonator. However, this is not a suitable solution if full integration is desired. With respect to this particular aspect, microcavity-based solitons are a very promising solution since they are capable of directly generating wide-spanning combs. Figure 8.9 shows how (Brasch et al. 2016), by carefully designing the waveguide dispersion and exciting the microring resonator at a single resonance with relatively high power (continuous wave (CW), $P_{in} = 2$ W in this case), the resonator emitted a comb spanning two thirds of an octave.

The spectrum further shows a distinct peak at 155 THz which arises from Cherenkov radiation due to normal group velocity dispersion (GVD) whereas the soliton is generated due to anomalous dispersion, see Figure 8.10.

This effect of the Cherenkov radiation arises from higher-order dispersion terms and results in radiative tails of the soliton due to a change of the shape and velocity of the stationary solitons. As a result, the spectrum becomes asymmetric (Soliton peak area in Figure 8.10) and an additional, local maximum is formed (dispersive wave). Although the generated comb did not quite span a complete octave, it extended the

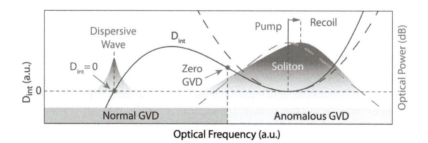

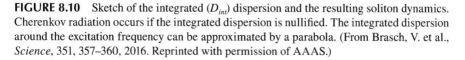

FIGURE 8.10 Sketch of the integrated (D_{int}) dispersion and the resulting soliton dynamics. Cherenkov radiation occurs if the integrated dispersion is nullified. The integrated dispersion around the excitation frequency can be approximated by a parabola. (From Brasch, V. et al., *Science*, 351, 357–360, 2016. Reprinted with permission of AAAS.)

bandwidth enough to enable a number of new applications of microcavity-based combs, particularly in spectroscopy and metrology.

8.3   QUANTUM PHOTONICS IN MICROCAVITIES

The field of quantum optics has become one of the most active avenues of fundamental research, as well as for practical applications (Zeilinger et al. 2005). Based on wide array of approaches for the generation (Kwiat et al. 1995), manipulation (Politi et al. 2008), and detection (Hadfield 2009) of single photons, quantum optics has enabled a wide range of fundamental experiments, such as the Einstein-Podolsky-Rosen paradox (Einstein et al. 1935; Howell et al. 2004) and tests of non-locality (Gröblacher et al. 2007) including loophole-free Bell inequality violations (Giustina et al. 2015; Shalm et al. 2015), as well as the first realization of experimental measurement-based quantum computation (Walther et al. 2005). Moreover, secure communications (Gisin et al. 2002), imaging (Lloyd 2008), microscopy (Israel et al. 2014), and spectroscopy (Kira et al. 2011) applications can also be improved by exploiting the quantum properties of light. The unique properties of nonlinear microcavities enable them to efficiently generate entangled (i.e., quantum correlated) photon pairs (Caspani et al. 2017) and many other functions. Microresonators are also a very efficient and elegant way to realize quantum experiments and applications such as optical quantum information processing (QIP), and quantum key distribution (QKD) (see (Caspani et al. 2017), (Kues et al. 2019) and references therein) with Kues, Michael, Cristian Reimer, Joseph M. Lukens, William J Munro, Andrew M. Weiner, David J. Moss, and Roberto Morandotti. 2019. "Quantum optical microcombs." Nature Photonics 13 (3). Springer US: 170–79. Doi: 10.1038/s41566-019-0363-0). Solutions based on microcavity resonators have many advantages over other approaches such as those based on free-space optics. These include: (i) the use of low pump powers (sub-mW regime) which allows easier filtering; (ii) the generation of resolvable and individually addressable modes (which allow individual mode control); (iii) resonance linewidths that are compatible with quantum memories; and (iv) cavity mode spacings that are in the radiofrequency (RF)-range. Furthermore, the systems are compatible with off-the-shelf telecommunication components which allow easier implementation in telecommunications systems than for comparable bulk optic setups (Reimer et al. 2018).

8.3.1   PHOTON-PAIR GENERATION

One of the most common techniques to generate optical quantum states is based on exploiting nonlinear optical interactions. While individual quantum emitters such as molecules (Siyushev et al. 2014), quantum dots (Buckley et al. 2012) and defect centers in crystals (Togan et al. 2010) can be used to generate single photons (and can be combined to form multi-photon states), nonlinear optical sources are commonly used to directly generate photon pairs. Microcavities based on different material platforms (e.g., Si (Clemmen et al. 2009), Si (Azzini et al. 2012), SiN (Reimer et al. 2014), AlN (Guo et al. 2016)) have played an increasing role in efficiently generating photon pairs. Parametric photon generation during a nonlinear interaction (e.g., FWM) is effectively instantaneous (on a time-scale given by the inverse of the phase-matching bandwidth) hence signal and idler photons are generated at the same time and place. For energy

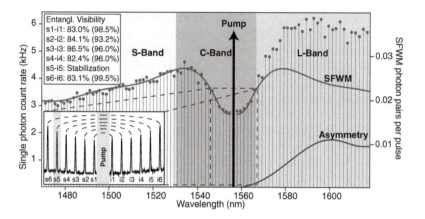

FIGURE 8.11   Single-photon spectrum (dots) emitted by the microring resonator, measured using a grating-based spectrometer and a high-resolution digital tunable filter in the C band (bottom inset). The upper solid curve indicates photons ideally generated through SFWM, whereas the bottom solid curve depicts a spectral asymmetry due to Raman scattering (noise). The bottom inset shows the channels used in the entanglement measurements. Further, the measured raw entanglement visibilities (background-corrected values in brackets) for the individual channel pairs are displayed in the top inset. (From Reimer, C. et al., *Science*, 351, 1176–1180, 2016. Reprinted with permission of AAAS.)

conversion, those photon pairs are generated spectrally symmetric to the incident pump frequency, as shown in Figure 8.11 (Reimer et al. 2016).

There generally exist two schemes for the generation of photon pairs using a microcavity: continuous wave (CW) and pulsed excitation. CW signals can be coupled effectively into a resonance despite the presence of thermal nonlinearities (Carmon et al. 2004) yet, typically require active feedback. However, for pump lasers with linewidths smaller than the cavity resonance, this results in the generation of low purity spectrally entangled photon pairs (Helt et al. 2010). This is detrimental, since pure single-photon sources are required for high-visibility quantum interference (for e.g., linear quantum computing (Kok et al. 2007)), **with** Kok, Pieter, William J. Munro, Kae Nemoto, Timothy C. Ralph, Jonathan P. Dowling, and Gerard J. Milburn. 2007. "Linear Optical Quantum Computing with Photonic Qubits." Reviews of Modern Physics 79 (1): 135–74. doi: 10.1103/RevModPhys.79.135.) and for scaling up state complexities (for, e.g., multi-photon states). On the other hand, pulsed excitation can generate high-purity photon pairs. As well, such schemes allow for simplified synchronization with the photon emission, and enable the reduction of detector noise counts by means of temporal gating or post-selection. Unfortunately, large bandwidth pulses are difficult to couple into narrow (hundreds of MHz) resonances. As well, the generation rate of photons is typically limited by available pulsed laser sources (and is not easily increased). Toward this end, recent work has targeted the development of schemes merging self-locked laser architectures with microcavities, such that the oscillating field is passively locked to and filtered by the microcavity resonance (Reimer et al. 2015). Furthermore, by employing harmonic mode-locking while maintaining adequate pulse powers, the same work has shown that the photon pair generation rate can be increased without sacrificing either photon pair Coincidence-to-Accidental Rate (CAR, the coincidence measurement equivalent of the signal-to-noise ratio) or purity (Roztocki et al. 2017; MacLellan et al. 2018).

As previously discussed schemes have shown, microcavity-based sources are ideal to generate entangled quantum states, where the sources can be designed to emit photon pairs (Barreiro et al. 2005). Further, due to the conservation of energy, momentum and spin in the nonlinear generation process (Boyd 2008), photon pair sources can directly generate a large variety of different entangled states, where the two photons are entangled in, e.g., energy (Franson 1989), polarization (Kwiat et al. 1995), orbital angular momentum (Dada et al. 2011), etc. (for a discussion on entanglement see the following section), and are therefore widely used for applications relying on entanglement as a quantum resource.

8.3.2 TWO-PHOTON ENTANGLEMENT

Integrated sources have managed to produce different types of entangled quantum states such as path (Fulconis et al. 2007; Silverstone et al. 2014, 2015; Solntsev and Sukhorukov 2017), polarization (Matsuda et al. 2012; Horn et al. 2013; Olislager et al. 2013), as well as continuous-variable energy-time entanglement (Grassani et al. 2015). Most sources have mainly relied on simple waveguides (Sharping et al. 2006), photonic crystals (Collins et al. 2013), as well as microring resonators (Engin et al. 2013). While microring resonators have been used for the generation of photon pairs (Engin et al. 2013), and recently also the generation of continuous energy-time entanglement (Grassani et al. 2015), the generation and manipulation of photons in multiple (high-dimensional) frequency modes is a very new field (Kues, Reimer, Roztocki, et al. 2017; Imany et al. 2018).

If a stable CW laser is used to generate a spectrally broadband photon pair, the photons are generated in a coherent superposition of temporal and frequency modes, forming continuous-variable entangled states (Grassani et al. 2015), see Figure 8.12a.

However, since it is very difficult (or even impossible) to address the individual quantum modes of continuous-variable states, their implementation in many applications is very challenging (Su et al. 2013). For this reason, it is often preferred to work with discrete quantum states, especially for the implementation of quantum gates (Humphreys et al. 2013; Knill et al. 2001).

The discrete forms of energy-time entanglement (where discrete temporal and/or frequency modes are present) can be categorized into time-bin (Brendel et al. 1999) and frequency-bin (Olislager et al. 2010) entanglement, depending on which variable (temporal or frequency mode) can be controlled, see Figures 8.12b and c.

In particular, if photons are in a superposition of two (or more) distinct temporal modes, which do not overlap and can be manipulated independently (Brendel et al. 1999), the state is referred to as time-bin entangled. Since temporal modes are linked to the frequency domain through Fourier transformation, the presence of a minimum of two temporal modes also dictates the presence of at least two frequency modes. However, these frequency modes usually overlap within the spectral bandwidth of the photon and cannot be selected or controlled independently, see Figure 8.12b.

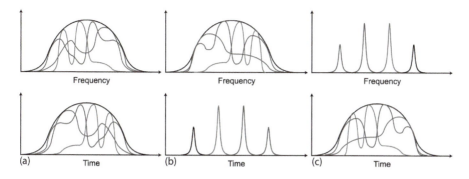

FIGURE 8.12 Schematic of the different types of energy-time entanglement: Temporal and frequency modes of (a) continuous-variable, (b) discrete time-bin, and (c) discrete frequency-bin entanglement. Note that this figure is a very simplified visualization of otherwise complex mode structures. The main point is that in the continuous case all temporal and frequency modes overlap and cannot be independently addressed, while in the time-bin and frequency-bin discrete forms, either the temporal or frequency modes can be manipulated independently, since they do not overlap.

If two (or more) frequency modes do not overlap and are spectrally separated far enough so that they can be selected and manipulated by means of optical filters (usually requiring a spectral separation in the GHz range), the quantum state is referred to as frequency-bin entangled (Olislager et al. 2010). For such states, the temporal modes overlap and cannot be addressed independently, see Figure 8.12c.

In addition to their potential on-chip generation, quantum state manipulation of discrete energy-time entanglement can in principle be implemented using on-chip devices, by means of interferometers, frequency filters, modulators, and phase shifters, all of which can be achieved by means of integrated photonics (Editorial 2010). For the above-stated reasons, the work presented in this section focuses on the realization of discrete energy-time entanglement using integrated microring resonator-based optical frequency combs. Many different approaches for quantum state generation and manipulation based on the exploitation of microcavities have been undertaken.

The first demonstration of time-energy entanglement in microcavities was shown by Grassani et al. in 2015. For their experiments, they used SOI microring resonators to show energy-time entanglement and compared the results to energy-time entangled photon pairs generated in second-order nonlinear materials. As a result, this work paved the road for time-energy entanglement using microcavity resonators in contrast to previously used techniques, as for example polarization entanglement, which are limited in scalability.

Another type of entanglement is time-bin entanglement. Here, the initial pump pulse is separated and delayed by using an unbalanced (fiber) interferometer ("Pump interferometer" in Figure 8.13) allowing for the preparation of a coherent pair of input pulses ($|S>$ and $|L>$) as shown in Figure 8.13.

Afterwards, the separated pulses enter the microcavity and created a signal-idler pair superposition for two different generation times, separated by the interferometer delay. The concept is of particular interest, as it is well-suited for use with standard

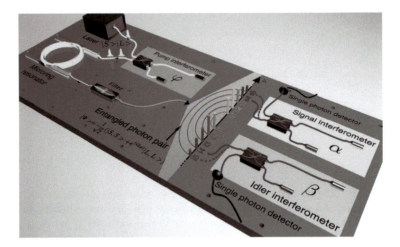

FIGURE 8.13 Setup for time-bin entanglement using fiber interferometers and microring resonators. (From Reimer, C. et al., *Science*, 351, 1176–1180, 2016. Reprinted with permission of AAAS.)

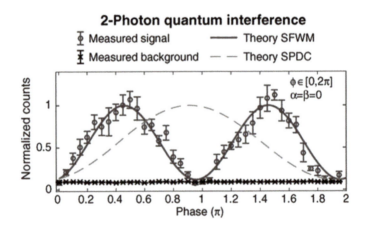

FIGURE 8.14 Measurement of the time-bin entanglement by varying the phase of the pump interferometer while keeping the signal and idler interferometer stable. Interference visibilities of more than ~0.71 prove a violation of Bell's inequality and are thus a measure of entanglement. (From Reimer, C. et al., *Science*, 351, 1176–1180, 2016. Reprinted with permission of AAAS.)

telecommunications/fiber as well as electronic infrastructures. For verifying the fidelity of the quantum states, correlation measurements between these time-bins can be performed (using another set of interferometers that coherently interfere such time-bins). These measurements allow the characterization of the generated states as illustrated in Figure 8.14 and thus prove entanglement by showing violations of Bell's inequality.

Another newly exploited type of entanglement easily accessible via microcavities is frequency-bin entanglement. This entanglement is based on the separation of entangled modes in the frequency domain rather than in the time domain as discussed before. This offers a wider variety of possibilities regarding the control and manipulation of the modes due to the exploitation of standard telecom components. Frequency-bin entanglement usually requires simpler setups than for time-bin entanglement (in general no pump interferometer is required). As well, frequency-bin entanglement is a suitable for the generation of high-dimensional states in a single spatial mode and manipulation via electronic components such as spectral pulse shapers and frequency modulators (Kues, Reimer, Roztocki, et al. 2017; Imany et al. 2018).

Here, the generated photon pairs are in a superposition of many different frequency modes and thus can be subsequently manipulated using programmable filters and a phase modulator. Since phase and frequency are directly linked together, the phase modulator allows shifting the frequency modes which are thus suitable for manipulation and processing of frequency encoded data.

8.3.3 Complex Quantum States

For powerful future applications, multi-photon and/or high-dimensional quantum states will be required to create larger Hilbert spaces (Lanyon et al. 2009). Polarization entanglement is intrinsically two-dimensional, and therefore incompatible with the generation of high-dimensional states. In addition, the realization of polarization-entangled multi-photon states requires elements such as directional couplers and polarization beam splitters, with equal dispersion for both polarization modes. Such devices have until now only been achieved in laser-written glass waveguides (Ciampini et al. 2016), but not yet in integrated on-chip devices due to strong polarization mode dispersion. However, there now exist different ways of achieving larger Hilbert spaces such as (i) additional spatial dimensions (i.e., waveguides); (ii) multi-photon approaches; (iii) higher dimensional photon states.

It is obvious that an increase in spatial modes is not an ideal solution toward augmented information storage, since the dimensions of the chip will quickly exceed the limits of practical manageability. The use of more photons is a good way of increasing the Hilbert space, but the disadvantage here is the use of more complex setups for its realization (Matthews et al. 2009). Additionally, multi-photon approaches show an exponential decrease of the coincidence detection probability. The exploitation of frequency-bin entanglement, as discussed in the previous section, is a very promising way to achieve larger Hilbert spaces. Moreover, various high-dimensional entanglement (e.g. frequency-bin or combination of multiple degrees of freedom, such as time- and frequency-bins (Reimer et al. 2019)), **with** Reimer, Christian, Stefania Sciara, Piotr Roztocki, Mehedi Islam, Luis Romero Cortés, Yanbing Zhang, Bennet Fischer, et al. 2019. "High-Dimensional One-Way Quantum Processing Implemented on d-Level Cluster States." Nature Physics 15 (2): 148–53. doi:10.1038/s41567-018-0347-x.) can be obtained through on-chip microcavities. Hence, microcavities are a very promising way to overcome challenges in quantum applications, as they make photonic-based

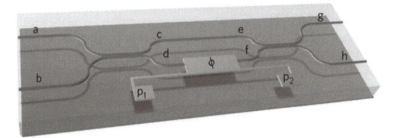

FIGURE 8.15 Waveguide structure used to perform the manipulation of multiphoton states consisting of two couplers which create an interferometer. By applying a voltage between the contact pads p_1 and p_2, the relative optical phase ϕ can be controlled. (With kind permission from Springer Science+Business Media: Matthews, J.C.F. et al., *Nat. Photonics*, Manipulation of Multiphoton Entanglement in Waveguide Quantum Circuits, 3, 2009, 346–350.)

chip solutions a favorable alternative to "classical" quantum approaches based on, for example, ultra-cold atoms and superconductors.

The first experiments on the exploitation of larger Hilbert spaces were recently conducted using multi-photon states and free-space generation of photon pairs (Matthews et al. 2009). They demonstrated the manipulation of free space generated multi-photon pairs by exploiting waveguide circuits with integrated phase modulators as shown in Figure 8.15.

Although on-chip manipulation was successfully demonstrated, the approach of generating multiple photons with free-space setups is not a very suitable solution for a scalable on-chip solution.

Other work has focused on exploiting the higher dimensions of a photon in order to increase the Hilbert space (Schaeff et al. 2015). Here, the focus relied on the manipulation of high-dimensional states (two-qutrit states), demonstrating manipulation of a 9-dimensional systems. As before, the manipulation was carried out by waveguides consisting of multiple integrated beam splitters. Although that work demonstrated the generation and manipulation of larger Hilbert spaces, it relied on bulk optic approaches for the realization of the photon pairs and hence this approach is not scalable.

The first work on the generation and manipulation of high-dimensional states on-chip was presented in 2017 (Kues, Reimer, Roztocki, et al. 2017). Figure 8.16 illustrates the related setup.

Here, frequency-bin entanglement is employed, which can generate up to $d = 100$ dimensions, as shown by the correlation measurements. Furthermore, entanglement measurements were performed for three different dimensionalities, see Figure 8.17. They indeed showed frequency manipulation of high-dimensional states using frequency mixing techniques.

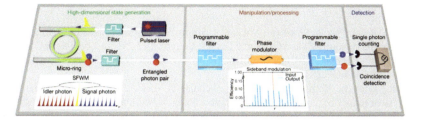

FIGURE 8.16 Setup for frequency-bin generation, manipulation, and characterization. On the left, the generation process is shown including a pulsed laser source which is spectrally filtered to pump one of the microring resonances. Afterwards, the generated photon-pairs are separated from the pump and subsequently sent to the manipulation stage. Such stage consists of two programmable filters and a phase modulator in between. On the far right, the detection module consists of two single-photon detectors. (With kind permission from Springer Science+Business Media: Kues, M. et al., *Nature*, On-Chip Generation of High-Dimensional Entangled Quantum States and Their Coherent Control, 546, 2017, 622–626.)

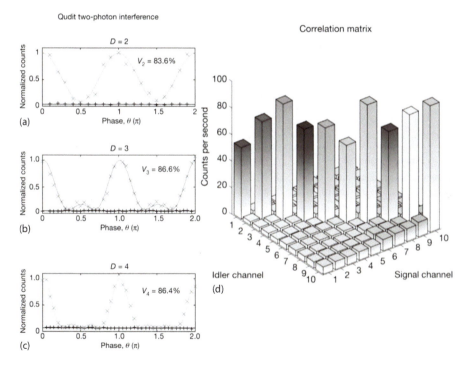

FIGURE 8.17 Results of the high-dimensional frequency entanglement. (a)–(c) shows the entanglement for different dimensionalities after projection measurement with frequency modes of different phases. By changing their relative phases, a variation of coincidence counts can be registered. (d) Correlation matrix between the generated signal and idler photons. The whole system has the capability to generate and manipulate up to $10 \times 10 = 100$ dimensions. (With kind permission from Springer Science+Business Media: Kues, M. et al., *Nature*, On-Chip Generation of High-Dimensional Entangled Quantum States and Their Coherent Control, 546, 2017, 622–626.)

REFERENCES

Azzini, Stefano, Davide Grassani, Michael J. Strain, Marc Sorel, Lukas G. Helt, J. E. Sipe, Marco Liscidini, Matteo Galli, and Daniele Bajoni. 2012. "Ultra-Low Power Generation of Twin Photons in a Compact Silicon Ring Resonator." *Optics Express* 20 (21): 23100. doi:10.1364/OE.20.023100.

Barreiro, Julio T., Nathan K. Langford, Nicholas A. Peters, and Paul G. Kwiat. 2005. "Generation of Hyperentangled Photon Pairs." *Physical Review Letters* 95 (26): 260501. doi:10.1103/PhysRevLett.95.260501.

Becker, Michael F., Dirk J. Kuizenga, and Anthony E. Siegman. 1972. "Harmonic Mode Locking of the Nd:YAG Laser." *IEEE Journal of Quantum Electronics* 8 (8): 687–93. doi:10.1109/JQE.1972.1077271.

Boyd, Robert W. 2008. *Nonlinear Optics.* Burlington, NJ, Elsevier.

Brasch, Victor, Michael Geiselmann, Tobias Herr, G. Lihachev, Martin H. P. Pfeiffer, Michael L. Gorodetsky, and Tobias J. Kippenberg. 2016. "Photonic Chip-Based Optical Frequency Comb Using Soliton Cherenkov Radiation." *Science* 351 (6271): 357–60. doi:10.1126/science.aad4811.

Brendel, Jürgen, Nicolas Gisin, Wolfgang Tittel, and Hugo Zbinden. 1999. "Pulsed Energy-Time Entangled Twin-Photon Source for Quantum Communication." *Physical Review Letters* 82 (12): 2594–97. doi:10.1103/PhysRevLett.82.2594.

Buckley, Sonia, Kelley Rivoire, and Jelena Vučković. 2012. "Engineered Quantum Dot Single-Photon Sources." *Reports on Progress in Physics* 75 (12): 126503. doi:10.1088/0034-4885/75/12/126503.

Butterworth, Stuart D., Sylvain Girard, and David C. Hanna. 1996. "A Simple Technique to Achieve Active Cavity-Length Stabilisation in a Synchronously Pumped Optical Parametric Oscillator." *Optics Communications* 123 (4–6): 577–82. doi:10.1016/0030-4018(95)00533-1.

Carmon, Tal, Lan Yang, and Kerry J. Vahala. 2004. "Dynamical Thermal Behavior and Thermal Self-Stability of Microcavities." *Optics Express* 12 (20): 4742. doi:10.1364/OPEX.12.004742.

Caspani, Lucia, Christian Reimer, Michael Kues, Piotr Roztocki, Matteo Clerici, Benjamin Wetzel, Yoann Jestin, et al. 2016. "Multifrequency Sources of Quantum Correlated Photon Pairs On-Chip: A Path toward Integrated Quantum Frequency Combs." *Nanophotonics* 5 (2). doi:10.1515/nanoph-2016-0029.

Caspani, Lucia, Chunle Xiong, Benjamin J. Eggleton, Daniele Bajoni, Marco Liscidini, Matteo Galli, Roberto Morandotti, and David J. Moss. 2017. "Integrated Sources of Photon Quantum States Based on Nonlinear Optics." *Light: Science & Applications* 6 (11): e17100. doi:10.1038/lsa.2017.100.

Chu, Sai T., Brent E. Little, W. Pan, T. Kaneko, S. Sato, and Y. Kokubun. 1999. "An Eight-Channel Add-Drop Filter Using Vertically Coupled Microring Resonators over a Cross Grid." *IEEE Photonics Technology Letters* 11 (6): 691–93. doi:10.1109/68.766787.

Ciampini, Mario Arnolfo, Adeline Orieux, Stefano Paesani, Fabio Sciarrino, Giacomo Corrielli, Andrea Crespi, Roberta Ramponi, Roberto Osellame, and Paolo Mataloni. 2016. "Path-Polarization Hyperentangled and Cluster States of Photons on a Chip." *Light: Science & Applications* 5 (4): e16064–e16064. doi:10.1038/lsa.2016.64.

Clemmen, Stéphane, Kien Phan Huy, Wim Bogaerts, Roel Baets, Philippe Emplit, and Serge Massar. 2009. "Continuous Wave Photon Pair Generation in Silicon-on-Insulator Waveguides and Ring Resonators." *Optics Express* 17 (19): 16558. doi:10.1364/OE.17.016558.

Cole, Daniel C., Erin S. Lamb, Pascal Del'Haye, Scott A. Diddams, and Scott B. Papp. 2017. "Soliton Crystals in Kerr Resonators." *Nature Photonics* 11 (10). Springer US: 671–76. doi:10.1038/s41566-017-0009-z.

Collins, Matthew J., Chunle Xiong, Isabella H. Rey, Trung D. Vo, Jiakun He, Shayan Shahnia, Cristopher Reardon, et al. 2013. "Integrated Spatial Multiplexing of Heralded Single-Photon Sources." *Nature Communications* 4 (1): 2582. doi:10.1038/ncomms3582.

Corcoran, Bill, Christelle Monat, Christian Grillet, David J. Moss, Benjamin J. Eggleton, Thomas P. White, Liam O'Faolain, and Thomas F. Krauss. 2009. "Green Light Emission in Silicon through Slow-Light Enhanced Third-Harmonic Generation in Photonic-Crystal Waveguides." *Nature Photonics* 3 (4): 206–10. doi:10.1038/nphoton.2009.28.

Dada, Adetunmise C., Jonathan Leach, Gerald S. Buller, Miles J. Padgett, and Erika Andersson. 2011. "Experimental High-Dimensional Two-Photon Entanglement and Violations of Generalized Bell Inequalities." *Nature Physics* 7 (9): 677–80. doi:10.1038/nphys1996.

Del'Haye, Pascal, A. Schliesser, O. Arcizet, T. Wilken, R. Holzwarth, and Tobias J. Kippenberg. 2007. "Optical Frequency Comb Generation from a Monolithic Microresonator." *Nature* 450 (7173): 1214–17. doi:10.1038/nature06401.

Del'Haye, Pascal, O. Arcizet, A. Schliesser, R. Holzwarth, and Tobias J. Kippenberg. 2008. "Full Stabilization of a Microresonator-Based Optical Frequency Comb." *Physical Review Letters* 101 (5): 053903. doi:10.1103/PhysRevLett.101.053903.

Del'Haye, Pascal, Tobias Herr, E. Gavartin, Michael L. Gorodetsky, R. Holzwarth, and Tobias J. Kippenberg. 2011. "Octave Spanning Tunable Frequency Comb from a Microresonator." *Physical Review Letters* 107 (6): 063901. doi:10.1103/PhysRevLett.107.063901.

Editorial. 2010. "Simply Silicon." *Nature Photonics* 4 (8): 491–491. doi:10.1038/nphoton.2010.190.

Einstein, Albert, Boris Podolsky, and Nathan Rosen. 1935. "Can Quantum-Mechanical Description of Physical Reality Be Considered Complete?" *Physical Review* 47 (10): 777–80. doi:10.1103/PhysRev.47.777.

Engin, Erman, Damien Bonneau, Chandra M. Natarajan, Alex S. Clark, Michael G. Tanner, Robert H. Hadfield, Sanders N. Dorenbos, et al. 2013. "Photon Pair Generation in a Silicon Micro-Ring Resonator with Reverse Bias Enhancement." *Optics Express* 21 (23): 27826. doi:10.1364/OE.21.027826.

Ferrera, Marcello, Luca Razzari, David Duchesne, Roberto Morandotti, Zhenshan Yang, Marco Liscidini, J. E. Sipe, Sai T. Chu, Brent E. Little, and David J. Moss. 2008. "Low-Power Continuous-Wave Nonlinear Optics in Doped Silica Glass Integrated Waveguide Structures." *Nature Photonics* 2 (12): 737–40. doi:10.1038/nphoton.2008.228.

Ferrera, Marcello, Yongwoo Park, Luca Razzari, Brent E. Little, Sai T. Chu, Roberto Morandotti, David J. Moss, and José Azaña. 2010. "On-Chip CMOS-Compatible All-Optical Integrator." *Nature Communications* 1 (3): 1–5. doi:10.1038/ncomms1028.

Ferrera, Marcello, Yongwoo Park, Luca Razzari, Brent E. Little, Sai T. Chu, Roberto Morandotti, David J. Moss, and José Azaña. 2011. "All-Optical 1st and 2nd Order Integration on a Chip." *Optics Express* 19 (23): 23153. doi:10.1364/OE.19.023153.

Franson, James D. 1989. "Bell Inequality for Position and Time." *Physical Review Letters* 62 (19): 2205–8. doi:10.1103/PhysRevLett.62.2205.

Fulconis, Jérémie, Olivier Alibart, Jeremy L. O'Brien, William J. Wadsworth, and John G. Rarity. 2007. "Nonclassical Interference and Entanglement Generation Using a Photonic Crystal Fiber Pair Photon Source." *Physical Review Letters* 99 (12): 120501. doi:10.1103/PhysRevLett.99.120501.

Gisin, Nicolas, Grégoire Ribordy, Wolfgang Tittel, and Hugo Zbinden. 2002. "Quantum Cryptography." *Reviews of Modern Physics* 74 (1): 145–95. doi:10.1103/RevModPhys.74.145.

Giustina, Marissa, Marijn A. M. Versteegh, Sören Wengerowsky, Johannes Handsteiner, Armin Hochrainer, Kevin Phelan, Fabian Steinlechner, et al. 2015. "Significant-Loophole-Free Test of Bell's Theorem with Entangled Photons." *Physical Review Letters* 115 (25): 250401. doi:10.1103/PhysRevLett.115.250401.

Grassani, Davide, Stefano Azzini, Marco Liscidini, Matteo Galli, Michael J. Strain, Marc Sorel, J. E. Sipe, and Daniele Bajoni. 2015. "Micrometer-Scale Integrated Silicon Source of Time-Energy Entangled Photons." *Optica* 2 (2): 88. doi:10.1364/OPTICA.2.000088.

Gröblacher, Simon, Tomasz Paterek, Rainer Kaltenbaek, Časlav Brukner, Marek Żukowski, Markus Aspelmeyer, and Anton Zeilinger. 2007. "An Experimental Test of Non-Local Realism." *Nature* 446 (7138): 871–75. doi:10.1038/nature05677.

Guo, Xiang, Chang-ling Zou, Carsten Schuck, Hojoong Jung, Risheng Cheng, and Hong X. Tang. 2016. "Parametric Down-Conversion Photon-Pair Source on a Nanophotonic Chip." *Light: Science & Applications* 6 (5): e16249. doi:10.1038/lsa.2016.249.

Hadfield, Robert H. 2009. "Single-Photon Detectors for Optical Quantum Information Applications." *Nature Photonics* 3 (12): 696–705. doi:10.1038/nphoton.2009.230.

Haken, Hermann, and H. Sauermann. 1963. "Nonlinear Interaction of Laser Modes." *Zeitschrift Fuer Physik* 173 (3): 261–75. doi:10.1007/BF01377828.

Helt, Lukas G., Zhenshan Yang, Marco Liscidini, and J. E. Sipe. 2010. "Spontaneous Four-Wave Mixing in Microring Resonators." *Optics Letters* 35 (18): 3006. doi:10.1364/OL.35.003006.

Henry, Charles H., Rudolph F. Kazarinov, Hyung J. Lee, Kenneth J. Orlowsky, and L. E. Katz. 1987. "Low Loss Si3N4–SiO2 Optical Waveguides on Si." *Applied Optics* 26 (13): 2621. doi:10.1364/AO.26.002621.

Herr, Tobias, Victor Brasch, John D. Jost, C. Y. Wang, N. M. Kondratiev, Michael L. Gorodetsky, and Tobias J. Kippenberg. 2014. "Temporal Solitons in Optical Microresonators." *Nature Photonics* 8 (2): 145–52. doi:10.1038/nphoton.2013.343.

Horn, Rolf T., Piotr Kolenderski, Dongpeng Kang, Payam Abolghasem, Carmelo Scarcella, Adriano Della Frera, Alberto Tosi, et al. 2013. "Inherent Polarization Entanglement Generated from a Monolithic Semiconductor Chip." *Scientific Reports* 3 (1): 2314. doi:10.1038/srep02314.

Howell, John C., Ryan S. Bennink, Sean J. Bentley, and Robert W. Boyd. 2004. "Realization of the Einstein-Podolsky-Rosen Paradox Using Momentum- and Position-Entangled Photons from Spontaneous Parametric Down Conversion." *Physical Review Letters* 92 (21): 210403. doi:10.1103/PhysRevLett.92.210403.

Humphreys, Peter C., Benjamin J. Metcalf, Justin B. Spring, Merritt Moore, Xian-Min Jin, Marco Barbieri, W. Steven Kolthammer, and Ian A. Walmsley. 2013. "Linear Optical Quantum Computing in a Single Spatial Mode." *Physical Review Letters* 111 (15): 150501. doi:10.1103/PhysRevLett.111.150501.

Ikeda, Kazuhiro, Robert E. Saperstein, Nikola Alic, and Yeshaiahu Fainman. 2008. "Thermal and Kerr Nonlinear Properties of Plasma-Deposited Silicon Nitride/ Silicon Dioxide Waveguides." *Optics Express* 16 (17): 12987. doi:10.1364/OE.16.012987.

Imany, Poolad, Jose A. Jaramillo-Villegas, Ogaga D. Odele, Kyunghun Han, Daniel E. Leaird, Joseph M. Lukens, Pavel Lougovski, Minghao Qi, and Andrew M. Weiner. 2018. "50-GHz-Spaced Comb of High-Dimensional Frequency-Bin Entangled Photons from an On-Chip Silicon Nitride Microresonator." *Optics Express* 26 (2): 1825. doi:10.1364/OE.26.001825.

Ippen, Erich P. 1994. "Principles of Passive Mode Locking." *Applied Physics B Laser and Optics* 58 (3): 159–70. doi:10.1007/BF01081309.

Israel, Yonatan, Shamir Rosen, and Yaron Silberberg. 2014. "Supersensitive Polarization Microscopy Using NOON States of Light." *Physical Review Letters* 112 (10): 103604. doi:10.1103/PhysRevLett.112.103604.

Ji, Xingchen, Felippe A. S. Barbosa, Samantha P. Roberts, Avik Dutt, Jaime Cardenas, Yoshitomo Okawachi, Alex Bryant, Alexander L. Gaeta, and Michal Lipson. 2016. "Breaking the Loss Limitation of On-Chip High-Confinement Resonators." *Optica* 4 (6): 619. doi:10.1364/OPTICA.4.000619.

Jones, David J., Scott A. Diddams, Jinendra K. Ranka, Andrew Stentz, Robert S. Windeler, John L. Hall, and Steven T. Cundiff. 2000. "Carrier-Envelope Phase Control of Femtosecond Mode-Locked Lasers and Direct Optical Frequency Synthesis." *Science* 288 (5466): 635–39. doi:10.1126/science.288.5466.635.

Jung, Hojoong, Rebecca Stoll, Xiang Guo, Debra Fischer, and Hong X. Tang. 2014. "Green, Red, and IR Frequency Comb Line Generation from Single IR Pump in AlN Microring Resonator." *Optica* 1 (6): 396. doi:10.1364/OPTICA.1.000396.

Kaminov, Ivan, Tingye Li, and Alan E. Willner, eds. 2013. "Optical Fiber Telecommunications VIB." In *Optical Fiber Telecommunications*, iii. Elsevier. doi:10.1016/B978-0-12-396960-6.00025-0.

Kelly, Stephen M.J. 1989.. "Mode-Locking Dynamics of a Laser Coupled to an Empty External Cavity." *Optics Communications* 70 (6): 495–501. doi:10.1016/0030-4018(89)90372-6.

Kippenberg, Tobias J., R. Holzwarth, and Scott A. Diddams. 2011. "Microresonator-Based Optical Frequency Combs." *Science* 332 (6029): 555–59. doi:10.1126/science.1193968.

Kira, Mackillo, Stephan W. Koch, Ryan P. Smith, Andrew E. Hunter, and Steven T. Cundiff. 2011. "Quantum Spectroscopy with Schrödinger-Cat States." *Nature Physics* 7 (10): 799–804. doi:10.1038/nphys2091.

Knill, Emanuel, R. Laflamme, and G. J. Milburn. 2001. "A Scheme for Efficient Quantum Computation with Linear Optics." *Nature* 409 (6816): 46–52. doi:10.1038/35051009.

Kues, Michael, Christian Reimer, Benjamin Wetzel, Piotr Roztocki, Brent E. Little, Sai T. Chu, Tobias Hansson, Evgeny A. Viktorov, David J. Moss, and Roberto Morandotti. 2017. "Passively Mode-Locked Laser with an Ultra-Narrow Spectral Width." *Nature Photonics* 11 (3): 159–62. doi:10.1038/nphoton.2016.271.

Kues, Michael, Christian Reimer, Piotr Roztocki, Luis Romero Cortés, Stefania Sciara, Benjamin Wetzel, Yanbing Zhang, et al. 2017. "On-Chip Generation of High-Dimensional Entangled Quantum States and Their Coherent Control." *Nature* 546 (7660): 622–26. doi:10.1038/nature22986.

Kwiat, Paul G., Klaus Mattle, Harald Weinfurter, Anton Zeilinger, Alexander V. Sergienko, and Yanhua Shih. 1995. "New High-Intensity Source of Polarization-Entangled Photon Pairs." *Physical Review Letters* 75 (24): 4337–41. doi:10.1103/PhysRevLett.75.4337.

Lanyon, Benjamin P., Marco Barbieri, Marcelo P. Almeida, Thomas Jennewein, Timothy C. Ralph, Kevin J. Resch, Geoff J. Pryde, Jeremy L. O'Brien, Alexei Gilchrist, and Andrew G. White. 2009. "Simplifying Quantum Logic Using Higher-Dimensional Hilbert Spaces." *Nature Physics* 5 (2): 134–40. doi:10.1038/nphys1150.

Leo, François, Jassem Safioui, Bart Kuyken, Gunther Roelkens, and Simon-Pierre Gorza. 2014. "Generation of Coherent Supercontinuum in A-Si:H Waveguides: Experiment and Modeling Based on Measured Dispersion Profile." *Optics Express* 22 (23): 28997. doi:10.1364/OE.22.028997.

Leuthold, Juerg, Christian Koos, and Wolfgang Freude. 2010. "Nonlinear Silicon Photonics." *Nature Photonics* 4 (8): 535–44. doi:10.1038/nphoton.2010.185.

Levy, Jacob S., Alexander Gondarenko, Mark A. Foster, Amy C. Turner-Foster, Alexander L. Gaeta, and Michal Lipson. 2010. "CMOS-Compatible Multiple-Wavelength Oscillator for On-Chip Optical Interconnects." *Nature Photonics* 4 (1): 37–40. doi:10.1038/nphoton.2009.259.

Li, Jiang, Hansuek Lee, Tong Chen, and Kerry J. Vahala. 2012. "Chip-Based Frequency Comb with Microwave Repetition Rate." In *Conference on Lasers and Electro-Optics 2012*, CTh3A.3. Washington, DC.: OSA. doi:10.1364/CLEO_SI.2012.CTh3A.3.

Liang, Di, Gunther Roelkens, Roel Baets, and John Bowers. 2010. "Hybrid Integrated Platforms for Silicon Photonics." *Materials* 3 (3): 1782–1802. doi:10.3390/ma3031782.

Liang, Tak-Keung, and Hon K. Tsang. 2004. "Role of Free Carriers from Two-Photon Absorption in Raman Amplification in Silicon-on-Insulator Waveguides." *Applied Physics Letters* 84 (15): 2745–47. doi:10.1063/1.1702133.

Lloyd, Seth. 2008. "Enhanced Sensitivity of Photodetection via Quantum Illumination." *Science* 321 (5895): 1463–65. doi:10.1126/science.1160627.

Lomsadze, Bachana, and Steven T. Cundiff. 2017. "Frequency Combs Enable Rapid and High-Resolution Multidimensional Coherent Spectroscopy." *Science* 357 (6358): 1389–91. doi:10.1126/science.aao1090.

Lu, Xiyuan, Jonathan Y. Lee, Philip X.-L. Feng, and Qiang Lin. 2014. "High Q Silicon Carbide Microdisk Resonator." *Applied Physics Letters* 104 (18): 181103. doi:10.1063/1.4875707.

MacLellan, Benjamin, Piotr Roztocki, Michael Kues, Christian Reimer, Luis Romero Cortés, Yanbing Zhang, Stefania Sciara, et al. 2018. "Generation and Coherent Control of Pulsed Quantum Frequency Combs." *Journal of Visualized Experiments*, no. 136 (June). doi:10.3791/57517.

Mandon, Julien, Guy Guelachvili, and Nathalie Picqué. 2009. "Fourier Transform Spectroscopy with a Laser Frequency Comb." *Nature Photonics* 3 (2): 99–102. doi:10.1038/nphoton.2008.293.

Matsuda, Nobuyuki, Hanna Le Jeannic, Hiroshi Fukuda, Tai Tsuchizawa, William John Munro, Kaoru Shimizu, Koji Yamada, Yasuhiro Tokura, and Hiroki Takesue. 2012. "A Monolithically Integrated Polarization Entangled Photon Pair Source on a Silicon Chip." *Scientific Reports* 2 (1): 817. doi:10.1038/srep00817.

Matthews, Jonathan C. F., Alberto Politi, André Stefanov, and Jeremy L. O'Brien. 2009. "Manipulation of Multiphoton Entanglement in Waveguide Quantum Circuits." *Nature Photonics* 3 (6): 346–50. doi:10.1038/nphoton.2009.93.

Miller, Steven A., Mengjie Yu, Xingchen Ji, Austin G. Griffith, Jaime Cardenas, Alexander L. Gaeta, and Michal Lipson. 2017. "Low-Loss Silicon Platform for Broadband Mid-Infrared Photonics." *Optica* 4 (7): 707. doi:10.1364/OPTICA.4.000707.

Moss, David J., Roberto Morandotti, Alexander L. Gaeta, and Michal Lipson. 2013. "New CMOS-Compatible Platforms Based on Silicon Nitride and Hydex for Nonlinear Optics." *Nature Photonics* 7 (8): 597–607. doi:10.1038/nphoton.2013.183.

Okawachi, Yoshitomo, Kasturi Saha, Jacob S. Levy, Y. Henry Wen, Michal Lipson, and Alexander L. Gaeta. 2011. "Octave-Spanning Frequency Comb Generation in a Silicon Nitride Chip." *Optics Letters* 36 (17): 3398. doi:10.1364/OL.36.003398.

Olislager, Laurent, J. Cussey, A. T. Nguyen, Philippe Emplit, Serge Massar, J.-M. Merolla, and Kien Phan Huy. 2010. "Frequency-Bin Entangled Photons." *Physical Review A* 82 (1): 013804. doi:10.1103/PhysRevA.82.013804.

Olislager, Laurent, Jassem Safioui, Stéphane Clemmen, Kien Phan Huy, Wim Bogaerts, Roel Baets, Philippe Emplit, and Serge Massar. 2013. "Silicon-on-Insulator Integrated Source of Polarization-Entangled Photons." *Optics Letters* 38 (11): 1960. doi:10.1364/OL.38.001960.

Orieux, Adeline, Marijn A. M. Versteegh, Klaus D. Jöns, and Sara Ducci. 2017. "Semiconductor Devices for Entangled Photon Pair Generation: A Review." *Reports on Progress in Physics* 80 (7): 076001. doi:10.1088/1361-6633/aa6955.

Parson, William W. 2015. *Modern Optical Spectroscopy*. Berlin, Germany: Springer Berlin Heidelberg. doi:10.1007/978-3-662-46777-0.

Pasquazi, Alessia, Lucia Caspani, Marco Peccianti, Matteo Clerici, Marcello Ferrera, Luca Razzari, David Duchesne, et al. 2013. "Self-Locked Optical Parametric Oscillation in a CMOS Compatible Microring Resonator: A Route to Robust Optical Frequency Comb Generation on a Chip." *Optics Express* 21 (11): 13333. doi:10.1364/OE.21.013333.

Pasquazi, Alessia, Marco Peccianti, Brent E. Little, Sai T. Chu, David J. Moss, and Roberto Morandotti. 2012. "Stable, Dual Mode, High Repetition Rate Mode-Locked Laser Based on a Microring Resonator." *Optics Express* 20 (24): 27355. doi:10.1364/OE.20.027355.

Pasquazi, Alessia, Marco Peccianti, Luca Razzari, David J. Moss, Stéphane Coen, Miro Erkintalo, Yanne K. Chembo, et al. 2018. "Micro-Combs: A Novel Generation of Optical Sources." *Physics Reports* 729 (January): 1–81. doi:10.1016/j.physrep.2017.08.004.

Pasquazi, Alessia, Raja Ahmad, Martin Rochette, Michael Lamont, Brent E. Little, Sai T. Chu, Roberto Morandotti, and David J. Moss. 2010. "All-Optical Wavelength Conversion in an Integrated Ring Resonator." *Optics Express* 18 (4): 3858. doi:10.1364/OE.18.003858.

Paul, Pierre-Marie, Elena S. Toma, Pierre Breger, Geneviève M. Mullot, Frédérika Augé, Philippe Balcou, Harm G. Müller, and Pierre Agostini. 2001. "Observation of a Train of Attosecond Pulses from High Harmonic Generation." *Science* 292 (5522): 1689–92. doi:10.1126/science.1059413.

Peccianti, Marco, Alessia Pasquazi, Yongwoo Park, Brent E. Little, Sai T. Chu, David J. Moss, and Roberto Morandotti. 2012. "Demonstration of a Stable Ultrafast Laser Based on a Nonlinear Microcavity." *Nature Communications* 3 (1): 765. doi:10.1038/ncomms1762.

Pfeiffer, Martin H. P., Arne Kordts, Victor Brasch, Michael Zervas, Michael Geiselmann, John D. Jost, and Tobias J. Kippenberg. 2016. "Photonic Damascene Process for Integrated High-Q Microresonator Based Nonlinear Photonics." *Optica* 3 (1): 20. doi:10.1364/OPTICA.3.000020.

Politi, Alberto, M. J. Cryan, John G. Rarity, S. Yu, and Jeremy L. O'Brien. 2008. "Silica-on-Silicon Waveguide Quantum Circuits." *Science* 320 (5876): 646–49. doi:10.1126/science.1155441.

Pysher, Matthew, Yoshichika Miwa, Reihaneh Shahrokhshahi, Russell Bloomer, and Olivier Pfister. 2011. "Parallel Generation of Quadripartite Cluster Entanglement in the Optical Frequency Comb." *Physical Review Letters* 107 (3): 030505. doi:10.1103/PhysRevLett.107.030505.

Razzari, Luca, David Duchesne, Marcello Ferrera, Roberto Morandotti, Sai T. Chu, Brent E. Little, and David J. Moss. 2010. "CMOS-Compatible Integrated Optical Hyper-Parametric Oscillator." *Nature Photonics* 4 (1): 41–45. doi:10.1038/nphoton.2009.236.

Reichert, Joerg, Ronald Holzwarth, Thomas Udem, and Theodor W. Hänsch. 1999. "Measuring the Frequency of Light with Mode-Locked Lasers." *Optics Communications* 172 (1–6): 59–68. doi:10.1016/S0030-4018(99)00491-5.

Reimer, Christian, Lucia Caspani, Matteo Clerici, Marcello Ferrera, Michael Kues, Marco Peccianti, Alessia Pasquazi, et al. 2014. "Integrated Frequency Comb Source of Heralded Single Photons." *Optics Express* 22 (6): 6535. doi:10.1364/OE.22.006535.

Reimer, Christian, Michael Kues, Lucia Caspani, Benjamin Wetzel, Piotr Roztocki, Matteo Clerici, Yoann Jestin, et al. 2015. "Cross-Polarized Photon-Pair Generation and Bi-Chromatically Pumped Optical Parametric Oscillation on a Chip." *Nature Communications* 6 (1): 8236. doi:10.1038/ncomms9236.

Reimer, Christian, Michael Kues, Piotr Roztocki, Benjamin Wetzel, Fabio Grazioso, Brent E. Little, Sai T. Chu, et al. 2016. "Generation of Multiphoton Entangled Quantum States by Means of Integrated Frequency Combs." *Science* 351 (6278): 1176–80. doi:10.1126/science.aad8532.

Reimer, Christian, Yanbing Zhang, Piotr Roztocki, Stefania Sciara, Luis Romero Cortés, Mehedi Islam, Bennet Fischer, et al. 2018. "On-Chip Frequency Combs and Telecommunications Signal Processing Meet Quantum Optics." *Frontiers of Optoelectronics* 11 (2): 134–47. doi:10.1007/s12200-018-0814-0.

Rong, Haisheng, Richard Jones, Ansheng Liu, Oded Cohen, Dani Hak, Alexander Fang, and Mario Paniccia. 2005. "A Continuous-Wave Raman Silicon Laser." *Nature* 433 (7027): 725–28. doi:10.1038/nature03346.

Roztocki, Piotr, Michael Kues, Christian Reimer, Benjamin Wetzel, Stefania Sciara, Yanbing Zhang, Alfonso Cino, et al. 2017. "Practical System for the Generation of Pulsed Quantum Frequency Combs." *Optics Express* 25 (16): 18940. doi:10.1364/OE.25.018940.

Savchenkov, Anatoliy A., Andrey B. Matsko, Vladimir S. Ilchenko, Iouri Solomatine, David Seidel, and Lute Maleki. 2008. "Tunable Optical Frequency Comb with a Crystalline Whispering Gallery Mode Resonator." *Physical Review Letters* 101 (9): 093902. doi:10.1103/PhysRevLett.101.093902.

Schaeff, Christoph, Robert Polster, Marcus Huber, Sven Ramelow, and Anton Zeilinger. 2015. "Experimental Access to Higher-Dimensional Entangled Quantum Systems Using Integrated Optics." *Optica* 2 (6): 523. doi:10.1364/OPTICA.2.000523.

Schawlow, Arthur L., and Charles H. Townes. 1958. "Infrared and Optical Masers." *Physical Review* 112 (6): 1940–49. doi:10.1103/PhysRev.112.1940.

Schindl, Andreas, Martin J. Schindl, Heidemarie Pernerstorfer-Schön, and Liesbeth Schindl. 2000. "Low-Intensity Laser Therapy: A Review." *Journal of Investigative Medicine : The Official Publication of the American Federation for Clinical Research* 48 (5): 312–26. http://www.ncbi.nlm.nih.gov/pubmed/10979236.

Schwelb, Otto. 2008. "A Decade of Progress in Microring and Microdisk Based Photonic Circuits: A Personal Selection." In, edited by Alexis V. Kudryashov, Alan H. Paxton, and Vladimir S. Ilchenko, 68720H. doi:10.1117/12.772046.

Shalm, Lynden K., Evan Meyer-Scott, Bradley G. Christensen, Peter Bierhorst, Michael A. Wayne, Martin J. Stevens, Thomas Gerrits, et al. 2015. "Strong Loophole-Free Test of Local Realism." *Physical Review Letters* 115 (25): 250402. doi:10.1103/PhysRevLett.115.250402.

Sharping, Jay E., Kim F. Lee, Mark A. Foster, Amy C. Turner, Bradley S. Schmidt, Michal Lipson, Alexander L. Gaeta, and Prem Kumar. 2006. "Generation of Correlated Photons in Nanoscale Silicon Waveguides." *Optics Express* 14 (25): 12388. doi:10.1364/OE.14.012388.

Silverstone, Joshua W., Damien Bonneau, Kazuya Ohira, Nobuo Suzuki, Haruhiko Yoshida, Norio Iizuka, Mizunori Ezaki, et al. 2014. "On-Chip Quantum Interference between Silicon Photon-Pair Sources." *Nature Photonics* 8 (2): 104–8. doi:10.1038/nphoton.2013.339.

Silverstone, Joshua W., Raffaele Santagati, Damien Bonneau, Michael J. Strain, Marc Sorel, Jeremy L. O'Brien, and Mark G. Thompson. 2015. "Qubit Entanglement between Ring-Resonator Photon-Pair Sources on a Silicon Chip." *Nature Communications* 6 (1): 7948. doi:10.1038/ncomms8948.

Singh, Neetesh, Darren D. Hudson, Yi Yu, Christian Grillet, Stuart D. Jackson, Alvaro Casas-Bedoya, Andrew Read, et al. 2015. "Midinfrared Supercontinuum Generation from 2 to 6 Mm in a Silicon Nanowire." *Optica* 2 (9): 797. doi:10.1364/OPTICA.2.000797.

Siyushev, Petr, Guilherme Stein, Jörg Wrachtrup, and Ilja Gerhardt. 2014. "Molecular Photons Interfaced with Alkali Atoms." *Nature* 509 (7498): 66–70. doi:10.1038/nature13191.

Solntsev, Alexander S., and Andrey A. Sukhorukov. 2017. "Path-Entangled Photon Sources on Nonlinear Chips." *Reviews in Physics* 2 (November): 19–31. doi:10.1016/j.revip.2016.11.003.

Soltani, Mohammad, Siva Yegnanarayanan, and Ali Adibi. 2007. "Ultra-High Q Planar Silicon Microdisk Resonators for Chip-Scale Silicon Photonics." *Optics Express* 15 (8): 4694. doi:10.1364/OE.15.004694.

Strekalov, Dmitry V, Christoph Marquardt, Andrey B. Matsko, Harald G L Schwefel, and Gerd Leuchs. 2016. "Nonlinear and Quantum Optics with Whispering Gallery Resonators." *Journal of Optics* 18 (12): 123002. doi:10.1088/2040-8978/18/12/123002.

Su, Xiaolong, Shuhong Hao, Xiaowei Deng, Lingyu Ma, Meihong Wang, Xiaojun Jia, Changde Xie, and Kunchi Peng. 2013. "Gate Sequence for Continuous Variable One-Way Quantum Computation." *Nature Communications* 4 (November). doi:10.1038/ncomms3828.

Tan, Dawn T. H., Doris K. T. Ng, Ting Wang, Siu-Kit Ng, Yeow-Teck Toh, Qian Wang, George F. R. Chen, and Arthur K. L. Chee. 2016. "Ultra-Silicon-Rich Nitride Devices for CMOS Nonlinear Optics." *SPIE Newsroom*, April. doi:10.1117/2.1201603.006326.

Telle, Harald R., Günter D. Steinmeyer, Amy E. Dunlop, Joern Stenger, Dirk H. Sutter, and Ursula Keller. 1999. "Carrier-Envelope Offset Phase Control: A Novel Concept for Absolute Optical Frequency Measurement and Ultrashort Pulse Generation." *Applied Physics B* 69 (4): 327–32. doi:10.1007/s003400050813.

The Royal Swedish Academy of Sciences. 2005. "Quantum-Mechanical Theory of Optical Coherence Laser-Based Precision Spectroscopy and Optical Frequency Comb Techniques Quantum-Mechanical Theory of Optical Coherence." *October*, no. October: 1–14.

Thorpe, Michael J., Kevin D. Moll, Ronald Jason Jones, Benjamin R. Safdi, and Jun Ye. 2006. "Broadband Cavity Ringdown Spectroscopy for Sensitive and Rapid Molecular Detection." *Science* 311 (5767): 1595–99. doi:10.1126/science.1123921.

Togan, Emre, Yiwen Chu, Alexei S. Trifonov, Liang Jiang, Jerónimo R. Maze, Lillian I. Childress, M. V. Gurudev Dutt, et al. 2010. "Quantum Entanglement between an Optical Photon and a Solid-State Spin Qubit." *Nature* 466 (7307): 730–34. doi:10.1038/nature09256.

Vahala, Kerry J. 2003. "Optical Microcavities." *Nature* 424 (6950): 839–46. doi:10.1038/nature01939.

Walther, Philip, Kevin J. Resch, Terry G. Rudolph, Emmanuel Schenck, Harald Weinfurter, Vlatko Vedral, Markus Aspelmeyer, and Anton Zeilinger. 2005. "Experimental One-Way Quantum Computing." *Nature* 434 (7030): 169–76. doi:10.1038/nature03347.

Wang, Ting, Doris K. T. Ng, Siu-Kit Ng, Yeow-Teck Toh, Arthur K. L. Chee, George F. R. Chen, Qian Wang, and Dawn T. H. Tan. 2015. "Supercontinuum Generation in Bandgap Engineered, Back-End CMOS Compatible Silicon Rich Nitride Waveguides." *Laser & Photonics Reviews* 9 (5): 498–506. doi:10.1002/lpor.201500054.

Weiner, Andrew M. 2009. *Ultrafast Optics*. Hoboken, NJ: John Wiley & Sons, Inc. doi:10.1002/9780470473467.

Ycas, Gabriel G., Franklyn Quinlan, Scott A. Diddams, Steve Osterman, Suvrath Mahadevan, Stephen Redman, Ryan Terrien, et al. 2012. "Demonstration of On-Sky Calibration of Astronomical Spectra Using a 25 GHz near-IR Laser Frequency Comb." *Optics Express* 20 (6): 6631. doi:10.1364/OE.20.006631.

Yi, Xu, Qi-Fan Yang, Ki Youl Yang, and Kerry J. Vahala. 2016. "Active Capture and Stabilization of Temporal Solitons in Microresonators." *Optics Letters* 41 (9): 2037. doi:10.1364/OL.41.002037.

Yoshida, Eiji, and Masataka Nakazawa. 1997. "Low-Threshold 115-GHz Continuous-Wave Modulational-Instability Erbium-Doped Fiber Laser." *Optics Letters* 22 (18): 1409. doi:10.1364/OL.22.001409.

Zeilinger, Anton, Gregor Weihs, Thomas Jennewein, and Markus Aspelmeyer. 2005. "Happy Centenary, Photon." *Nature* 433 (7023): 230–38. doi:10.1038/nature03280.

Zhang, Mian, Cheng Wang, Rebecca Cheng, Amirhassan Shams-Ansari, and Marko Lončar. 2017. "Monolithic Ultra-High-Q Lithium Niobate Microring Resonator." *Optica* 4 (12): 1536. doi:10.1364/OPTICA.4.001536.

Zhang, Yanbing, Christian Reimer, Jenny Wu, Piotr Roztocki, Benjamin Wetzel, Brent E. Little, Sai T. Chu, et al. 2017. "Multichannel Phase-Sensitive Amplification in a Low-Loss CMOS-Compatible Spiral Waveguide." *Optics Letters* 42 (21): 4391. doi:10.1364/OL.42.004391.

9 Harmonic Generation with Mie Resonant Nanostructures

Costantino De Angelis, Luca Carletti, Davide Rocco, Andrea Locatelli, Lavinia Ghirardini, Marco Finazzi, Michele Celebrano, Lei Xu, and Andrey Miroshnichenko

CONTENTS

9.1 INTRODUCTION

Light manipulation by means of nanoscale structures has attracted enormous interest due to its impact on modern photonic technologies. Nanoscale objects not only might facilitate the integration of various linear and nonlinear functionalities in nanophotonic circuits, but they also allow to replace bulky components by tremendously improving miniaturization [1]. Recently, high-refractive-index nanostructures with large intrinsic optical nonlinearities have been demonstrated to provide a powerful and competitive alternative to nonlinear nanoplasmonics [2,3]. Fascinating results have been predicted and obtained for second- [4–10] and third- [11–13] harmonic generation from dielectric nanoparticles; these results set their roots in the unique linear optical properties of nonmetallic nanoresonators, which have recently attracted incredible attention for the possibility that they offer to simultaneously engineer electric and magnetic resonances [14–17]. Without a doubt, we can say that all-dielectric nanoantennas and metasurfaces are very promising building blocks that are promoting a paradigm shift for nonlinear optics at the nanoscale [18].

Recent theoretical predictions, as well as experimental realizations, demonstrated the huge impact that nonlinear optics in nanostructured materials can have in applications ranging from high-throughput sensing to small-footprint quantum optics. High-index dielectric nanostructures have demonstrated second-harmonic generation (SHG) [5–10] and third-harmonic generation (THG) [11–13,19] from an optical pump beam either at visible (VIS) or near infrared (NIR) frequencies with unprecedented record high conversion efficiencies in proportion to the structure footprint. Key points to achieve enhanced nonlinear optical processes at the nanoscale are material transparency at both the pump and emission wavelengths, large susceptibilities ($\chi^{(2)}$, $\chi^{(3)}$) values, and the possibility to sustain not only electric but also magnetic resonances. As far as $\chi^{(2)}$ is concerned, this is in contrast to plasmonic nanostructures, where the nonlinear optical response is mainly associated with the surface currents [20–22], high-index dielectric resonators may display a strong nonlinear response from the material volume thanks to their large bulk nonlinear coefficients and the sizable electric fields attained inside the structure. Magnetic resonances played a pivotal role in this context due to the strong electric field enhancement provided within the nanoparticle volume. Thus, the design of nanoparticles for enhanced harmonic generation sets its basis on the analysis of their basic linear optical properties such as resonant frequencies, bandwidth, field enhancement and radiation diagrams at fundamental and harmonic frequencies [3]. Engineering of the nanoparticle geometry has emerged as a very important aspect to increase efficiency; different strategies can be employed in this approach, but due to the close link with the observed physical features, multipole decomposition of the electric fields and currents is widely adopted [23,24]. This framework revealed that the interplay between strong electric and magnetic Mie-type resonances can result not only in constructive or destructive interference leading to beam shaping in the far-field, but also to resonant enhancement of electromagnetic fields in the nanoparticles, which opens up new possibilities for efficient nonlinear processes [25]. Along with this research line, recently it was suggested [26] that subwavelength nanoscale resonators can support localized states with high-Q factors (on the order of 10^2 in the visible and near-infrared spectrum), provided that their parameters are closely matched to a bound state in the continuum (BIC) [27] formed via destructive interference of two similar leaky modes [28,29]. Obviously, the BIC is a mathematical abstraction and its realization demands infinite size of the structure or either zero or infinite permittivity [30,31]; however, using this approach, high-index dielectric nanoparticles can exhibit high-Q resonances associated with the so-called supercavity modes [32] and these large Q factors can dramatically enhance nonlinear effects at the nanoscale [2,3,9].

Before discussing in detail harmonic generation in high-index dielectric nanostructures, it is useful to introduce the definition of the conversion efficiency of such processes. In the literature we find in fact different definitions and, depending on the adopted viewpoint, different record efficiencies were claimed at the same time on different devices. When we are interested in the power efficiency, it is customary to define the conversion efficiency for n-th harmonic generation as

$$\eta_n = \frac{P_n}{P_{FF}} \tag{9.1}$$

where P_n is the power emitted at the n-th harmonic (i.e., *SH* for $n = 2$ or *TH* for $n = 3$), and P_{FF} is the pump power incident on the structure. This efficiency parameter depends both on the properties of the nanoparticle and on the value of the pump power which is used in the particular experiment. Thus, when this parameter is used to evaluate the efficiency, one should always report the pump power used in the experiment. The main drawback of this definition is that, since the power of the n-th harmonic is proportional to P_{FF}^n by changing the pump power one can dramatically vary the efficiency; therefore a nonlinear parameter that reflects solely the properties of the nano-object independently of the incident power selected is strongly preferable. For this reason, another common parameter to evaluate harmonic efficiency is the so-called nonlinear coefficient that is defined as:

$$\tilde{n}_n = \frac{P_n}{P_{FF}^n} \tag{9.2}$$

where the n-th power of P_{FF} is taken to sort out the dependence on the pump power level, thus introducing a way to characterize the nanoparticle. Strictly speaking, the nonlinear coefficient as defined in (2) still bears a possible ambiguity related to the definition of pump power P_{FF} and power of the n-th harmonic P_n: to define a parameter depending only on the intrinsic nanoparticle properties one should consider the fraction of the pump power coupled to the nanoparticle and the total radiated power of the n-th harmonic, respectively [9]. It is also worth mentioning here that, since nonlinear processes in nanoscale objects are always induced by ultrafast pump sources, to address the pure nonlinear response of a nanostructures and be able to compare between different experimental outcomes, one should consider the peak powers/intensities and, hence, rule out the properties of the impinging pulses [20].

In this chapter we review the young history of second- and third-harmonic generation in all-dielectric nanostructure exhibiting Mie-type resonances at visible and near-infrared wavelengths which, despite being still in its infancy, already promises to greatly contribute to the emerging field of nonlinear optics at the nanoscale [33–35] as well as to a huge variety of applications [36–42]. Due also to a more accessible fabrication technology, silicon has been the first material platform used in this field; as a consequence of its centrosymmetric crystalline structure, bulk second-order nonlinear processes in Si are forbidden at first order and thus the dominant nonlinear contribution is coming from $\chi^{(3)}$. For this reason, third-harmonic generation has been the first nonlinear process to be demonstrated in all-dielectric nanoparticles [11]. Only more recently the research line of second-harmonic generation in all-dielectric nanoantennas and metasurfaces was developed in different material platforms (mainly GaAs and GaP [5–8]).

In Section 9.2, we will deal with second-harmonic generation while in Section 9.3 we will consider third-harmonic generation; the chapter ends with a section devoted to perspectives.

9.2   SECOND-HARMONIC GENERATION FROM DIELECTRIC NANOSTRUCTURES

Frequency conversion phenomena driven by second-order optical nonlinearities (including optical rectification (OR), second-harmonic generation (SHG), sum frequency generation (SFG) and difference frequency generation (DFG)) can be modeled in the frequency domain by means of the following simple formula:

$$P_i^{\omega_i} = \varepsilon_0 \chi_{ijk}^{(2)}\left(\omega_i; \omega_j, \omega_k\right) E_j^{\omega_j} E_k^{\omega_k} \tag{9.3}$$

where ε_0 is the dielectric permittivity of vacuum, $E_{j(k)}^{\omega_{j(k)}}$ is the $j(k)$-component of the electric field at the $\omega_{j(k)}$ frequency, $P_i^{\omega_i}$ is the nonlinear polarization at the frequency ω_i driven by second-order nonlinearities, and $\chi_{ijk}^{(2)}$ is one of the 27 elements of the second-order nonlinear susceptibility tensor.

By 1965, optical rectification, second-harmonic generation, sum frequency generation and difference frequency generation had been demonstrated in bulk nonlinear optics and rapidly found many applications including optical parametric amplification [43]; however, the interaction of light with small objects is not just a simple replica of what happens in bulk media, thus nanoscale devices for frequency conversion using second-order nonlinearities still represent a big challenge as new physical phenomena need to be understood. As already mentioned in the introduction, contrary to what happens in plasmonics, in all-dielectric nanoparticles third-harmonic generation has been first observed and only later on second-harmonic generation has been reported. The material platform where second-order nonlinear nanophotonics in all-dielectric nanoantennas has been first studied is gallium arsenide; GaAs is associated with a mature fabrication technology and its noncentrosymmetric crystalline structure gives rise to nonzero second-order nonlinear susceptibility. Second-order guided wave nonlinear optics in GaAs is, in fact, a well-established and still active research topic [44]; however the idea of exploiting GaAs as a constituent material in metasurfaces for second-order nonlinearities was at first not considered as a viable route; in fact, it is well known that, due to its zinc-blende crystal symmetry, a flat GaAs surface excited by a pump beam at normal incidence cannot produce second-harmonic in the far-field [45,46].

A completely different scenario, however, appears if the surface is decorated with nanoresonators whose natural modes are excited by the pump beam giving rise to a fundamental frequency which, as a consequence, has electric field components orthogonal to the surface, which in turn can generate harmonic field lines tangent to the surface contributing to far-field radiation.

Pursuing this idea using $Al_{0.18}Ga_{0.82}As$ and GaAs nanoparticles led to remarkable achievements. In the first pioneering theoretical study on magnetic-driven SHG in high-index dielectric nanoparticles, Carletti and co-workers [4] showed SHG from $Al_{0.18}Ga_{0.82}As$ nanodisks with a conversion efficiency, η_{SH}, of about 0.1% using a pump intensity of 1 GW/cm^2. For these conceptual demonstrations, AlGaAs disk-shaped resonators in a uniform air background were designed to exhibit a magnetic dipole (MD) resonance at the pump wavelengths between 1600 and 1650 nm. The design

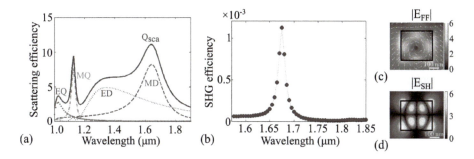

FIGURE 9.1 (a) Scattering efficiency Q_{sca} of an AlGaAs disk in air, decomposed in magnetic dipole (MD), electric dipole (ED), magnetic quadrupole (MQ), and electric quadrupole (EQ) contributions, as a function of wavelength calculated for $r = 225$ nm and $h = 400$ nm. (b) SHG efficiency as a function of pump wavelength for a AlGaAs cylinder with $r = 225$ nm, $h = 400$ nm suspended in air and pump intensity $I_0 = 1$ GW/cm². (c) Normalized $|E_{FF}|$ field amplitude in the x-z plane through the axis of the AlGaAs cylinder at a wavelength of 1680 nm. The arrows represent the direction of the electric field E in the same x-z plane. (d) Normalized second-harmonic field amplitude $|E_{SH}|$ on a cross section in the x-z plane at the center of the cylinder generated from a pump at a wavelength of 1680 nm.

approach exploited the potential of multipole decomposition of the scattered electric field to characterize the optical resonances of the nanodisks.

As it can be seen from Figure 9.1a, the strong resonance peak at a wavelength of about 1640 nm is due to a magnetic dipole resonance, while other multipolar resonances contribute to scattering peaks observed at lower wavelengths. The electric fields induced in the nanodisks are then used to determine the nonlinear currents generated at the *SH* by the second-order nonlinear susceptibility [47]. As the investigated nanostructures present a high volume to surface ratio, only volume optical nonlinearities are considered in the model. In the case of AlGaAs, the material crystalline structure exhibits a cubic symmetry and it belongs to the *3m* symmetry group. For these materials, the second-order nonlinear susceptibility tensor has non-vanishing elements of the type $\chi_{ijk}^{(2)}$ where $i \neq j \neq k$ are crystalline axes. Exploiting Kleinman symmetry condition and introducing the 3×6 *d* matrix formulation [48], the only relevant term is d_{36}, which for AlGaAs is 50 pm/V [49]. Thus, the nonlinear currents at the *SH* frequency can be expressed as

$$J_i^{SH} = i\omega_{SH} 4\varepsilon_0 d_{36} E_j^\omega E_k^\omega$$

where ε_0 is the dielectric permittivity of vacuum, ω_{SH} is the *SH* angular frequency, and $E_{j(k)}^\omega$ is the $j(k)$-component of the electric field at the pump frequency. Using such currents as optical sources for numerical simulations at the *SH* frequency, and assuming an undepleted pump regime, it is possible to predict the nonlinear *SH* scattering of the nanodisks.

The SHG efficiency (Figure 9.1b) varies strongly when the *FF* frequency is on or off the MD resonance with a sharp peak to above 10^{-3} (incident pump intensity 1 GW/cm²). Apart from the MD resonance, also the coupling between the generated nonlinear currents and the optical modes of the nanodisk at the *SH* frequencies influences η_{SH}. Indeed, the wavelength of the η_{SH} peak is slightly shifted from the MD peak. A way to describe, at least qualitatively and in the absence of absorption, the phenomenology of SHG in such nanostructures can be given by evaluating the parameter v that is proportional to η_{SH}:

$$v = \left(Q_{FF}\right)^2 \times Q_{SH} \times \gamma$$

where Q_{FF} and Q_{SH} are the scattering efficiencies of the resonator at *FF* and *SH* frequencies, respectively, and γ is the SHG nonlinear overlap integral between the electric fields at the *FF* and *SH*, i.e.:

$$\gamma = \sum_{i,j,k} \frac{\left| \int E_i^{2\omega} E_j^{\omega} E_k^{\omega} dV \right|^2}{\int \left| E_i^{2\omega} \right|^2 dV \int \left| E_j^{\omega} \right|^2 dV \int \left| E_k^{\omega} \right|^2 dV}$$

calculated over the indexes (i, j, k) for which $\chi_{ijk}^{(2)} \neq 0$. Thus, the conversion efficiency increases proportionally to $\left(Q_{FF}\right)^2$ since the electric field enhancement inside the resonator scales as the scattering efficiency and the *SH* response depends on the square of the electric field. Moreover, η_{SH} is proportional to γ and Q_{SH} as one needs an efficient coupling between the nonlinear currents and the resonator modes and an efficient radiation of these local fields. A similar behavior and enhancement is found also when the nanodisk radius is varied between 180 and 230 nm, while the corresponding pump wavelength red-shifts continuously as expected when resonator dimensions increase (Figure 9.2).

This work prompted the first experimental demonstration of magnetic-driven SHG in high-index nanoresonators based on the key enabling technology of the

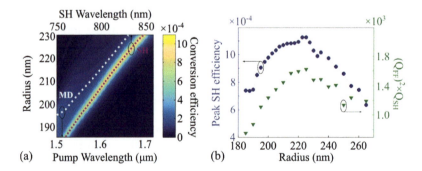

FIGURE 9.2   (a) SHG efficiency as a function of pump wavelength and nanodisk radius. The magnetic dipole resonance (MD) and the resonant mode at the emission wavelength (SH) are outlined with a dotted white and magenta line respectively. (b) Maximum SHG efficiency (circles) and $(Q_{FF})^2 \times Q_{SH}$ (triangles) as a function of nanodisk radius. The nanodisks have constant height of 400 nm and are suspended in air. The pump intensity is 1 GW/cm².

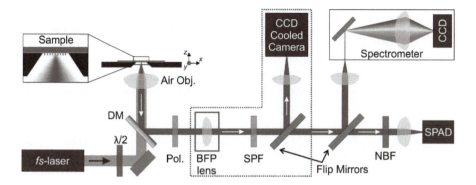

FIGURE 9.3 Schematic of the experimental setup for nonlinear spectroscopy and BFP imaging. $\lambda/2$: half-wavelength plate, DM: dichroic mirror, Pol.: polarizer, BFP: back-focal plane lens, SPF: short-pass filter, NBF: narrow-band filter, SPAD: single-photon avalanche detector.

AlGaAs-on-AlOx material system [5]. The typical experimental setup employed to address the nonlinear optical properties of these nanostructures is a confocal microscope that allows recording diffraction-limited nonlinear spatial maps by raster-scanning the sample as well as polar plots and polarization-resolved back-focal plane images by exciting each individual nanoantenna. Figure 9.3 shows a typical experimental arrangement that combines all these functionalities in a single optical setup. The excitation light is a femtosecond pump laser working at telecom wavelengths, usually a linearly polarized ultrafast (150 fs pulse length) Erbium-doped fiber laser centered at 1550 nm, whose polarization is precisely controlled thanks to a $\lambda/2$ plate before entering the objective. In order to acquire wavelength-dependent responses, one could also employ a tunable ultrafast laser source, which is often constituted by an optical parametric oscillator (OPO) or amplifier (OPA) working in the MHz repetition-rate range to avoid sample damage. The excitation beam is focused onto the sample through a high-NA air objective (typically $NA = 0.7/0.85$), and the emission from the individual nanostructures is characterized by recording confocal maps while raster-scanning the sample with a piezoelectric stage. The light emitted by the sample is collected through the same optical path, and sent to the detection path by a polarization-maintaining dichroic mirror (DM), which also partially rejects the reflected excitation light. A short-pass filter (SPF) and a narrow-band filter further clean the nonlinear signal from the excitation light and from any possible two-photon photoluminescence (TPL), while a linear polarizer (Pol.) allows selecting the light components to be analyzed and recorded. The filtered signal is then detected with a single-photon avalanche detector (SPAD), while spectra of the unfiltered emitted radiation can be collected with a visible/near-infrared grating-based spectrometer. This nonlinear microscopy setup allows attaining a lateral resolution of about 700 nm in the nonlinear regime, which enables the resolution of the emission from single particles.

Together with intensity, polarization distribution, and spectral response, a full characterization of the emission properties of nanoresonators includes also the angular distribution or radiation pattern. This measurement is carried out by the so-called back focal plane (BFP) or Fourier imaging. The BFP is that plane where all the light rays collected from an objective are parallel to each other. As such, an image of the radiation distribution on this plane, performed by inserting a Bertrand lens (BFP lens) that focuses at the BFP of the objective in the collection path, is an image of the radiation directionality of the emitter located at the focus of the objective (see the area confined in the pink dashed line in Figure 9.3). The imaging process is then realized with a CCD cooled camera.

In this experimental realization, the nanodisk is situated on a low refractive index substrate of alumina ($n \approx 1.6$) and surrounded by air. In [5] the resonator design is adjusted to achieve MD resonance at a wavelength of 1550 nm. Measurements of the light emitted at the *SH* frequency show that the power is proportional to the square of the pump power, as expected for second-order nonlinear phenomena. A ratio of $\sqrt{2}$ between the linewidth of the *SH* and *FF* spectra confirmed that the *SH* signal originates from a coherent two-photon process. Using an intensity of the pump beam of 1.6 GW/cm², the SHG conversion efficiency, η_{SH}, in this system was on the order of 10^{-4}. Collection of the *SH* light through an objective with a *NA* of 0.85 diminishes the measured efficiency to about 5×10^{-5}. The collected SHG intensity as a function of the nanodisk radius is shown in Figure 9.4a. From the acquired intensity, and accounting for the optical throughput of the collection path (see [5]), we estimate a conversion efficiency of the order of 10^{-5} for the most efficient nanodisk. The numerical calculations, which are also shown in Figure 9.4, are in good qualitative and quantitative agreement with the measurements. Small discrepancies between the numerical and experimental results could be, at least, partially ascribed to fabrication uncertainties. The results in Figure 9.4 indicate that the theoretical model used for the SHG provides a comprehensive picture of the physical phenomena at play. In particular, this confirms that the observed nonlinear response is due to the bulk nonlinearity while the contribution of the surface nonlinearities can be considered negligible.

From the measurements shown in Figure 9.4, three different peaks can be identified at approximately 195, 205, and 220 nm radii. Analyzing the optical response of

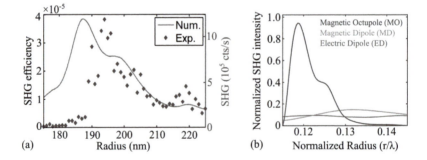

FIGURE 9.4 (a) Numerical (continuous line) and experimental (diamonds) dependences of SHG on the radius of the AlGaAs nanoantennas for a fundamental wavelength of 1550 nm. The background SHG from the AlOx substrate is about three orders of magnitude below the SHG signal from the nanoantenna and was subtracted in the plot. (b) Multipole decomposition of the fields emitted at the *SH* wavelength as a function of r/λ.

the nanodisk at the *SH* frequencies revealed that these three peaks are due to three different modes, confirming the fundamental role of the nanoresonator modes in the SHG process. While the peaks at 190 and 205 nm are due to MO resonances, the peak at 220 nm is due to MD and ED modes.

The strong potential of AlGaAs nanodisks for efficient SHG has been demonstrated also on the AlGaAs-in-insulator material system [7,50]. In analogy with previously described work, the resonator shape was designed to sustain an MD resonance at the pump wavelength. Using pump intensities of about 7 GW/cm^2, [7] *SH* conversion efficiency of about 10^{-4} was reported. Always in the NIR spectral range, efficient SHG has been demonstrated using GaAs nanodisks on an alumina substrate using an optical pump at a wavelength of about 1 μm [51]. With a pump intensity of 3.4 GW/cm^2, η_{SH} was found to be about 10^{-5}. In the visible range, GaAs (but also AlGaAs) is no longer transparent, therefore pumping such structures at high optical intensities may induce damages and re-absorption of the emitted light, thus effectively limiting the *SH* generation, in analogy with plasmonic nanoantennas.

Therefore, as the transparency window of both AlGaAs and GaAs is limited to wavelengths above 750 nm, efficient SHG in a negligible loss regime requires to adopt materials with larger values of the corresponding bandgap energy. In this context, a promising work based on GaP as the nonlinear material was conducted by Cambiasso et al. [8]. Here, GaP nanodisk corrugations have been fabricated and characterized for SHG using a pump wavelength of about 900 nm. A strong enhancement of the conversion efficiency was predicted at the magnetic quadrupole resonance for the pump wavelength due to a strong electric field enhancement induced in the GaP nanodisk. From measurements, the estimated conversion efficiency, η_{SH}, was about 2×10^{-4} obtained by using a pump beam intensity of 200 GW/cm^2 (Figure 9.5).

Another interesting class of promising materials for nanoscale SHG are Perovskites such as $BaTiO_3$. A recent demonstration of $BaTiO_3$ nanospheres reported a magnetic-driven SHG in the UV spectral range with record high conversion efficiency at a pump beam intensity of 870 GW/cm^2 [52]. As it can be seen from Figure 9.6, this enhancement was strictly related to the electric field enhancement induced in the resonators by a magnetic quadrupole (MQ) mode at wavelengths of about 375 nm.

Very recently, Makarov et al. performed an interesting study of SHG driven Mie-type resonances in crystalline Si nanospheres [10]. This demonstration, at variance with the others presented before, exploits the surface nonlinearity of Si, since the bulk second-order nonlinear susceptibility is zero due to the centrosymmetric structure of Si crystal lattice. The conversion efficiency estimated in the experiments is about 10^{-6} using a pump wavelength of 1050 nm and an intensity of 30 GW/cm^2. Interestingly, the achieved conversion efficiency is comparable to that of works performed using noncentrosymmetric materials, showing a good potential for all-dielectric nanoantennas supporting Mie-type resonances to exploit both volume and surface nonlinear effects.

As far as the efficiency of the second-harmonic generation process is concerned, Table 9.1 summarizes the results reported in the last three years in the literature. However, efficiency is not the only parameter worth optimizing for practical applications of SHG at the nanoscale. As in any radiating system, directionality can be used at the advantage of specific needs [53–59]; interestingly, the use of subwavelength

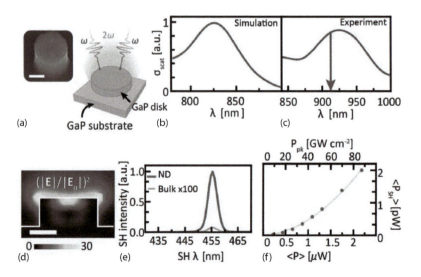

FIGURE 9.5 (a) Scanning electron micrograph of a 200 nm radius GaP disk (scale bar, 200 nm), and schematic view of the experimental setup for enhanced SHG from a GaP disk. (b, c) Simulated (b) and experimental (c) scattering cross-section spectrum of a single 200 nm thick GaP disk with $R = 200$ nm. The vertical arrow in (c) represents the chosen excitation wavelength band for the nonlinear characterization. (d) XZ cross section of the predicted electric field intensity distribution computed at the simulated scattering maximum. Scale bar: 200 nm. (e) *SH* spectra of the disk and bulk GaP substrate for an excitation wavelength of 910 nm. (f) Dependence of the average *SH* power $\langle P_{SH} \rangle$ on the average excitation power $\langle P \rangle$ for the GaP disk (corresponding pulse peak power $\langle P_{pk} \rangle$ on top axis). The solid line is a fit of the data considering the expected quadratic dependence of $\langle P_{SH} \rangle$ with $\langle P \rangle$.

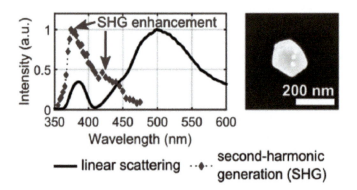

FIGURE 9.6 SHG spectra of BaTiO3 nanoparticles. The continuous line represents the normalized backscattering linear spectrum. The diamond line represents the normalized SHG signal values for the same nanoparticle. The inset show the SEM micrographs of the measured BaTiO3 nanoparticles. The arrows highlight the positions of the peaks in the SHG spectrum.

TABLE 9.1

Overview of SHG Experiments in All-Dielectric Nanoparticles

	$\chi^{(2)}$ (pm/V)	Fundamental Wavelength (nm)	Conversion Efficiency	Peak Pump Intensity (GW/cm²)	P_{SH}/P_{FW}^2 (1/W)
Gili et al. [5]	AlGaAs ~ 200	1550	1×10^{-5}	1.6	1.5×10^{-7}
Liu et al. [6]	GaAs ~ 400	1020	2×10^{-5}	3.4	1.5×10^{-8}
Camacho Morales et al. [7]	AlGaAs ~ 200	1550	8.5×10^{-5}	7	6.4×10^{-6}
Cambiasso et al. [8]	GaP ~ 300	900	2×10^{-6}	200	4×10^{-9}
Makarov et al. [10]	Si	1050	1×10^{-6}	30	4×10^{-8}

structures for nonlinear frequency mixing constitutes a major advantage over bulk crystals for controlling the properties of the emitted photons. Since many wavelengths of material are needed to achieve appreciable nonlinear optical responses in bulk crystals, phase-matching constraints and physical dimensions hinder spatial control of the emitted photons. In contrast, nanostructures exhibit high optical nonlinear responses with subwavelength dimensions and circumvent phase-matching problems. For these reasons, a fine control of the emitted photons in both near and far-field regions is possible using nanoantennas. Moreover, compared to plasmonic nanostructures, high-index dielectric nanostructures can exploit interference between magnetic and electric response to shape and control the properties of the generated photons and higher conversion efficiencies can also be achieved due to much higher damage threshold powers.

In this context, it has been recognized that both the polarization properties and the radiation pattern of the *SH* signal can be modified by changing the polarization and/or the angle of incidence of the *FF* pump impinging on AlGaAs nanodisks [60]. Due to the axial symmetry of the structure and incident fields, and the form of the $\chi^{(2)}$ tensor, radiation profiles of modes excited at the *SH* by nonlinear currents will always feature a null in the normal direction for nanodisks fabricated on the (100) surface normal pump incidence. This property has also been observed experimentally [7]. In recent works, it was first shown that this condition can be circumvented by using an angled pump beam [60] and effectively breaking the symmetry of the system [61]. As shown in Figures 9.7 and 9.8, for an s-polarized pump beam, as the angle of incidence of the pump increases, the generated nonlinear currents excite more and more efficiently an electric dipole mode at the *SH*. Consequently, the radiation pattern features a main lobe in the normal direction (see Figure 9.7 for the theoretical results [61] and Figure 9.8 for the experimental ones [62]).

Experiments on AlGaAs nanodisks also revealed the complex dependence between pump beam, nanostructure modes, and polarization properties of the *SH* photons. Polarization-resolved *SH* measurements have been conducted using monolithic AlGaAs resonators on alumina substrates [62]. These measurements, obtained

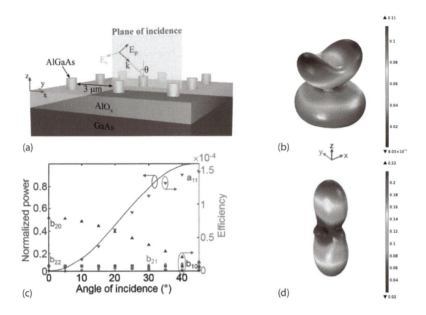

FIGURE 9.7 (a) Schematic representation of the fabricated AlGaAs-on-AlOx nanodisk array and the illumination geometry. The wavevector, k, and the electric field vectors for s, E_s, and p, E_p, polarizations on the plane of incidence are shown. Radiation patterns of SHG for tilted pump beams: (b) normal incidence, (d) incidence at 45° with respect to the z-axis and s-polarized light. (c) shows the multipolar decomposition of the radiation pattern as a function of the angle of incidence, with a monotonic increase of the electric dipole contribution a_{11}.

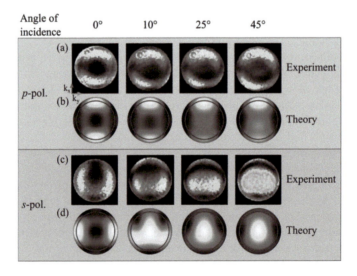

FIGURE 9.8 Back-focal plane images of the SH radiation emitted in the backward direction for different angles of incidence, ϑ, of the fundamental for (a, b) p-polarized and (c, d) s-polarized excitation: (a, c) Experimental measurements and (b, d) numerical calculations. The inner circle in (b, d) represents the experimental $NA = 0.85$.

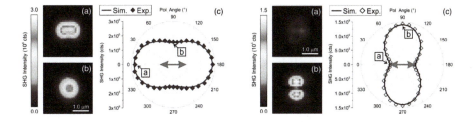

FIGURE 9.9   Left: Confocal SHG maps acquired on a $r/\lambda \approx 0.123$ nanodisk for emission polarization parallel [co-polarized, (a)] and perpendicular [cross-polarized, (b)] to the pump field. (c) Simulated (solid line) and experimentally acquired (diamonds) polar plots of the emitted *SH* from the nanostructure. Red arrow: pump polarization direction. Right: Confocal SHG maps acquired on an $r/\lambda \approx 0.140$ nanodisk for emission polarization parallel [co-polarized, (a)] and perpendicular [cross-polarized, (b)] to the pump field. (c) Simulated (solid line) and experimentally acquired (diamonds) polar plots of the emitted *SH* from the nanostructure. Red arrow: pump polarization direction.

using a confocal setup, are reported in Figure 9.9. As it can be seen, for two nanodisks that correspond to the first and third resonance peak in Figure 9.4, the *SH* emission exhibit a $\pi/2$ polarization change despite the *FF* beam has been left unaltered and is parallel to one of the (001) in-plane crystalline axis [63]. This phenomenon is somehow surprising by itself as, by considering the nonlinear susceptibility tensor of unstructured AlGaAs, *SH* currents are aligned in the direction orthogonal to the pump beam polarization. Comparison between numerical results and measurements confirmed the fundamental role of the nonlinear coupling to the nanoresonator modes at the *SH* for the SHG process. Indeed, the two nanodisks support modes of different nature at the *SH* frequency that effectively change the dominant polarization component of the emitted *SH* [63,64]. The excellent agreement between experiments and simulations, which are based solely on the bulk $\chi^{(2)}$ of the material, allow one to rule out the role of the surface for SHG in these nanopillars.

   A nontrivial dependence of the *SH* radiation pattern and pump polarization has also been shown on the AlGaAs-in-insulator material system. Exploiting the possibility of collecting the *SH* signal in both transmission and reflection configuration, the SHG backward-to-forward ratio was characterized for the first time in high-index dielectric nanostructures. Figure 9.10 shows that this ratio strongly depends on the

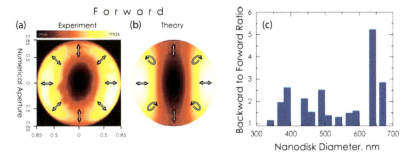

FIGURE 9.10   Directionality diagrams in the forward direction of the *SH* signal. (a) Experiment. (b) Theory. (c) Backward-to-forward ratio of the *SH* as a function of the disk diameter.

nanodisk diameter and, in general, the forward emission is dominant over backward emission. The radiation pattern at the *SH* was also characterized in [7] using a back-focal plane imaging setup. It was observed that despite the simple pump polarization (i.e., linear polarization) the *SH* radiation presents a vector-beam nature. In particular, the polarization of the emitted light is radial. It is also interesting to observe that the intensity radiation pattern displays a toroidal shape with a zero at the center.

Interestingly, a radially polarized *SH* beam is obtained from a Gaussian pump at the *FF*. As later investigated by Kruk et al. [50], interference between the electric and magnetic nonlinear response can manifest at the *SH* by exhibiting dominantly radially polarized or azimuthally polarized radiation patterns. As shown in Figure 9.11, depending on the orientation of the pump electric field polarization with respect to one of the in-plane (001) crystalline axes of AlGaAs, the *SH* radiation may result either radially or azimuthally polarized. This is motivated by the excitation of either the magnetic or the electric nonlinear response [50,64]. An excitation with E_{FF} parallel to one of the (001) crystal axis couples with magnetic multipole resonances. The characteristic of these modes is that their radiation pattern is dominated by radially polarized electric fields. Conversely, a pump with E_{FF} on the bisector of the two in-plane crystal axes, couples to electric multipole resonances that are characterized by azimuthally polarized emission (Figure 9.12).

A very interesting prospect of the research in metasurfaces is the access to dynamic control of the optical response. Optical modulation has been demonstrated both in the terahertz and in the optical regimes [65,66], where free carrier modulation with sub-picosecond and femtosecond dynamics was obtained in aluminum-doped zinc oxide and hydrogenated amorphous silicon. In particular, using magnetic dipolar resonances, 60% modulation of the metasurface transmittance and a 6 nm resonance shift were achieved by pumping each nanoantenna at picojoule energy [66]. The methods to achieve optical tunability consist in inducing refractive index modifications by linear or nonlinear absorption of an intense control beam (Figure 9.13) [67,68]. Using pump-probe setups, the tunability of the SH signal generated by AlGaAs nanodisks has been demonstrated [69]. The efficiency of the proposed methods to control the radiation pattern of the devised metasurfaces needs now to be systematically characterized.

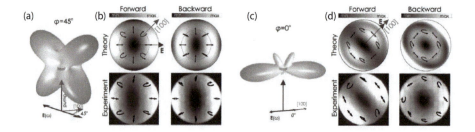

FIGURE 9.11 Directionality and polarization diagrams of the *SH* signal. (a, c) Calculated radiation patterns of *SH* for the pump polarization at 45° and 0° angles, respectively. (b, d) Polarization inclinations of *SH* in the forward and backward directions both experimentally measured and numerically calculated at 45° and 0° angles correspondingly. Numerical aperture of experimental forward and backward collection is 0.9 and 0.85. Calculations in (b, d) show full numerical apertures of 1.44 (inner circles show experimentally captured portion).

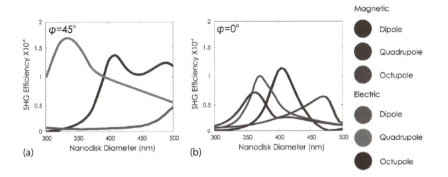

FIGURE 9.12 Multipolar decomposition of second-harmonic radiation as a function of the disk diameter for the orientation angles (a) $\varphi = 0°$, and (b) $\varphi = 45°$. Panel a is dominated by electric multipoles, whereas panel b is dominated by magnetic multipoles.

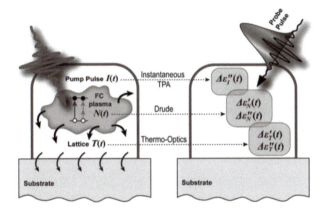

FIGURE 9.13 Illustration of transient optical processes in a nanoresonator. The interaction with a monochromatic pump beam (left panel) gives rise to a modification of the nanoresonator permittivity experienced by the broadband probe pulse (right panel), according to three different physical mechanisms: instantaneous TPA, Drude plasma response and thermo-optical effect.

The metasurface static re-configurability upon absorption of a control beam has to be addressed and is expected to be a real breakthrough for all-dielectric resonant nanophotonics as a platform for ultrafast optical devices, and opens the way to the possibility for ultrafast polarization multiplexed displays and polarization rotators.

9.3 THIRD-HARMONIC GENERATION FROM DIELECTRIC NANOSTRUCTURES

Similarly to what stated for second-order optical nonlinearities, all the frequency conversion phenomena driven by third-order optical nonlinearities can be modeled in the frequency domain by means of the following simple formula:

$$P_i^{\omega_i} = \varepsilon_0 \chi_{ijkl}^{(3)}(\omega_i; \omega_j, \omega_k, \omega_l) E_j^{\omega_j} E_k^{\omega_k} E_l^{\omega_l} \tag{9.4}$$

where ε_0 is the dielectric permittivity of vacuum, ω is the pump angular frequency, $E_{j(k,l)}^{\omega j(k,l)}$ is the $j(k,l)$-component of the electric field at the pump frequency, P_i^3 is the nonlinear polarization driven by third-order optical nonlinearities and $\chi_{ijkl}^{(3)}$ is one of the 81 elements of the third-order nonlinear susceptibility tensor. High-refractive-index dielectric nanostructures with large intrinsic third-order susceptibility, such as silicon, germanium etc., have recently emerged to provide a powerful platform to study the nonlinear process through the exploration of the associated multipolar optical responses of both electric and magnetic nature. By exciting and engineering the multipolar resonances supported by high-index dielectric nanostructures, the electromagnetic energy can be confined to a very small volume and penetrate inside the structure, not limited to the surface. This further enables downscaling of the required optical power during the third-harmonic generation (THG) process as the intensity of THG scales with the sixth power of the fundamental field strength. Concurrently, the phase-matching requirements, which are usually crucial for harmonic generation in bulk structures, can be greatly relaxed in the case of nanostructures, due to the efficient nonlinear light-matter interactions in a subwavelength area.

Silicon has been widely used in nanophotonics due to its CMOS compatibility, low cost, and well-established fabrication technologies. Silicon possesses strong bulk third-order optical susceptibility nearly 200 times larger than that of silica along with a high refractive index [70–74] making it a highly attractive choice to explore the third-order nonlinear process. For cubic materials like silicon, germanium etc., the nonlinear polarization at the third-harmonic (TH) frequency as a function of the pump field can be written as

$$P_i^{(3\omega)} = \varepsilon_0 \left\{ 3\chi_{1212}^{(3)} E_i \left(\overline{E} \cdot \overline{E} \right) + \left[\chi_{1111}^{(3)} - 3\chi_{1212}^{(3)} \right] E_i E_i E_i \right\} \qquad (9.5)$$

in the crystal reference frame [74]. The second term is responsible for the anisotropy which, in the general case, is a complex-valued frequency-dependent function. In crystalline silicon, the two components of the third-order susceptibility can be related by $\chi_{1111}^{(3)} = 2.36\chi_{1212}^{(3)}$ for low photon energies $\hbar\omega \leq 0.6\,\text{eV}$ [74]; in amorphous silicon, we have $\chi_{1111}^{(3)} = 3\chi_{1212}^{(3)}$ with no anisotropy.

As can be seen from Eq. (9.5), for a TH process, the induced polarization $P_i^{(3\omega)}$ scales as the cube of the local electric field. Thus, when pumping near the resonance of the dielectric nanostructure to induce a strong near-field enhancement, one could expect to obtain a large induced polarization, and, in turn, a high overall conversion efficiency.

M. Shcherbakov et al. have experimentally demonstrated a strong nonlinear response from individual silicon resonators near the magnetic dipole resonance [11]. In the experiment, the silicon nanodisks were fabricated on a 2-μm-thick SiO_2 layer situated on a bulk silicon substrate, and then pumped by an intense femtosecond laser pulse train with the frequency close to the magnetic dipole resonance. The results have shown that, by resonantly exciting the magnetic dipole mode from such silicon nanodisks to get locally enhanced electric fields, it is possible to get a two orders of magnitude enhancement of THG with respect to bulk silicon (Figure 9.14).

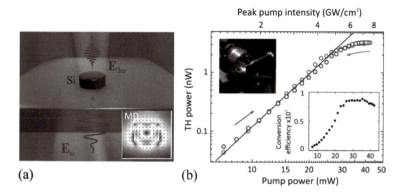

FIGURE 9.14 (a) Illustration of THG from an individual Si nanodisk at the magnetic dipole resonance excited by normal pump incidence at the fundamental frequency. Superimposed is the electric near-field distribution at MD resonance. (b) The measured power dependence and conversion efficiency of the resonant THG process in Si nanodisks. The THG power dependence upon increasing and decreasing pump power is denoted by full dots, while the empty circles denote the reverse procedure. In inset, a photographic image of the sample irradiated with the visible *IR* beam impinging from the back side of the sample. (Images are adapted from Shcherbakov, MR, et al., *Nano Lett.*, 14, 6488, 2014.)

An experimentally measured efficiency around 10^{-7} is achieved by using a pump power of 30 mW (corresponding to peak intensity ≈ 5.5 GW/cm^2).

By enhancing the near-fields at the pump frequency in dielectric nanostructures, researchers have been exploring various ways to further propel the nonlinear light-matter interactions and third-order nonlinear process. Constructive interference of different multipoles offers further possibilities to excite high-quality collective modes associated with enhanced fields inside the nanoparticles. It was demonstrated that high-quality Fano resonances can be supported by various nanostructures [75]. Fano resonance originates from the interference of two scattering channels, one of which is a resonant mode, and another is a nonresonant one with the same far-field symmetry. It usually appears as a resonant suppression of the light scattering at a specific point, corresponding to strong near-field enhancement; it can thus potentially boost the non-linear process in the nanoscale regime. In dielectric nanostructures, Fano resonances can lead to a much larger near-field enhancement compared to the magnetic dipole resonance supported by a single nanoparticle. Utilizing the large near-field enhancements near a Fano resonance, Y. Yang et al. have demonstrated highly efficient visible THG [76]. Here, the enhancement of THG is driven by a high-quality Fano resonance which stems from collective "bright" and "dark" modes of the silicon-based metasurface consisting of coupled rectangular silicon bars and disks, as shown in Figure 9.15. The nanobar supports an electric dipole resonance which can be excited by a normally incident pump with E-polarization parallel to the long axis, and the disk supports a "dark" out-of-plane magnetic dipole resonance which cannot be directly excited by normally incident light but can instead be excited through the coupling with the electric dipole resonance supported by the disk bar (Figure 9.15a). A sharp peak in the linear transmittance spectrum is observed with an experimental Q-factor

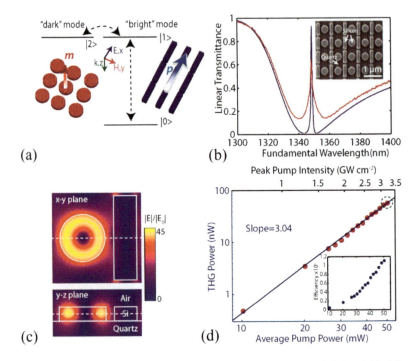

FIGURE 9.15 (a) Schematic of the Fano interference between the "bright" and "dark" mode resonators. (b) Simulated (blue) and experimentally measured (red) transmittance spectra of the Fano resonant metasurface. The inset shows an SEM image of the fabricated structure. (c) Simulated electric field amplitude at the Fano resonance peak wavelength of 1350 nm. (d) The measured THG power and conversion efficiency as a function of the pump power and the peak pump intensity. (Images are adapted from Yang, Z.-J. et al., *Phys. Rep.*, 701, 1, 2017.)

of 466 (Figure 9.15b). Near the peak position, a large near-field enhancement within the disk, associated with the excitation of the magnetic Fano resonance, is obtained (Figure 9.15c), leading to a THG enhancement factor of 1.5×10^5 from this metasurface as compared with unpatterned Si film. The result experimentally demonstrates a *TH* conversion efficiency of 1.2×10^{-6} under a peak pump intensity of 3.2 GW / cm^2.

Combining dielectric nanoparticles into oligomers can lead to the formation of collective modes with enhanced near-fields inside the nanoparticles, which is beneficial for nonlinear processes. In [13], by performing THG spectroscopy measurements, A. Shorokhov et al. observed a multifold enhancement of THG in dielectric quadrumers of silicon nanodisks supporting high-quality collective modes associated with the magnetic Fano resonance originating from the interplay between the collective optically induced magnetic responses of quadrumers and the individual magnetic responses of their constituent dielectric nanodisks. Nontrivial wavelength and angular dependencies of the generated harmonic signal were observed experimentally, featuring a multifold THG enhancement in such dielectric systems (Figure 9.16).

Besides, Fano-assisted THG enhancement has also been demonstrated in silicon nanodisks arranged in the form of trimer oligomers [12]. By varying the distance

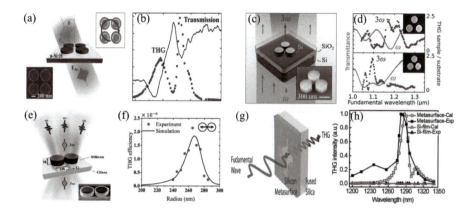

FIGURE 9.16 (a, b) Schematic of the resonant THG in silicon quadrumers and the measured THG spectroscopy. (From Shcherbakov, M.R. et al., *ACS Photonics*, 2, 578, 2015.) (c, d) Schematic of the THG process in silicon nanodisk trimers and the measured TH spectral responses with a different gap between its constituent elements. (From Shcherbakov, M.R. et al., *ACS Photonics*, 2, 578, 2015.) (e, f) Schematic of THG process from a silicon dimer structure, and the corresponding THG conversion efficiency as a function of the disk size. (From Wang, L. et al., *Nanoscale*, 9, 2201, 2017.) (g, h) Schematic of THG from the complementary silicon metasurface, and the experimentally measured and calculated THG spectra. (From Chen, S. et al., *ACS Photonics*, 5, 1671, 2018.)

between the nanoparticles, strong reshaping of the *TH* emission was observed due to the interference of the nonlinearly generated optical waves. The effect of hybridized modes supported by dielectric dimers allows for a substantial enhancement of both electric and magnetic fields, promoting the THG process [77]. S. Chen et al. [78] have designed a complementary silicon metasurface composed of nanoapertures of cross-like shape in the silicon film. Under normal pump incidence, such hole-array metasurface can excite different multipolar modes, and the interference between the guided and localized modes in the structure further results in a Fano-type resonant features with strongly enhanced near-field distribution and a large mode volume, allowing for an enhanced THG emission with an efficiency of a factor 100 higher than that from a planar silicon film of the same thickness.

High-index dielectric nanostructures also support other interesting modes which are endowed interesting features in view of enhancing light-matter interactions in both linear and nonlinear regimes. For instance, it was demonstrated that high-index dielectric nanostructures can support nonradiating anapole excitations. The anapole originates from the overlap of co-excited electric and toroidal dipole modes which have the same scattering magnitude but are out of phase. Thus, they can cancel the scattering of each other in the far-field and meanwhile exhibit strong near-field enhancement, providing a powerful approach to facilitate the nonlinear effects. The anapole can be excited for a disk with a low-enough aspect ratio where the leading contribution is provided by the electric dipole with all the other modes being strongly suppressed, making the nanostructures usually thicker than magnetic dipole resonators [79]. In [80], G. Grinblat et al. have demonstrated that the THG can be boosted by exciting

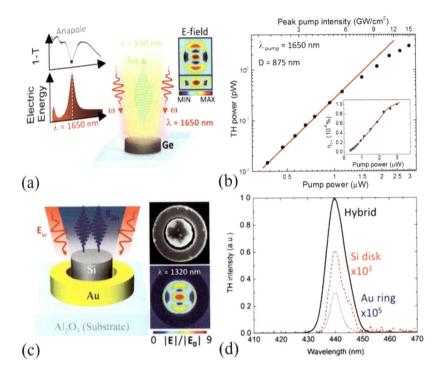

FIGURE 9.17 (a, b) Illustration of the THG process for a 100-nm-thick germanium nanodisk on glass pumped with near infrared light and the measured TH power versus pump power at the anapole resonance. (From Grinblat, G. et al., *Nano Lett.*, 16, 4635–4640, 2016.) The cubic dependence of TH emission on the pump power is shown by the red line in (b). (c, d) THG enhancement through a metal-dielectric hybrid nanostructure supporting an anapole mode. (From Shibanuma, T. et al., *Nano Lett.*, 17, 2647, 2017.) (c) Left: Schematic of the metal-dielectric nanostructure. Right upside: SEM micrograph. Right downside: The simulated electric field distribution at the anapole resonance. (d) gives the measured TH spectra for Hybrid (black), Si disk (red), and Au ring (blue) structures, respectively.

a germanium nanodisk in the vicinity of the nonradiative anapole mode under normal incident laser excitation (Figure 9.17a and b). The *TH* intensity is determined to be about one order of magnitude larger than that corresponding to excitation at electric dipolar resonances where the fields are mainly confined near the surfaces. The measured THG conversion efficiency reaches 0.0001% at an excitation wavelength of 1650 nm under peak pump intensity around 15 GW/cm², which is four orders of magnitude greater than the one obtained using an unstructured germanium film.

The anapole mode supported by a single free-standing nanodisk is usually formed by the spectral overlap between the off-resonantly excited electric and toroidal dipole modes. It is expected that the near-field can be enhanced efficiently if these two modes are resonantly excited and overlapped with each other. However, intrinsic drawbacks of the anapole mode are (i) its weak coupling to the far-field and (ii) the rather limited values of the overlap integral of the E field in the nanoparticle volume (see Figure 9.17c). To overcome these issues, the same group, suggested to add an Au nanoring outside the

Si nanodisk to enhance the electric near-field through the coupling between individual components (Figure 9.17c and d). The experimental results show a *TH* conversion efficiency reaching values of up to 0.007% near the anapole resonance.

A novel mechanism has been suggested by Xu et al. to enhance the excitation of the anapole mode [82]. They introduce a metallic mirror underneath the Si nanodisk supporting an anapole mode, as shown in Figure 9.18a. In this nanosystem, the Cartesian electric and toroidal dipole modes are both resonantly excited with the same scattering magnitude and spatially overlapped, leading to a significantly enhanced electric near-field inside the disk (Figure 9.18b). Meanwhile, during the nonlinear process, the mirror creates an image nonlinear source below the interface and coherently amplifies the nonlinear responses. This results in a nearly 100 times enhancement of the experimentally observed *TH* emission as compared to the silicon anapole-resonator-on-glass configuration (Figure 9.18c). The authors experimentally obtained a high *TH* conversion efficiency of 0.01% with peak pump intensity as low as 3.0 GW/cm^2 (Figure 9.18d).

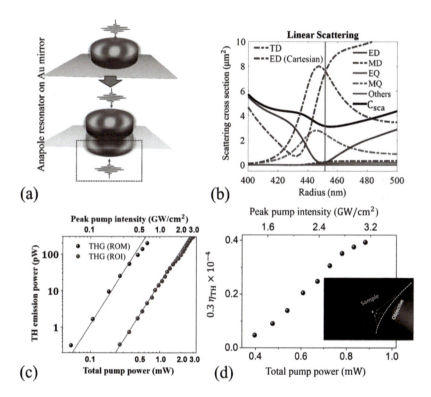

FIGURE 9.18 (a) Schematic illustration of an anapole resonator on an Au mirror surface. (b) Multipolar decomposition for the anapole-resonator-on-mirror nanostructure. (c) Experimentally measured TH power versus the pump power, where the cubic dependence is given by the red lines. (d) The measured THG conversion efficiency for different pump powers. The inset shows a photographic image of the TH emission from the sample. (Images are adapted from Xu, L. et al., *Light Sci. Appl.*, 7, 44, 2018.)

TABLE 9.2

THG Conversion Efficiencies Reported in Several Nanoscale Platforms Using Mie Resonators

THG	(m^2/V^2)	Fundamental Wavelength (nm)	Efficiency	Peak Pump Intensity (GW/cm^2)	Peak Pump Power (W)	P_{TH}/P_{FW}^3
Shcherbakov et al. [11]	Si ~ 10^{-19}	1260	$8\times10^{-6}\%$	5	1974	2.1×10^{-14}
Yang et al. [76]	Poly-Si ~ 10^{-18}	1350	1.2×10^{-6}	3.2	2500	1.9×10^{-13}
Shorokhov et al. [13]	Si ~ 10^{-19}	1340	~ 10^{-6}	5	2632	1.4×10^{-13}
Grinblat et al. [80]	Ge~ 5.65×10^{-19}	1650	0.0001%	15	166.7	3.6×10^{-11}
Wang et al. [77]	Poly-Si ~ 10^{-18}	1556	2×10^{-6}	8.49	300	2.2×10^{-11}
Melik-Gaykazyan et al. [83]	aSi ~ 10^{-19}	1550	4×10^{-8}	2.2	103.3	3.7×10^{-12}
Shibanuma et al. [81]	Si	1320	0.007%	25	138.9	3.6×10^{-9}
Chen et al. [78]	Silicon ~ 10^{-19}	1280	1.76×10^{-7}	1.04	1628	6.6×10^{-14}
Xu et al. [82]	aSi ~ 10^{-19}	1550	0.01%	3.0	17.8	3.2×10^{-7}

In Table 9.2 we have listed the experimentally obtained THG conversion efficiency and the *TH* nonlinear coefficient from several recently reported results at the time of the writing. Similarly to the *SH* nonlinear coefficient, the *TH* nonlinear coefficient is introduced to sort out the dependence on the pump power level during the THG process, which is defined as the ratio between the *TH* emission power and the cubic pump power $\tilde{n}_3 = P_{3}/P_{FF}^3$ based on the generated *TH* emission and peak pump power.

By tuning the arrangement of the geometry of nanoparticles and the input pump, one can excite different types of multipoles at the harmonic frequencies. The non-linear multipolar interference effect provides flexibilities for shaping the radiation patterns of the THG emission. The *TH* directivity can be tailored through the inter-ferences between the nonlinearly generated multipoles [84]: high directionality can be achieved for a silicon nanodisk by adjusting the aspect ratio to overlap the nonlin-early generated magnetic and electric dipole resonances (Figure 9.19a). In addition, by placing a nanoparticle near a silicon dimer to control the linear and nonlinear multipolar excitation, it is possible to achieve high directivity for both the linear and the nonlinear scatterings simultaneously [3]. Nonlinear optical processes exploiting multipolar Mie resonances enable novel functionalities, such as ultrafast all-optical switching [3], generation of nonlinear structured beams of radial or azimuthal polar-ization states [83], etc. Finally, exploiting the large even-order and odd-order optical

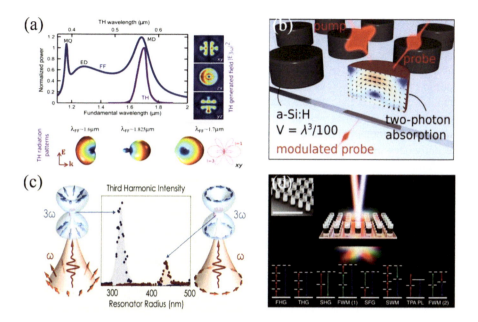

FIGURE 9.19 (a) Transformations of the TH radiation pattern with switching directionality. (From Smirnova, D. and Kivshar, Y.S., *Optica*, 3, 1241, 2016.) (b) Ultrafast all-optical switching in subwavelength nonlinear silicon nanostructures. (From Shcherbakov, M.R. et al., *Nano Lett.*, 15, 6985, 2015.) (c) Nonlinear structured light generation through silicon nanoparticles. (From Melik-Gaykazyan, E.V. et al., *ACS Photonics*, 5, 728, 2018.) (d) A broadband optical frequency mixer based on GaAs metasurfaces. (From Liu, S. et al., *Nat. Commun.*, 9, 2507, 2018.)

nonlinearities of noncentrosymmetric resonant GaAs metasurfaces, researchers have recently been able to demonstrate seven different nonlinear optical processes simultaneously through GaAs metasurfaces, realizing ultra-compact flat optical mixers [19].

9.4 PERSPECTIVES

Thanks to multipolar interference effects, all-dielectric nanostructures and metasurfaces have already shown the ability to achieve all four quadrants of electromagnetic responses ($\varepsilon > 0$, $\mu > 0$; $\varepsilon < 0$, $\mu > 0$; $\varepsilon > 0$, $\mu < 0$; $\varepsilon < 0$, $\mu < 0$), yielding unique possibilities for both linear and nonlinear light manipulation, including the phase, amplitude, and polarization states of light with subwavelength resolution. These features will further trigger advances in various applications including nonlinear holograms, nonlinear mirrors, quantum sources, etc.

Harmonic generation in high-index dielectric nanostructures has just begun to unveil the potential of all-dielectric platforms for nonlinear nanophotonics. The first milestone of establishing record high efficiencies being reached, we now witness new approaches to improve the state of the art. In fact, multipolar resonances such as MD or MQ are characterized by relatively low Q factors and ideas to achieve high-Q modes in high-index dielectric resonators are being explored [26]. Techniques based on collective Fano resonances have already been demonstrated as effective solutions to enhance third-harmonic generation in Si structures [13,85]. A particularly interesting phenomenon arises when such cavities can sustain optical modes that destructively interfere in the far-field. Such radiation-less modes can be a powerful tool to enhance cavity Q-factor by eliminating one radiation loss mechanism. One popular example is the anapole case that was used to enhance both third-harmonic generation in Ge [80] as well as second-harmonic generation in AlGaAs nanostructures [86]. Another class is constituted by bound states in the continuum that have recently been unveiled also for high-index dielectric nanostructures [9].

Numerical simulations have shown to be capable of predicting such features with quite a high accuracy. This is of fundamental importance in the perspective of the development of nonlinear metasurfaces featuring flexible and accurate control of the properties of generated photons.

Currently, frequency generation from second-order nonlinear effects is focused on SHG due to its simpler dynamics. However, this is just the preliminary step toward the investigation of other interesting second-order nonlinear processes such as SFG and DFG in high-index dielectric nanostructures. The first theoretical and experimental demonstrations have just appeared at the time of this writing [87,88]. The range of potential applications based on such nonlinear phenomena is extremely rich and broad. A not completely exhaustive list would include nonlinear imaging, holography, optical communications, parametric amplification, optical sensing, and quantum technologies.

Considering the second- or third-order dependence of the nonlinear signal on the incident light intensity through the nonlinear conversion process, the harmonic spectrum is expected to be spectrally narrower than the corresponding linear resonance. If properly combined with high nonlinear conversion efficiency, dielectric nanostructures provide a perfect tool for developing new generations of on-chip nonlinear biosensors with enhanced performance based on high-finesse optical resonances [35].

REFERENCES

1. D. Pile, P. Berini, Nanophotonics is big, *Nature Photonics*, 8 (2014) 878.
2. M. Kauranen, A.V. Zayats, Nonlinear plasmonics, *Nature Photonics*, 6 (2012) 737.
3. D. Smirnova, Y.S. Kivshar, Multipolar nonlinear nanophotonics, *Optica*, 3 (2016) 1241.
4. L. Carletti, A. Locatelli, O. Stepanenko, G. Leo, C. De Angelis, Enhanced second-harmonic generation from magnetic resonance in AlGaAs nanoantennas, *Optics Express*, 23 (2015) 26544.
5. V.F. Gili, L. Carletti, A. Locatelli, D. Rocco, M. Finazzi, L. Ghirardini, I. Favero, et al., Monolithic AlGaAs second-harmonic nanoantennas, *Optics Express*, 24 (2016) 15965.
6. S. Liu, M.B. Sinclair, S. Saravi, G.A. Keeler, Y. Yang, J. Reno, G.M. Peake, et al., Resonantly enhanced second-harmonic generation using III–V semiconductor all-dielectric metasurfaces, *Nano Letters*, 16 (2016) 5426.
7. R. Camacho-Morales, M. Rahmani, S. Kruk, L. Wang, L. Xu, D.A. Smirnova, A.S. Solntsev, et al., Nonlinear generation of vector beams from AlGaAs nanoantennas, *Nano Letters*, (2016) 7191. doi:10.1021/acs.nanolett.6b03525.
8. J. Cambiasso, G. Grinblat, Y. Li, A. Rakovich, E. Cortés, S.A. Maier, Bridging the gap between dielectric nanophotonics and the visible regime with effectively lossless gallium phosphide antennas, *Nano Letters*, 17 (2017) 1219.

9. L. Carletti, K. Koshelev, C. De Angelis, Y. Kivshar, Giant nonlinear response at the nanoscale driven by bound states in the continuum, *Physical Review Letters*, 121 (2018) 033903.

10. S.V. Makarov, M.I. Petrov, U. Zywietz, V. Milichko, D. Zuev, N. Lopanitsyna, A. Kuksin, et al., Efficient second-harmonic generation in nanocrystalline silicon nanoparticles, *Nano Letters*, 17 (2017) 3047.

11. M.R. Shcherbakov, D.N. Neshev, B. Hopkins, A.S. Shorokhov, I. Staude, E.V. Melik-Gaykazyan, M. Decker, et al., Enhanced third-harmonic generation in silicon nanoparticles driven by magnetic response, *Nano Letters*, 14 (2014) 6488.

12. M.R. Shcherbakov, A.S. Shorokhov, D.N. Neshev, B. Hopkins, I. Staude, E.V. Melik-Gaykazyan, A.A. Ezhov, et al., Nonlinear interference and tailorable third-harmonic generation from dielectric oligomers, *ACS Photonics*, 2 (2015) 578.

13. A.S. Shorokhov, E.V. Melik-Gaykazyan, D.A. Smirnova, B. Hopkins, K.E. Chong, D.-Y. Choi, M.R. Shcherbakov, et al., Multifold enhancement of third-harmonic generation in dielectric nanoparticles driven by magnetic fano resonances, *Nano Letters*, 16 (2016) 4857.

14. A.I. Kuznetsov, A.E. Miroshnichenko, Y.H. Fu, J. Zhang, B. Luk'yanchuk, Magnetic light, *Scientific Reports*, 2 (2012) 492.

15. A.I. Kuznetsov, A.E. Miroshnichenko, M.L. Brongersma, Y.S. Kivshar, B. Lukyanchuk, Optically resonant dielectric nanostructures, *Science*, 354 (2016) aag2472-aag2472.

16. A.E. Minovich, A.E. Miroshnichenko, A.Y. Bykov, T.V. Murzina, D.N. Neshev, Y.S. Kivshar, Functional and nonlinear optical metasurfaces: Optical metasurfaces, *Laser & Photonics Reviews*, 9 (2015) 195.

17. M. Decker, I. Staude, Resonant dielectric nanostructures: A low-loss platform for functional nanophotonics, *Journal of Optics*, 18 (2016) 103001.

18. A. Krasnok, M. Tymchenko, A. Alù, Nonlinear metasurfaces: A paradigm shift in nonlinear optics, *Materials Today*, 21 (2018) 8.

19. S. Liu, P.P. Vabishchevich, A. Vaskin, J.L. Reno, G.A. Keeler, M.B. Sinclair, I. Staude, I. Brener, An all-dielectric metasurface as a broadband optical frequency mixer, *Nature Communications*, 9 (2018) 2507.

20. M. Celebrano, X.F. Wu, M. Baselli, S. Grossmann, P. Biagioni, A. Locatelli, C. De Angelis, et al., Mode matching in multiresonant plasmonic nanoantennas for enhanced second harmonic generation, *Nature Nanotechnology* 10 (2015), 412.

21. C. Ciracì, E. Poutrina, M. Scalora, D.R. Smith, Second-harmonic generation in metallic nanoparticles: Clarification of the role of the surface, *Physical Review B*, 86 (2012) 115451.

22. D. Timbrell, J.W. You, Y.S. Kivshar, N.C. Panoiu, A comparative analysis of surface and bulk contributions to second-harmonic generation in centrosymmetric nanoparticles, *Scientific Reports*, 8 (2018) 3586.

23. A.B. Evlyukhin, C. Reinhardt, B.N. Chichkov, Multipole light scattering by nonspherical nanoparticles in the discrete dipole approximation, *Physical Review B*, 84 (2011) 235429.

24. P. Grahn, A. Shevchenko, M. Kaivola, Electromagnetic multipole theory for optical nanomaterials, *New Journal of Physics*, 14 (2012) 093033.

25. S. Long, M. McAllister, S. Liang, The resonant cylindrical dielectric cavity antenna, *IEEE Transactions on Antennas and Propagation*, 31 (1983) 406.

26. M.V. Rybin, K.L. Koshelev, Z.F. Sadrieva, K.B. Samusev, A.A. Bogdanov, M.F. Limonov, Y.S. Kivshar, High-Q supercavity modes in subwavelength dielectric resonators, *Physical Review Letters*, 119 (2017) 243901.

27. C.W. Hsu, B. Zhen, A.D. Stone, J.D. Joannopoulos, M. Soljačić, Bound states in the continuum, *Nature Reviews Materials*, 1 (2016) 16048.

28. H. Friedrich, D. Wintgen, Interfering resonances and bound states in the continuum, *Physical Review A*, 32 (1985) 3231.

29. A.F. Sadreev, E.N. Bulgakov, I. Rotter, Bound states in the continuum in open quantum billiards with a variable shape, *Physical Review B*, 73 (2006) 235342.

30. C.W. Hsu, B. Zhen, J. Lee, S.-L. Chua, S.G. Johnson, J.D. Joannopoulos, M. Soljačić, Observation of trapped light within the radiation continuum, *Nature*, 499 (2013) 188.
31. F. Monticone, A. Alù, Embedded photonic eigenvalues in 3D nanostructures, *Physical Review Letters*, 112 (2014) 213903.
32. M. Rybin, Y. Kivshar, Supercavity lasing, *Nature*, 541 (2017) 164.
33. C. De Angelis, A. Locatelli, L. Carletti, D. Rocco, O. Stepanenko, G. Leo, I. Favero, et al., Enhanced second-harmonic generation from magnetic resonance in AlGaAs nanoantennas, SPIE OPTO, SPIE, 2016, pp. 8.
34. S.S. Kruk, Y.S. Kivshar, Functional meta-optics and nanophotonics govern by Mie resonances, *ACS Photonics*, (2017) 2638. doi:10.1021/acsphotonics.7b01038.
35. D.G. Baranov, D.A. Zuev, S.I. Lepeshov, O.V. Kotov, A.E. Krasnok, A.B. Evlyukhin, B.N. Chichkov, All-dielectric nanophotonics: The quest for better materials and fabrication techniques, *Optica*, 4 (2017) 814.
36. S. Keren-Zur, L. Michaeli, H. Suchowski, T. Ellenbogen, Shaping light with nonlinear metasurfaces, *Advanced Optical and Photonics*, 10 (2018) 309.
37. C. Schlickriede, N. Waterman, B. Reineke, P. Georgi, G. Li, S. Zhang, T. Zentgraf, Imaging through nonlinear metalens using second harmonic generation, *Advanced Materials*, 30 (2018) 1703843.
38. A. Arbabi, Y. Horie, M. Bagheri, A. Faraon, Dielectric metasurfaces for complete control of phase and polarization with subwavelength spatial resolution and high transmission, *Nature Nanotechnology*, 10 (2015) 937.
39. E. Arbabi, A. Arbabi, S.M. Kamali, Y. Horie, A. Faraon, Multiwavelength polarization-insensitive lenses based on dielectric metasurfaces with meta-molecules, *Optica*, 3 (2016) 628.
40. S.A. Maier, Dielectric and low-dimensional-materials nanocavities for non-linear nanophotonics and sensing, Advanced Photonics 2018 (BGPP, IPR, NP, NOMA, Sensors, Networks, SPPCom, SOF), Optical Society of America, Zurich, 2018, pp. SeW2E.4.
41. O. Yavas, M. Svedendahl, P. Dobosz, V. Sanz, R. Quidant, On-a-chip biosensing based on all-dielectric Nanoresonators, *Nano Letters*, 17 (2017) 4421.
42. J. Cambiasso, M. König, E. Cortés, S. Schlücker, S.A. Maier, Surface-enhanced spectroscopies of a molecular monolayer in an all-dielectric nanoantenna, *ACS Photonics*, 5 (2018) 1546–1557.
43. C. Manzoni, G. Cerullo, Design criteria for ultrafast optical parametric amplifiers, *Journal of Optics*, 18 (2016) 103501.
44. L. Chang, X. Guo, D.T. Spencer, J. Chiles, A. Kowligy, N. Nader, D. Hickstein, et al., A gallium arsenide nonlinear platform on silicon, *Conference on Lasers and Electro-Optics*, Optical Society of America, San Jose, California, 2018, pp. STu3F.5.
45. M. Finazzi, P. Biagioni, M. Celebrano, L. Duò, Selection rules for second-harmonic generation in nanoparticles, *Physical Review B*, 76 (2007) 125414.
46. J.I. Dadap, Optical second-harmonic scattering from cylindrical particles, *Physical Review B*, 78 (2008) 205322.
47. D. de Ceglia, M.A. Vincenti, C. De Angelis, A. Locatelli, J.W. Haus, M. Scalora, Role of antenna modes and field enhancement in second harmonic generation from dipole nanoantennas, *Optics Express*, 23 (2015) 1715.
48. R.W. Boyd, *Nonlinear Optics*, Academic Press, 2008.
49. Z. Yang, P. Chak, A.D. Bristow, H.M. van Driel, R. Iyer, J.S. Aitchison, A.L. Smirl, J.E. Sipe, Enhanced second-harmonic generation in AlGaAs microring resonators, *Optics Letters*, 32 (2007) 826.
50. S.S. Kruk, R. Camacho-Morales, L. Xu, M. Rahmani, D.A. Smirnova, L. Wang, H.H. Tan, et al., Nonlinear optical magnetism revealed by second-harmonic generation in nanoantennas, *Nano Letters*, 17 (2017) 3914.

51. S. Liu, G.A. Keeler, J.L. Reno, M.B. Sinclair, I. Brener, Efficient second harmonic generation from GaAs all-dielectric metasurfaces, in *Conference on Lasers and Electro-Optics*, OSA Technical Digest (online) (Optical Society of America, 2016), paper FM2D.6.

52. F. Timpu, A. Sergeyev, N.R. Hendricks, R. Grange, Second-harmonic enhancement with Mie resonances in perovskite nanoparticles, *ACS Photonics*, 4 (2017) 76.

53. K.B. Crozier, A. Sundaramurthy, G.S. Kino, C.F. Quate, Optical antennas: Resonators for local field enhancement, *Journal of Applied Physics*, 94 (2003) 4632.

54. P. Mühlschlegel, H.J. Eisler, O.J.F. Martin, B. Hecht, D.W. Pohl, Resonant optical antennas, *Science*, 308 (2005) 1607.

55. A. Locatelli, C. De Angelis, D. Modotto, S. Boscolo, F. Sacchetto, M. Midrio, A.-D. Capobianco, F.M. Pigozzo, C.G. Someda, Modeling of enhanced field confinement and scattering by optical wire antennas, *Optics Express*, 17 (2009) 16792.

56. P. Bharadwaj, B. Deutsch, L. Novotny, Optical antennas, *Advanced Optical and Photonics*, 1 (2009) 438–483.

57. C. De Angelis, A. Locatelli, D. Modotto, S. Boscolo, M. Midrio, A.-D. Capobianco, Frequency addressing of nano-objects by electrical tuning of optical antennas, *JOSA B*, 27 (2010) 997–1001.

58. L. Novotny, N. van Hulst, Antennas for light, *Nature Photonics*, 5 (2011) 83.

59. B. Paolo, H. Jer-Shing, H. Bert, Nanoantennas for visible and infrared radiation, *Reports on Progress in Physics*, 75 (2012) 024402.

60. L. Carletti, D. Rocco, A. Locatelli, C. De Angelis, V.F. Gili, M. Ravaro, I. Favero, et al., Controlling second-harmonic generation at the nanoscale with monolithic AlGaAs-on-AlOx antennas, *Nanotechnology*, 28 (2017) 114005.

61. L. Carletti, A. Locatelli, D. Neshev, C. De Angelis, Shaping the radiation pattern of second-harmonic generation from AlGaAs dielectric nanoantennas, *ACS Photonics*, (2016) 1500. doi:10.1021/acsphotonics.6b00050.

62. L. Carletti, G. Marino, L. Ghirardini, V.F. Gili, D. Rocco, I. Favero, A. Locatelli, et al., Nonlinear goniometry by second-harmonic generation in AlGaAs nanoantennas, *ACS Photonics*, (2018) 4386.

63. L. Ghirardini, L. Carletti, V. Gili, G. Pellegrini, L. Duò, M. Finazzi, D. Rocco, et al., Polarization properties of second-harmonic generation in AlGaAs optical nanoantennas, *Optics Letters*, 42 (2017) 559.

64. M. Guasoni, L. Carletti, D. Neshev, C. De Angelis, Theoretical model for pattern engineering of harmonic generation in all-dielectric nanoantennas, *IEEE Journal of Quantum Electronics*, 53 (2017) 1.

65. A.M. Shaltout, A.V. Kildishev, V.M. Shalaev, Evolution of photonic metasurfaces: From static to dynamic, *Journal of the Optical Society of America B*, 33 (2016) 501.

66. M.R. Shcherbakov, P.P. Vabishchevich, A.S. Shorokhov, K.E. Chong, D.-Y. Choi, I. Staude, A.E. Miroshnichenko, et al., Ultrafast all-optical switching with magnetic resonances in nonlinear dielectric nanostructures, *Nano Letters*, 15 (2015) 6985.

67. D.G. Baranov, S.V. Makarov, A.E. Krasnok, P.A. Belov, A. Alù, Tuning of near-and far-field properties of all-dielectric dimer nanoantennas via ultrafast electron-hole plasma photoexcitation, *Laser & Photonics Reviews*, 10 (2016) 1009.

68. S.V. Makarov, A.S. Zalogina, M. Tajik, D.A. Zuev, M.V. Rybin, A.A. Kuchmizhak, S. Juodkazis, Y. Kivshar, Light-induced tuning and reconfiguration of nanophotonic structures, *Laser & Photonics Reviews*, 11 (2017) 1700108.

69. L. Ghirardini, L. Carletti, V. Gili, G. Pellegrini, L. Duò, M. Finazzi, D. Rocco, et al., Optical switching of the second harmonic generation in AlGaAs nanoantennas, *Advanced Photonics* 2018 (BGPP, IPR, NP, NOMA, Sensors, Networks, SPPCom, SOF), Optical Society of America, Zurich, 2018, pp. NpW3C.8.

70. Q. Lin, J. Zhang, G. Piredda, R.W. Boyd, P.M. Fauchet, G.P. Agrawal, Dispersion of silicon nonlinearities in the near infrared region, *Applied Physics Letters*, 91 (2007) 021111.

71. A.D. Bristow, N. Rotenberg, H.M. van Driel, Two-photon absorption and Kerr coefficients of silicon for 850–2200 nm, *Applied Physics Letters*, 90 (2007) 191104.

72. D.J. Moss, E. Ghahramani, J.E. Sipe, H.M. van Driel, Band-structure calculation of dispersion and anisotropy in $\chi^{(3)}$ for third-harmonic generation in Si, Ge, and GaAs, *Physical Review B*, 41 (1990) 1542.

73. D.J. Moss, H.M. van Driel, J.E. Sipe, Dispersion in the anisotropy of optical third-harmonic generation in silicon, *Optics Letters*, 14 (1989) 57.

74. L. Vivien, L. Pavesi, *Handbook of Silicon Photonics*, CRC Press, Taylor & Francis Group, Boca Raton, FL, 2016.

75. A.E. Miroshnichenko, S. Flach, Y.S. Kivshar, Fano resonances in nanoscale structures, *Reviews of Modern Physics*, 82 (2010) 2257.

76. Y. Yang, W. Wang, A. Boulesbaa, I.I. Kravchenko, D.P. Briggs, A. Puretzky, D. Geohegan, J. Valentine, Nonlinear fano-resonant dielectric metasurfaces, *Nano Letters*, 15, (2015) 7388–7393.

77. L. Wang, S. Kruk, L. Xu, M. Rahmani, D. Smirnova, A. Solntsev, I. Kravchenko, D. Neshev, Y. Kivshar, Shaping the third-harmonic radiation from silicon nanodimers, *Nanoscale*, 9 (2017) 2201.

78. S. Chen, M. Rahmani, K.F. Li, A. Miroshnichenko, T. Zentgraf, G. Li, D. Neshev, S. Zhang, Third harmonic generation enhanced by multipolar interference in complementary silicon metasurfaces, *ACS Photonics*, 5 (2018) 1671.

79. A.E. Miroshnichenko, A.B. Evlyukhin, Y.F. Yu, R.M. Bakker, A. Chipouline, A.I. Kuznetsov, B. Luk'yanchuk, B.N. Chichkov, Y.S. Kivshar, Nonradiating anapole modes in dielectric nanoparticles, *Nature Communications*, 6 (2015) 8069.

80. G. Grinblat, Y. Li, M.P. Nielsen, R.F. Oulton, S.A. Maier, Enhanced third harmonic generation in single germanium nanodisks excited at the anapole mode, *Nano Letters*, 16 (2016) 4635–4640.

81. T. Shibanuma, G. Grinblat, P. Albella, S.A. Maier, Efficient third harmonic generation from metal–dielectric hybrid nanoantennas, *Nano Letters*, 17 (2017) 2647.

82. L. Xu, M. Rahmani, K. Zangeneh Kamali, A. Lamprianidis, L. Ghirardini, J. Sautter, R. Camacho-Morales, et al., Boosting third-harmonic generation by a mirror-enhanced anapole resonator, *Light: Science & Applications*, 7 (2018) 44.

83. E.V. Melik-Gaykazyan, S.S. Kruk, R. Camacho-Morales, L. Xu, M. Rahmani, K. Zangeneh Kamali, A. Lamprianidis, et al., Selective third-harmonic generation by structured light in Mie-resonant nanoparticles, *ACS Photonics*, 5 (2018) 728.

84. D.A. Smirnova, A.B. Khanikaev, L.A. Smirnov, Y.S. Kivshar, Multipolar third-harmonic generation driven by optically induced magnetic resonances, *ACS Photonics*, 3 (2016) 1468–1476.

85. G.F. Walsh, L. Dal Negro, Enhanced second harmonic generation by photonic–plasmonic fano-type coupling in nanoplasmonic arrays, *Nano Letters*, 13 (2013) 3111.

86. V.F. Gili, L. Ghirardini, D. Rocco, G. Marino, I. Favero, I. Roland, G. Pellegrini, et al., Metal–dielectric hybrid nanoantennas for efficient frequency conversion at the anapole mode, *Beilstein Journal of Nanotechnology*, 9 (2018) 2306–2314.

87. G. Marino, A.S. Solntsev, L. Xu, V. Gili, L. Carletti, A.N. Poddubny, D. Smirnova, et al., Sum-frequency generation and photon-pair creation in AlGaAs nano-scale resonators, *2017 Conference on Lasers and Electro-Optics (CLEO)*, 2017, pp. 1–1.

88. G. Marino, A.S. Solntsev, L. Xu, V.F. Gili, L. Carletti, A.N. Poddubny, et al., Sum-frequency- and photon-pair-generation in AlGaAs nano-disks, *Advanced Photonics* 2018 (BGPP, IPR, NP, NOMA, Sensors, Networks, SPPCom, SOF), Optical Society of America, Zurich, 2018, pp. NpM2I.2.

10 Future Perspectives

Giuseppe Marino, Carlo Gigli, and Aloyse Degiron

CONTENTS

In this chapter, we will discuss a number of developments that, in our view, would be interesting to consider for the future of nonlinear metasurfaces. In Section 10.1, we will lay out an approach to gain an exquisite control over the phase and polarization of second harmonic light using electron beam lithography semiconductor metasurfaces. In Section 10.2, we will examine strategies that may be followed to produce overlooked and exotic nonlinearities. We will examine nonlinear phenomena for optical beams propagating parallel to the metasurfaces, which is at variance with the existing literature that mostly deals with out-of-plane nonlinear processes, and also consider the case of the strong coupling regime. Finally, in Section 10.3, we will argue that metasurfaces made of colloidal nanoparticles have a strong and as-yet untapped potential for fulfilling many of these goals (and also to serve as a powerful alternative to existing implementations) due to their ease of fabrication and high versatility.

10.1   EXQUISITE CONTROL OF NONLINEARITIES WITH SEMICONDUCTOR METASURFACES

One key advantage of using metasurfaces in nanophotonics is their characteristic precise modulation of amplitude, polarization, and phase at the nanoscale. This is achieved through interactions between propagating and localized surface resonances. The control of the nonlinear wavefront is then possible for metasurfaces made of highly nonlinear materials. In the previous chapters the control over the polarization [1–5] and phase [6–12] of second harmonic (SH) light through metasurfaces has been shown.

In the following paragraph we will present further advances to encode the SH phase in the zero-diffraction order, and to control the polarization of SH light. To this end, we will take advantage of nonlocal surface modes, with preliminary results obtained with metasurfaces made of periodically arranged high refractive index meta-atoms. Nonlocal surface modes can take place in two different optical regimes: Mie-resonant metamaterials and photonic crystals. In the former case, $n \gg 1$, $p \ll \lambda$ and $\lambda / n = \lambda_{Mie} / n \approx L$, with n, L the refractive index, and characteristic size of the metasurface meta-atoms, p the periodicity and λ the first harmonic (FH) wavelength, thus light interacts with the metasurface as it was a dilute system. In the latter case, $\lambda / n_{eff} < \frac{\lambda_{Bragg}}{n_{eff}} = 2p$, with n_{eff} the effective index in the unit cell, thus the metasurface constituents are subject to Bragg scattering [13,14].

Here we focus on this latter regime and consider arrays of AlGaAs cylinders as an example (Figure 10.1a). The physical quantity that describes the coupling between neighbor meta-atoms is the array factor [16–18]:

$$S(\theta) = \sum e^{ikr} \left[\frac{(1-ikr)(3\cos^2\theta - 1)}{r^3} + \frac{k^2 \sin^2\theta}{r} \right], \qquad (10.1)$$

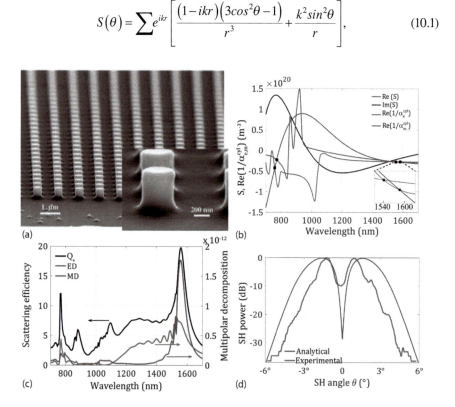

FIGURE 10.1 (a) SEM viewgraph of structure. (b) Re(S) (dark grey line with a broad resonance at around 950 nm) and Im(S) (black line) for a metasurface with R = 200 nm and p = 1025 nm. The two light grey lines represents the inverse of the electric (with two narrow resonances at around 900 nm) and magnetic (with a narrow resonance at around 750 nm) polarizability of the single cylinder. Inset: zoom near FW. Square or circular markers represent intersections between Re(S) and Re$\left(1/\alpha_e^{cyl}\right)$ or Re$\left(1/\alpha_m^{cyl}\right)$. (c) Scattering efficiency and multipolar decomposition associated to the metasurface (ED: electric dipole; MD: magnetic dipole). (d) SHG radiation pattern in the NA = 0.1 for the metasurface: analytically calculated (dark grey line) and experimental (light grey line) [15].

with r and θ the polar coordinates in the plane perpendicular to the metasurface (here we consider only the nearest neighbors, with $r = p$). Because nonlocal surface modes can constructively interfere with local Mie modes supported by the meta-atoms, the metasurface overall response can be described in the framework of the coupled-mode theory with effective electric and magnetic polarizabilities [16–18]:

$$\alpha_{e,m}^{meta} = \left[1 / \alpha_{e,m}^{cyl} - S \right]^{-1}, \tag{10.2}$$

where $\alpha_e^{cyl} = i\left(k^3 / 6\pi\right)^{-1} a_1$ and $\alpha_m^{cyl} = i\left(k^3 / 6\pi\right)^{-1} b_1$ are the electric and magnetic polarizabilities of an isolated meta-atom [19], with coefficients a_1 and b_1 defined in terms of Riccati-Bessel functions under Mie theory, and with $k = nk_0$ and k_0 the wave vectors in the medium embedding the meta-atoms and in vacuum (in our case, $n = n_{air} = 1$). The resonances of the metasurfaces correspond to the poles of the metasurface effective polarizability, i.e., to $\mathrm{Re}\left(1 / \alpha_{e,m}^{cyl} - S\right) = 0$. The quality factor of these resonances also depends on the radiative damping, i.e., on $\mathrm{Im}(S)$ at the effective polarizability poles. For example, the effective polarizability of a metasurface with $p = 1025$ nm, $R = 200$ nm and $h = 400$ nm has its poles at the bullet points of Figure 10.1b, where $\mathrm{Re}\left(S\right)$ intersects $\mathrm{Re}\left(1 / \alpha_e^{cyl}\right)$ around $\lambda = 1540$ nm and 760 nm, and $\mathrm{Re}\left(1 / \alpha_m^{cyl}\right)$ around $\lambda = 1570$ nm and 770 nm. Correspondingly, $\mathrm{Im}\left(S\right)$ turns out to be small at both telecom wavelengths and the near-infrared, thus low radiative damping is expected.

We now show how such nonlocal modes can be leveraged to generate polarized beams with narrow divergence and controlled orientation. The array factor represents a weak perturbation, so that the SH far-field from the metasurface can be approximated as:

$$I_{far}^{tot}\left(\theta\right) = \left|S\left(\theta\right)\right|^2 I_{far}\left(\theta\right) \tag{10.3}$$

i.e., the overlap between the radiation pattern of the isolated meta-atom and the array factor considering $N \times N$ meta-atoms, where N is a fit parameter. Eq. 10.3 is plotted in Figure 10.1d (light blue line) within a numerical aperture of 6 degrees from the backward direction and compared with experimentally measured SHG from the metasurface (red line). A good agreement between the two curves is achieved with two maxima at 1 degree from the backward direction. This zero-order phase encoded by the metasurface represents a big step forward with respect to the isolated meta-atom whose radiation pattern is maximum at 60° from the backward direction.

The measured SHG is plotted as a function of λ, R, and p in Figure 10.2 and compared with calculations. In agreement with the characteristic mode excited at the SH in Figure 10.1c, a Q factor of 50 has been demonstrated for a metasurface with $R = 200$ nm, $h = 400$ nm, and $p = 1025$ nm (Figure 10.2a). The measured dependence of SHG from the geometry (R) of the meta-atom is compared with numerical simulations of the Fourier transform of the SH near-field and with the analytical model approximating the diffraction at the nearest neighbors (solid blue line of Figure 10.2b). With regard to the period dependence, the SHG efficiency drops both for small

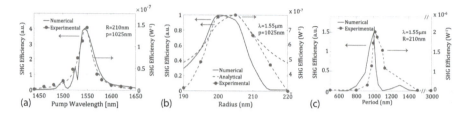

FIGURE 10.2 Zero-diffraction order SHG efficiency from metasurface in reflection geometry within 0.1 NA. Measurements (dashed line with markers) and FEM calculations (solid line) vs: (a) pump wavelength, for $R = 200$ nm, $h = 400$ nm and $p = 1025$ nm; (b) nanoantenna radius, for $l = 1550$ nm, $p = 1025$ nm; (c) period, for $l = 1550$ nm and $R = 200$ nm. In panel (b) the dashed line without markers represents the analytically calculated overlap between the single nanoantenna far-field and the array factor for $N = 15$, and both blue curves are normalized to their maximum. The maximum measured efficiency is 2×10^{-6} W^{-1} as reported in Figure 10.2c [15]. Importantly, given an input power of 1 GW/cm^2, SHG power in a 0.1 NA is 0.05 mW [20] from the isolated cylinder and 2.6 mW from the metasurface [15].

values of p of the order of the diffraction limit, where the zero-order grating lobes get sharper and narrower, and for values of p bigger than the FH wavelength, where the zero-order grating lobes get lower and wider. For values of p of the order of 1 μm, the SH far-field of the isolated meta-atom overlap constructively with the zero-order grating lobes associated to the array factor, thereby resulting in a 50 times enhancement with respect to the case of an isolated meta-atom.

So far, we have seen how to use the nonlocal modes to constructively diffract SH light from the meta-atoms into the zero order. However, the SH polarization is that of the isolated meta-atom, because of the weak perturbation of the meta-atom SH modes. To control the SH polarization, we have fabricated samples made of elliptical nanocylinders, so that their response varies as a function of their orientation with respect to the fixed input linear polarization. In the scheme shown in Figure 10.3a, this results in the excitation of identical nonlinear currents and SH modes solely orthogonal to each other according to the orientation of the elliptical basis (Figure 10.3b). The numerically calculated and measured SHG radiation diagram is reported in Figure 10.3c within an NA of 53 degree. As the maxima of the two SHG lobes are expected at 60 degree from the backward direction, only half of them appears in the diagram. Furthly, the two orientations of the elliptical-basis result in orthogonal polarizations, as reported in Figure 10.3d. However, this polarization control is not very practical for on-axis applications as a consequence of the high angular separation between the two lobes. This is why we have multiplexed the antenna in the form of a square array and exploited the metasurface constructive interference in the zero-order diffracted SH. The numerically calculated and measured SHG radiation diagram from the metasurface within an NA of 11.5 degree is reported in Figure 10.3e. In this case, the two maxima result at 6 degree from the backward direction. In both cases, the measured polarization conversion ratio H/V (Horizontal/Vertical) is 80%.

Taken together, these results show the potential of nonlocal modes in all-dielectric nonlinear metasurfaces. They offer extra degrees of freedom compared to individual

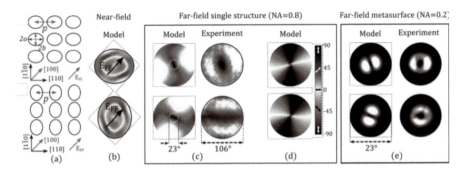

FIGURE 10.3 SH field distributions for horizontally (top) and vertically (bottom) oriented nanocylinders with elliptical base. (a) Control scheme with fixed FF pump polarization along and elliptical-basis meta-atoms rotated by 90°. The best result for polarization control was obtained with semi-major axis $a = 240$ nm and semi-minor axis $b = 200$ nm. Reference system, corresponding to AlGaAs crystal axes. (b) Near-field distribution of the mainly excited SH mode, with the indication of FF polarization. (c) SH radiation pattern of a single element within a solid angle set by $NA = 0.8$ (corresponding to 106° field of view) of the microscope objective. White dashes delimit a region with $NA = 0.2$ (corresponding to 23° field of view). (d) Inclination of SH linear polarization direction for the single element. (e) Back focal plane imaging of SH generated by a metasurface with the same constitutive elements as in c, period $p = 940$ nm, and collection restricted to NA = 0.2 [21].

Mie resonators that makes it possible to create highly polarized SHG along the vertical axis, which is promising for applications like on-axis imaging and free-space optical interconnects.

10.2 TOWARD EXOTIC AND/OR UNEXPLORED NONLINEARITES

10.2.1 IN-PLANE PROPAGATION: SOLITONS, INTEGRATED CIRCUITS

Most of the literature considers nonlinear metasurfaces as free-space elements, with electromagnetic waves coming in- and out-of-plane. However, the problem of nonlinear effects occurring within the plane of the metasurface also deserves our attention. A first category of nonlinear solutions that holds promises are solitons [22], which are modes observed with media exhibiting a positive Kerr nonlinearity (that is, a medium with index $n = n_0 + \alpha E^2$, with $\alpha > 0$). Under the right conditions, the term proportional to E^2 creates a self-focused waveguide that prevents the energy of the mode from diverging. Such solution is a spatial soliton and is also found in many other physical systems (such as the solitary waves in hydrodynamics [23]). Spatial solitons are not the only solutions of interest. The same self-focusing Kerr nonlinearity makes it also possible to counter the group velocity dispersion that affects all pulses in linear media because each optical frequency responds slightly differently to its environment. The resulting mode is a temporal soliton which propagates without modification of the pulse [24]. Finally, it is possible to combine the two sets of solutions to create spatiotemporal solitons which propagate without spreading nor distorting their pulse shape.

Nonlinear metasurfaces promise interesting new developments for these modes. In particular, they may prove a powerful tool to manipulate spatial solitons since it is possible to design complex gradients in their structure that will directly influence the path of such waves. This approach could be used, among other, to study new soliton trajectories beyond those already known (e.g., self-bending solutions), or to have a better control over the interactions between multiple solitons (which leads to, e.g., collisions, filamentation, and intertwined propagation [25]). In addition, the rapid developments of topological photonics [26] may lead to the observation of new families of modes, such as temporal solitons that are topologically protected and that act as spatiotemporal solitons as a result. Because topological photonics rely on artificially structured media to create protected modes, optical metasurfaces are a choice candidate to investigate these new families of solitons.

Besides solitons, and perhaps more importantly for immediate applications, is the urgency to develop and generalize studies for all types of nonlinearities in the context of integrated circuits. Including nonlinear effects in integrated circuits implies to leverage interactions with waves that propagate within optical waveguides. One solution to tackle this problem is to render the waveguides themselves nonlinear [27], but making the nonlinear parts of the circuit coexist with the linear parts poses significant production challenges. Another approach consists in replacing traditional photonic waveguides by plasmonic circuits [28,29]. In this case, a significant part of the energy is guided outside the metallic parts and the mode can thus interact with a nonlinear medium placed in its vicinity. Alternatively, it is also possible to use the same strategy for conventional dielectric waveguides, by placing the evanescent part of the mode that develops outside the high-index core within a nonlinear substance.

It is for these two latter cases, namely nonlinear plasmons and nonlinear interactions with the evanescent tails of dielectric waveguide modes, that metasurfaces may play a useful role. Rather than just surrounding the guides with a nonlinear substance, the hybridization with a metasurface would serve at tuning and enhancing the nonlinear interactions. Studies are needed to engineer metasurfaces specifically adapted to integrated photonics. By analogy with the developments in free space, showing that strong nonlinear effects can be achieved with a single metasurface placed in the path of the beam(s), it will be interesting to investigate whether the same can be achieved with guided modes encountering a single line of nonlinear meta-atoms placed perpendicular to the waveguide. Such a feat would alleviate the problem of phase matching that is critical for frequency generation and wave mixing, and also considerably reduce the footprint of the nonlinear functionalities within the photonic circuits.

10.2.2 THE ULTRA-STRONG COUPLING REGIME

At variance with the weak coupling interactions that are usually leveraged in nonlinear metasurfaces, exotic nonlinearities could be also achieved in the regime of (ultra-) strong coupling. At the heart of cavity quantum electrodynamics [30], strong coupling denotes a situation for which a material system (e.g., a quantum dot, a molecule or a quantum well) is coupled to an optical cavity with so little losses that energy is periodically exchanged between the two. The resulting eigenmodes of the structure are hybrid light-matter states, with energies distinct from those of the uncoupled system. The difference between the energies, known as Rabi splitting, is all the more pronounced that the interactions are strong. In the limit of ultra-strong coupling, the coupled-modes can have radically new properties, such as lifetimes that have no relationship with those of the modes of the uncoupled system. Recently, there has been a considerable interest in these so-called non-Markovian states [31], as it has been demonstrated by Ebbesen et al. and other groups that these states can be used for many purposes, ranging from enhancing the conductivity of molecular layers [32,33] to dramatically changing the kinetics of chemical reactions [34]. In other words, ultra-strong coupling opens a new realm in nanochemistry and material science, as it makes it possible to create states with no equivalent that can be manipulated and leveraged at will. Spectacularly, most of these demonstrations have been made in the dark, in the absence of real photons. These results prove that the phenomena behind the observations truly belong to the realm of cavity quantum electrodynamics, since experiments in the dark can only be interpreted by the fact that it is the quantum fluctuations of the vacuum that ensure the formation of the coupled states.

A first reason why (ultra-) strong coupling is relevant in the context of nonlinear optics is that the emitters involved in the interactions are oscillating in phase. In other words, these emitters form macroscopic states with spatial and temporal coherence [35], and therefore strong coupling offers new and uninvestigated degrees of freedom to optimize nonlinear processes requiring complex phase matching at different frequencies (e.g., harmonic generation, sum and difference frequency generation, or four-wave mixing). Even more tantalizing is the fact that in the ultra-strong coupling regime, the usual approximations used to predict and understand nonlinear phenomena break down [36]. There is therefore an almost entirely unchartered territory to explore, and in particular the fascinating possibility of reaching nonlinearities in the few photon limit—a regime already attained at deep cryogenic temperatures [37], which is completely at odds with conventional nonlinear optics relying on very intense and brief light pulses.

Resonant metasurfaces appear as a natural platform for making breakthroughs in this field because their unit cells can be used as miniature cavities that can be readily hybridized with a variety of material systems. In the visible range, for example, strong coupling between plasmonic metasurfaces and molecular layers coated onto them has been experimentally reported [38]. In the near- and mid-infrared, similar demonstrations have been made with metasurfaces patterned on top of multiple quantum wells [39,40]. Moreover, ultra-strong coupling has been observed for molecules interacting with nanohole arrays [32] and thin-film Fabry-Pérot cavities [41]. Importantly, these experiments have been made at room pressure and temperature and with relatively basic experimental setups, suggesting that the most exotic nonlinearities, such as those involving one or a handful of photons only, may find useful applications one day.

10.3   THE CASE FOR THE COLLOIDAL APPROACH

As is often the case in nanophotonics, one considerable roadblock in the path to actual products concerns the manufacturing process. Even though metasurfaces are easier to fabricate than bulk metamaterials, they consist of intricate patterns of subwavelength dimensions, which, in the visible and near-infrared range, implies feature sizes well below the micrometer scale. Up to now, the majority of nonlinear metasurfaces investigated in the optical regime have been realized with electron beam lithography [42,43], which is intrinsically slow and expensive—with the best systems currently available, it requires several hours to pattern centimeter square areas. There is, however, a number of alternative nanofabrication techniques at hand, from nanoimprint, where all the features of the metasurface are created at once using a reusable stamp, to approaches based on colloidal materials. Such fabrication processes have already been considered for linear structures [44] but their potential for nonlinear architectures appears to have been little discussed so far. In this section, we will specifically focus on the colloidal approach, as we shall see that it opens interesting prospects for the future of nonlinear metasurfaces.

The idea behind the colloidal approach is to synthetize the metasurface building blocks in solution (that is, under the form of metallic or semiconducting nanoparticles stabilized by organic ligands), and then assemble them into larger architectures. What makes this strategy technologically enabling is that it is now possible to fabricate the nanoparticles with a high degree of monodispersity, and that many of their structural and physical properties can be finely tuned with the ligands surrounding them. Even if the syntheses are sometimes difficult, this technique is scalable and compatible with a variety of platforms and substrates, including nonplanar and flexible ones.

Perhaps the most emblematic colloidal structures in this context are those based on noble metal nanoparticles—a central fixture in optics since the middle of the 19th century [45]. Over the years, tremendous progresses in chemical synthesis have allowed for an exquisite control of the size and shape of metallic nanoparticles. Not only it is possible to produce particles with a high degree of symmetry, such as spheres and pyramids, but also to fabricate particles with different aspect ratios, such as polygonal nanoplates and elongated nanorods [46]. The capability for self-organization are excellent for certain geometries [47], allowing the formation of patterned metasurfaces with a long range order. As is the case for their counterparts fabricated by top-down (e.g., lithography) techniques, such metallic nanoparticle assemblies are interesting for nonlinear applications for two reasons. First, as all nanostructures, they exhibit an intrinsic second-order nonlinearity due to their high surface over volume ratio that makes the electrons at their surface bask in an asymmetric potential. Second, the nonlinearities can be exacerbated by the fact that they sustain surface plasmon resonances associated with very strong electromagnetic fields [48]. The plasmon resonances can also be used to enhance the nonlinearities of a nearby dielectric substance. In this respect, we posit that colloidal metallic nanoparticles are especially suited for one class of nonlinear metasurfaces—one in which they are deposited on top of a nonlinear thin film that is itself coated on a ground plane. This geometry, which has already been investigated as passive metasurfaces [49] and at the single unit cell level as active light-emitting elements [50], appears very promising for studying nonlinear effects since the field is squeezed between the ground plane and the metal nanoparticles. It has led, among others, to record molecule photoluminescence enhancements due to the enormous electromagnetic field build-up between a ground plane and colloidal silver nanocubes [50]. There is therefore plenty of opportunities for exploiting such high field densities in the nonlinear regime. This is all the more true that colloidal unit cells, unlike those obtained by top-down fabrication techniques, are

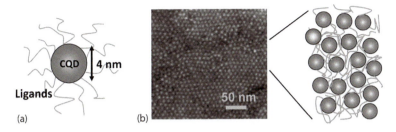

FIGURE 10.4   (a) Schematic of a PbS colloidal quantum dot (CQD). (b) Top surface of a PbS CQD-film as observed by a scanning electron microscope.

monocrystalline. The absorption losses by Joule effect are considerably smaller as a result [51], implying that the damage threshold, which is one of the biggest limitations of nonlinear metallic metasurfaces [42,43], is necessarily higher.

A second class of colloidal nanoparticles that are highly relevant for the future of nonlinear metasurfaces are semiconducting nanocrystals, or colloidal quantum dots (CQDs, Figure 10.4a). In terms of size, these objects are ten to a hundred times smaller than the plasmonic resonators discussed in the previous paragraph since they must be small enough to confine the electron wave functions in at least one (and generally three) dimensions. As a result, their semiconducting properties, and particularly their energy levels and energy gap, are not directly related to their chemical composition but rather defined by their size. In this field also, progresses in chemical synthesis have been staggering with the advent of nanocrystals with controlled size and shape [52,53], and, most importantly, doping [54,55]. This, in turn, has led to tremendous developments in solution-processed optoelectronic devices [56], in which thin solid films of CQDs replace standard semiconducting wafers (Figure 10.4b). The method for fabricating such granular semiconducting films is especially simple, as it suffices to spin-cast the colloidal solution with the desired concentration to obtain a compact granular film of one to several monolayers of quantum dots after evaporation of the solvent. Such granular fields are intrinsically disordered but more careful deposition techniques makes it possible to obtain nearly perfect regular arrangements of the individual nanocrystals [57]. The community explores all types of optoelectronic components, i.e., active sources [58,59], photodetectors [60,61], and solar cells [62,63], from the UV to the mid-infrared.

Up to now, the nonlinear properties of semiconducting colloidal nanocrystals have been the object of little attention, except for a few studies on optical limitation using quantum dots embedded in dielectric matrices [64] and the demonstration of the Kerr effect in isolated PbS nanocrystals [65]. In particular, it would be highly interesting to study the nonlinearities that one can obtain with the compact nanocrystal assemblies (i.e., without dielectric matrix) used in the optoelectronic community discussed in the previous paragraph, since it would add valuable functionalities and would open new opportunities in this application-oriented field. Being made of inorganic materials, they are intrinsically more resistant to high power damage than other solution-processed technologies (organic layers) although it remains to be seen whether high energy pulses would not lead to partial sintering and changes in their optical and electronic properties.

As a natural first step, it would be highly relevant to study the susceptibilities of the third order associated with the Kerr effect and three-photon generation. They should be readily observed in a wide variety of nanocrystal assemblies since they do not require noncentrosymmetric materials. A central question in our opinion is the role played by the interstitial gaps between the nanocrystals forming the granular medium—are they going to enhance the nonlinear susceptibilities, by becoming hot spots for the electromagnetic field, or are they going to dilute the nonlinear response of the system? It is likely that the answer to this question depends in part of the ligands surrounding the nanocrystals since the latter define the average center-to-center distance between the nanoparticles as well as the optical properties of the interstitial gaps.

Second, the rapid emergence of optoelectronic devices operating with semiconducting perovskite nanocrystals [66] opens the door to colloidal-based architectures with second-order nonlinearities because perovskites can be made noncentrosymmetric. In fact, SH generation and four-wave mixing have already been demonstrated with larger perovskite crystals that do not confine the electron wavefunctions [67]. Compared to a homogeneous layer, whose crystalline structure imposes an anisotropic second-order susceptibility, a granular layer with randomly oriented nanocrystals will likely exhibit an isotropic response. This prospect will considerably relax the selection rules that are required for SH generation and other second-order nonlinear phenomena—but it will weaken the strength of the nonlinearities for a given polarization since they will be averaged over the three dimensions of space.

Finally, another promising avenue to endow nanocrystal assemblies with second-order nonlinearities is based on nanoplatelets building blocks. Nanoplatelets are colloidal crystals that confine the electron wavefunctions in one dimension only [52,53]. Because one can control the doping of these objects and even synthetize complex heterostructures acting as colloidal quantum wells [68], it should be possible to use colloidal nanoplatelets to build the equivalent of epitaxial multiple quantum wells, which are among the structures exhibiting the strongest second- and third-order nonlinearities [69], by simply spin-coating different colloidal solutions onto the sample of interest.

To conclude this part, it is worth noting that the architectures based on noble metal nanoparticles or those based on semiconducting nanocrystals are certainly not the only ones with a promising potential for nonlinear meta-optics. If the literature related to the linear regime is any guide, it is likely that interesting new phenomena will emerge by hybridizing colloidal nano-objects with other nanostructures, or by blending several families of colloidal materials, or by synthetizing complex hybrid colloidal objects mixing different materials, shape, and aspect ratio [70,71].

10.4 CONCLUSION

In conclusion, we have elaborated on several aspects of nonlinear meta-optics that may be worth investigating in the coming years. It would be of course preposterous to predict which, if any, of these points will actually lead to fruitful scientific and/or technological advances. This is all the more true that the field is still at an early stage of development. Nevertheless, it is quite likely that many of these directions will be simultaneously explored in the foreseeable future and that they will cross the path of other highly active fields such as low-dimensionality systems, topological photonics, and few photon optics.

REFERENCES

1. F. Ding, Z. Wang, S. He, V. M. Shalaev, and A. V. Kildishev, *ACS Nano* **9**, 4111 (2015).
2. B. K. Canfield, S. Kujala, K. Jefimovs, T. Vallius, J. Turunen, and M. Kauranen, *J. Opt. Pure Appl. Opt.* **7**, S110 (2005).
3. A. Shaltout, J. Liu, A. Kildishev, and V. Shalaev, *Optica* **2**, 860 (2015).
4. G. Li, S. Chen, N. Pholchai, B. Reineke, P. W. H. Wong, E. Y. B. Pun, K. W. Cheah, T. Zentgraf, and S. Zhang, *Nat. Mater.* **14**, 607 (2015).
5. G. Marino, P. Segovia, A. V. Krasavin, P. Ginzburg, N. Olivier, G. A. Wurtz, and A. V. Zayats, *Laser Photonics Rev.* **12**, 1700189 (2018).
6. F. Aieta, P. Genevet, M. A. Kats, N. Yu, R. Blanchard, Z. Gaburro, and F. Capasso, *Nano Lett.* **12**, 4932 (2012).
7. N. Yu, P. Genevet, M. A. Kats, F. Aieta, J.-P. Tetienne, F. Capasso, and Z. Gaburro, *Science* **334**, 333 (2011).
8. X. Ni, A. V. Kildishev, and V. M. Shalaev, *Nat. Commun.* **4**, 2807 (2013).
9. M. S. Seghilani, M. Myara, M. Sellahi, L. Legratiet, I. Sagnes, G. Beaudoin, P. Lalanne, and A. Garnache, *Sci. Rep.* **6**, 38156 (2016).
10. S. Keren-Zur, O. Avayu, L. Michaeli, and T. Ellenbogen, *ACS Photonics* **3**, 117 (2015).
11. F. J. Löchner, A. N. Fedotova, S. Liu, G. A. Keeler, G. M. Peake, S. Saravi, M. R. Shcherbakov, S. Burger, A. A. Fedyanin, I. Brener, T. Pertsch. *ACS Photonics*, 5, 1786–1793 (2018).
12. L. Wang, S. S. Kruk, K. L. Koshelev, I. I. Kravchenko, B. Luther-Davies, and Y. S. Kivshar, *Nano Lett.* 18, 3978–3984 (2018).
13. M. V. Rybin, D. S. Filonov, K. B. Samusev, P. A. Belov, Y. S. Kivshar, and M. F. Limonov, *Nat. Commun.* **6**, 10102 (2015).
14. S. V. Li, Y. S. Kivshar, and M. V. Rybin, *ACS Photonics* 5, 4751–4757 (2018).
15. G. Marino, C. Gigli, D. Rocco, A. Lemaître, I. Favero, C. De Angelis, and G. Leo, *ACS Photonics* **6**, 1226–1231 (2019).
16. B. Auguié, and W. L. Barnes, *Phys. Rev. Lett.* **101**, 143902 (2008).
17. S. Zou, and G. C. Schatz, *J. Chem. Phys.* **121**, 12606 (2004).
18. V. A. Markel, and A. K. Sarychev, *Phys. Rev. B* **75**, 085426 (2007).
19. P. Grahn, A. Shevchenko, and M. Kaivola, *New J. Phys.* **14**, 093033 (2012).
20. L. Ghirardini, G. Marino, V. F. Gili, I. Favero, D. Rocco, L. Carletti, C. De Angelis, M. Finazzi, and M. Celebrano, *Nano Lett.* **18**, 6750 (2018).
21. C. Gigli, G. Marino, S. Suffit, G. Patriarche, G. Beaudoin, K. Pantzas, I. Sagnes, I. Favero, and G. Leo, *Josa B* **36**, E55–E64 (2019).
22. M. Wadati, *Pramana* **57**, 841 (2001).
23. J. W. Miles, *Annu. Rev. Fluid Mech.* **12**, 11 (1980).
24. A. Hasegawa, and F. Tappert, *Appl. Phys. Lett.* **23**, 142 (1973).
25. G. I. Stegeman, *Science* **286**, 1518 (1999).
26. L. Lu, J. D. Joannopoulos, and M. Soljačić, *Nat. Photonics* **8**, 821 (2014).
27. S. M. Hendrickson, A. C. Foster, R. M. Camacho, and B. D. Clader, *J. Opt. Soc. Am. B* **31**, 3193 (2014).
28. A. R. Davoyan, I. V. Shadrivov, and Y. S. Kivshar, *Opt. Express* **16**, 21209 (2008).
29. A. Degiron, and D. R. Smith, *Phys. Rev. A* **82**, 033812 (2010).
30. S. Haroche, and J.-M. Raimond, *Sci. Am.* **268**, 54 (1993).

31. A. Canaguier-Durand, C. Genet, A. Lambrecht, T. W. Ebbesen, and S. Reynaud, *Eur. Phys. J. D* **69**, (2015).
32. E. Orgiu, J. George, J. A. Hutchison, E. Devaux, J. F. Dayen, B. Doudin, F. Stellacci, C. Genet, J. Schachenmayer, C. Genes, G. Pupillo, P. Samorì, and T. W. Ebbesen, *Nat. Mater.* **14**, 1123 (2015).
33. N. Bartolo, and C. Ciuti, *Cond-Mat Physicsquant-Ph.* ArXiv180502623 (2018).
34. A. Thomas, J. George, A. Shalabney, M. Dryzhakov, S. J. Varma, J. Moran, T. Chervy, X. Zhong, E. Devaux, C. Genet, J. A. Hutchison, and T. W. Ebbesen, Angew. *Chem. Int. Ed.* **55**, 11462 (2016).
35. S. Aberra Guebrou, C. Symonds, E. Homeyer, J. C. Plenet, Yu. N. Gartstein, V. M. Agranovich, and J. Bellessa, *Phys. Rev. Lett.* **108**, 066401 (2012).
36. E. Sánchez-Burillo, J. García-Ripoll, L. Martín-Moreno, and D. Zueco, *Faraday Discuss.* **178**, 335 (2015).
37. T. Peyronel, O. Firstenberg, Q.-Y. Liang, S. Hofferberth, A. V. Gorshkov, T. Pohl, M. D. Lukin, and V. Vuletić, *Nature* **488**, 57 (2012).
38. S. R. K. Rodriguez, and J. G. Rivas, *Opt. Express* **21**, 27411 (2013).
39. A. Benz, S. Campione, M. W. Moseley, J. J. Wierer, A. A. Allerman, J. R. Wendt, and I. Brener, *ACS Photonics* **1**, 906 (2014).
40. S. Campione, S. Liu, A. Benz, J. F. Klem, M. B. Sinclair, and I. Brener, *Phys. Rev. Appl.* **4**, (2015).
41. T. Schwartz, J. A. Hutchison, C. Genet, and T. W. Ebbesen, *Phys. Rev. Lett.* **106**, (2011).
42. G. Li, S. Zhang, and T. Zentgraf, *Nat. Rev. Mater.* **2**, 17010 (2017).
43. A. Krasnok, M. Tymchenko, and A. Alù, Mater. *Today* **21**, 8 (2018).
44. V.-C. Su, C. H. Chu, G. Sun, and D. P. Tsai, *Opt. Express* **26**, 13148 (2018).
45. M. Faraday, *Philos. Trans. R. Soc. Lond.* **147**, 145 (1857).
46. P. Zhao, N. Li, and D. Astruc, *Coord. Chem. Rev.* **257**, 638 (2013).
47. A. Tao, P. Sinsermsuksakul, and P. Yang, *Nat. Nanotechnol.* **2**, 435 (2007).
48. K. L. Kelly, E. Coronado, L. L. Zhao, and G. C. Schatz, *J. Phys. Chem. B* **107**, 668 (2003).
49. G. M. Akselrod, J. Huang, T. B. Hoang, P. T. Bowen, L. Su, D. R. Smith, and M. H. Mikkelsen, *Adv. Mater.* **27**, 8028 (2015).
50. G. M. Akselrod, C. Argyropoulos, T. B. Hoang, C. Ciracì, C. Fang, J. Huang, D. R. Smith, and M. H. Mikkelsen, *Nat. Photonics* **8**, 835 (2014).
51. H. Ditlbacher, A. Hohenau, D. Wagner, U. Kreibig, M. Rogers, F. Hofer, F. R. Aussenegg, and J. R. Krenn, *Phys. Rev. Lett.* **95**, (2005).
52. S. Ithurria, M. D. Tessier, B. Mahler, R. P. S. M. Lobo, B. Dubertret, and Al. L. Efros, *Nat. Mater.* **10**, 936 (2011).
53. M. Nasilowski, B. Mahler, E. Lhuillier, S. Ithurria, and B. Dubertret, *Chem. Rev.* **116**, 10934 (2016).
54. Z. Ning, D. Zhitomirsky, V. Adinolfi, B. Sutherland, J. Xu, O. Voznyy, P. Maraghechi, X. Lan, S. Hoogland, Y. Ren, and E. H. Sargent, *Adv. Mater.* **25**, 1719 (2013).
55. G. Shen, M. Chen, and P. Guyot-Sionnest, *J. Phys. Chem. Lett.* **8**, 2224 (2017).
56. C. R. Kagan, E. Lifshitz, E. H. Sargent, and D. V. Talapin, *Science* **353**, aac5523 (2016).
57. D. M. Balazs, B. M. Matysiak, J. Momand, A. G. Shulga, M. Ibáñez, M. V. Kovalenko, B. J. Kooi, and M. A. Loi, *Adv. Mater.* **30**, 1802265 (2018).
58. J. M. Caruge, J. E. Halpert, V. Wood, V. Bulović, and M. G. Bawendi, *Nat. Photonics* **2**, 247 (2008).
59. X. Yang, P. L. Hernandez-Martinez, C. Dang, E. Mutlugun, K. Zhang, H. V. Demir, and X. W. Sun, *Adv. Opt. Mater.* **3**, 1439 (2015).
60. S. Keuleyan, E. Lhuillier, V. Brajuskovic, and P. Guyot-Sionnest, *Nat. Photonics* **5**, 489 (2011).

61. S. L. Diedenhofen, D. Kufer, T. Lasanta, and G. Konstantatos, *Light Sci. Appl.* **4**, e234 (2015).
62. I. J. Kramer, and E. H. Sargent, *ACS Nano* **5**, 8506 (2011).
63. A. H. Ip, S. M. Thon, S. Hoogland, O. Voznyy, D. Zhitomirsky, R. Debnath, L. Levina, L. R. Rollny, G. H. Carey, A. Fischer, K. W. Kemp, I. J. Kramer, Z. Ning, A. J. Labelle, K. W. Chou, A. Amassian, and E. H. Sargent, *Nat. Nanotechnol.* **7**, 577 (2012).
64. A. M. Malyarevich, M. S. Gaponenko, V. G. Savitski, K. V. Yumashev, G. E. Rachkovskaya, and G. B. Zakharevich, *J. Non-Cryst. Solids* **353**, 1195 (2007).
65. H. Cheng, Y. Wang, H. Dai, J.-B. Han, and X. Li, *J. Phys. Chem. C* **119**, 3288 (2015).
66. Q. A. Akkerman, G. Rainò, M. V. Kovalenko, and L. Manna, *Nat. Mater.* **17**, 394 (2018).
67. F. Timpu, A. Sergeyev, N. R. Hendricks, and R. Grange, *ACS Photonics* **4**, 76 (2017).
68. E. Izquierdo, M. Dufour, A. Chu, C. Livache, B. Martinez, D. Amelot, G. Patriarche, N. Lequeux, E. Lhuillier, and S. Ithurria, *Chem. Mater.* **30**, 4065 (2018).
69. M. M. Fejer, S. J. B. Yoo, R. L. Byer, A. Harwit, and J. S. HarrisJr., *Phys. Rev. Lett.* **62**, 1041 (1989).
70. P. Barois, V. Ponsinet, A. Baron, and P. Richetti, *J. Phys. Conf. Ser.* **963**, 012007 (2018).
71. B. Ji, E. Giovanelli, B. Habert, P. Spinicelli, M. Nasilowski, X. Xu, N. Lequeux, J.-P. Hugonin, F. Marquier, J.-J. Greffet, and B. Dubertret, *Nat. Nanotechnol.* **10**, 170 (2015).

Index

Note: Page numbers in italic and bold refer to figures and tables, respectively.
Page number followed by n refer to footnote.